$$\frac{A}{IA}$$

A COURSE IN
ALGEBRA

A COURSE IN
ALGEBRA

Yun Fan
Q. Y. Xiong
Y. L. Zheng

Department of Mathematics
Wuhan University
P. R. China

World Scientific
Singapore • New Jersey • London • Hong Kong

Published by

World Scientific Publishing Co. Pte. Ltd.

P O Box 128, Farrer Road, Singapore 912805

USA office: Suite 1B, 1060 Main Street, River Edge, NJ 07661

UK office: 57 Shelton Street, Covent Garden, London WC2H 9HE

Library of Congress Cataloging-in-Publication Data
Fan, Yun.
    A course in algebra / Yun Fan, Q. Y. Xiong, Y. L. Zheng.
      p.   cm.
    Includes bibliographical references and index.
    ISBN 9810240619 (alk. paper)
    1. Algebra, Abstract.   I. Xiong, Q. Y.   II. Zheng, Y. L.   III. Title.
QA162.F36   2000
512'.02--dc21   99-053874

British Library Cataloguing-in-Publication Data
A catalogue record for this book is available from the British Library.

Printed in Singapore.

Preface

This textbook is based on the lecture notes of the authors at Wuhan University and Hubei University in courses of abstract algebra for several decades. This volume is prepared for first- or second-year students.

The algebraic courses in curriculums of universities are subject to variation; on the other hand, there are various choices of the materials as a course in abstract algebra, and there are various ways to treat the materials. The present textbook is intended to suit the variations. It contains the fundamental concepts and basic properties on groups, rings, modules and fields; including the interplay between them and other mathematical branches and applied aspects. For example, the residue classes of integers and the Chinese Remainder Theorem appear several times, as concrete models of the abstract notations, and as inspirations of the mathematical ideas as well. In this book there are not many materials which could be regarded as "structure theorems"; for example, in the part on groups only cyclic groups are cleared and classified completely. These are enough, we think, for a course of first stage in algebra. On the other hand, some attentions to training "mathematical maturity" are paid; one of them is that homomorphisms and homomorphism theorems are emphasized throughout. For example, instead of an elementary number-theoretic way (which is left as exercises), the structure of cyclic groups is deduced by means of the homomorphism theorems. Consistently, the structure of single extension fields is deduced in the same way. Though the latter case seems more complicated than the former one, their basic ideas are exactly the same.

v

For prerequisites of this book, only the fundamentals of linear algebra are presumed; in fact, readers will find that not only many notations are enlightened by linear algebra, and many lines of idea in this book come from linear algebra as well. Except the above, the book is self-contained. For example, even for residue classes of integers, we introduce them in detail. Indeed, most of the first three chapters and the first stages of the last two chapters of this book are explained meticulously, so that a beginner can read them by himself without difficulties. The necessary set-theoretical and logical preliminaries are stated in the first chapter, but in a somewhat intuitive way; this is because to get the precise mathematical formulation one has to go into a deep and abstract mathematical branch, called the Mathematical Foundation. However, the decomposition of maps is still explained carefully in that chapter, it is in fact the first version of the later homomorphism theorems.

There are a number of exercises following each sections. Some of the exercises marked with + are fundamental and quoted elsewhere. There are a few exercises marked with * which are somewhat difficult or require more background. Some exercises are followed by hints or remarks which show some keys for them or connections of them with other questions.

This book can be used in the following ways:

- All the materials in this book could serve a course of 144 hours.

- Leaving half of Chapter 4 and half of Chapter 5 as complementary reading, it could be a course of 108 hours.

- Most of Chapter 1, Chapter 2 and Chapter 3 and part of the last two chapters form a course of 72 hours.

Here are a few words about the numbering of definitions, theorems, propositions etc. In order to search easily and quote conveniently, all the statements and formulas are numbered on the left uniformly:

2.3.12 Definition. $\cdots$

which means: chapter 2, section 3, number 12, a definition; and it is quoted in the book as "$\cdots$ by Def.2.3.12 $\cdots$". On the other hand, when an exercise is quoted, "Exer.3" means exercise 3 of the same section; while "Exer.3.2.12(2)" means the exercise of subnumber 2 of number 12 of section 2 of chapter 3.

As usual, □ denotes the end of a proof, or that the proof is omitted.

The first author and the third author were both students of the second author, they were and are supervised by him, and benefited very much from him; they are greatly indebted to the second author.

Any comments and suggestions about this book will warmly be welcomed.

The authors
Luojia Shan, Wuhan
August, 1998.

Contents

A COURSE IN
ALGEBRA

Chapter 1

Preliminaries

1.1 Sets

A *set S* means a collection of some objects such that for any object a we can recognize whether or not a belongs to S. If a belongs to S, we write $a \in S$ and say that a is an *element* of S, or S contains a; otherwise, we write $a \notin S$ and say that a is not an element of S, or S does not contain a.

The set which contains nothing is called *empty set* and denoted by $\emptyset$.

If the elements of a set S can be listed, we can then describe it by listing its elements $S = \{a, b, c, \cdots\}$. A set consisting of finite number of elements can certainly be described in this way; another example which can be described like this is the set of all natural numbers $\mathbf{N} = \{1, 2, \cdots\}$.

However, not all sets can be listed. For example, the elements of the set $[0, 1]$ of all the real numbers r such that $0 \leq r \leq 1$ cannot be listed. To see this, we write any real number r as an infinite decimal; suppose the elements of $[0, 1]$ can be listed: $[0, 1] = \{r_1, r_2, \cdots\}$, then we can find a decimal $s = 0.d_1 d_2 \cdots$ such that the first digit d_1 of s is different from the first digit of r_1, and the second digit d_2 of s is different from the second digit of r_2, and so on; it is clear that $0 \leq s \leq 1$ hence $s \in [0, 1]$; but s is different from any of $r_1, r_2, \cdots$, i.e. $s \notin [0, 1]$; this is a contradiction.

Another way to describe a set S is to specify how to recognize the elements of S, i.e. write it as $S = \{a \mid a \text{ satisfies } ... \}$. For example, the above set $[0,1]$ can be written as

$$[0,1] = \{r \mid r \text{ is a real number and } 0 \le r \le 1\}.$$

The number of elements of a set S is called the *cardinal number* of S, or *cardinality* of S, and denoted by $|S|$. By ∞ we denote the *infinite cardinality*. We write $|S| < \infty$ and say that S is a *finite set* if S contains finitely many elements. In fact, the mathematical formulation of the cardinality of a set initiated the modern set theory; here we just introduce it intuitively.

As shown above, there are two distinct kinds of infinite cardinalities. We say that a set S is *countable* if the elements of S can be listed (i.e. $|S|$ is finite or countably infinite); otherwise, we say that S is *uncountable*. For example, $\mathbf{N}$ is countable; but $[0,1]$ is uncountable.

Let S and T be two sets. We say that S is a *subset* of T, and write $S \subset T$ or $S \subseteq T$, if every element of S is also an element of T. We write $S = T$ if both $S \subset T$ and $T \subset S$, or equivalently, $a \in S$ if and only if $a \in T$. We say that S is a *proper subset* of T and write it as $S \subsetneqq T$ if $S \subset T$ and $S \ne T$.

Let S and T be two sets again. The *union* of S and T is defined as $S \cup T = \{x \mid x \in S \text{ or } x \in T\}$. While the *intersection* of S and T is defined as: $S \cap T = \{x \mid x \in S \text{ and } x \in T\}$. If $S \cap T = \emptyset$, then we say that S and T are *disjoint* and the union $S \cup T$ is a *disjoint union*. Another set constructed from S and T is the difference: $S - T = \{x \mid x \in S \text{ but } x \notin T\}$.

For many sets S_i, $i \in I$, where I is an index set to indicate the sets, the union $\cup_{i \in I} S_i$ and the intersection $\cap_{i \in I} S_i$ can be defined similarly; and we say that the subsets S_i, $i \in I$ are disjoint and $\cup_{i \in I} S_i$ is a disjoint union if $S_i \cap S_j = \emptyset$ for any $i \ne j \in I$.

1.1.1 Proposition. *For any sets S, T and W the following hold:*

$$S \cup T = T \cup S, \quad S \cap T = T \cap S \qquad \text{(commutative law)}$$
$$(S \cup T) \cup W = S \cup (T \cup W)$$
$$(S \cap T) \cap W = S \cap (T \cap W) \qquad \text{(associative law)}$$
$$S \cup S = S, \qquad S \cap S = S \qquad \text{(idempotent law)}$$
$$S \cup (S \cap T) = S, \quad S \cap (S \cup T) = S \qquad \text{(absorption law)}$$

$$W \cup (S \cap T) = (W \cup S) \cap (W \cup T)$$
$$W \cap (S \cup T) = (W \cap S) \cup (W \cap T) \qquad \text{(distributive law)}$$
$$W - (S \cap T) = (W - S) \cup (W - T)$$
$$W - (S \cup T) = (W - S) \cap (W - T) \qquad \text{(De Morgan's law)}$$

Note that all the above formulas can be extended to more than three sets.

We prove the first formula of De Morgan's law as follows, the other formulas are left as exercises. For $a \in W - (S \cap T)$, we have $a \in W$ and $a \notin S \cap T$, the latter means that $a \notin S$ or $a \notin T$; so, $a \in W$ and $a \notin S$, or $a \in W$ and $a \notin T$; in other words, $a \in W - S$ or $a \in W - T$; i.e. $a \in (W - S) \cup (W - T)$. Conversely, for $a \in (W - S) \cup (W - T)$, we have $a \in W - S$ or $a \in W - T$; in the former, $a \in W$ and $a \notin S$ hence $a \notin S \cap T$, i.e. $a \in W - (S \cap T)$; in the latter, the same argument shows $a \in W - (S \cap T)$. $\qquad \square$

Consider a given set X. All the subsets of X form a set

$$\mathcal{P}(X) = \{S \mid S \subset X\},$$

which is called the *power set* of X. Note that $\mathcal{P}(X)$ contains the empty set $\emptyset$ as a member even if X is empty; thus, if $X \neq \emptyset$, then $\mathcal{P}(X)$ contains at least two elements $\emptyset$ and X which are called the trivial subsets of X. On $\mathcal{P}(X)$, besides the operations "$\cup$", "$\cap$", and "$-$" defined above, we have another operation as follows: for $S \in \mathcal{P}(X)$, i.e. $S \subset X$, we denote the difference set $\overline{S} = X - S \in \mathcal{P}$, and call it the *complement* of S (in X). Then for $S, T \in \mathcal{P}$ we have another version of the De Morgan's law:

$$\overline{(S \cup T)} = \overline{S} \cap \overline{T}, \qquad \overline{(S \cap T)} = \overline{S} \cup \overline{T}.$$

Let S and T be sets. The Cartesian product set $S \times T$ of S and T is defined as

$$S \times T = \{(s, t) \mid s \in S \text{ and } t \in T\},$$

note that here it is allowed that $S = T$. A typical example is the real plane $\mathbf{R} \times \mathbf{R} = \{(x, y) \mid x, y \in \mathbf{R}\}$. For sets $S_1, \cdots, S_n$, their Cartesian product set is defined similarly; besides $S_1 \times \cdots \times S_n$, for convenience we also denote it as

$$\prod_{i=1}^{n} S_i = \left\{(s_1, \cdots, s_n) \,\middle|\, s_i \in S_i \text{ for } i = 1, \cdots, n\right\}$$

and we say that S_j is the jth-component of $\prod_{i=1}^{n} S_i$ and also say that

s_j is the jth-component of an element $(s_1, \cdots, s_n)$.

Let S_i, $i \in I$, be sets for an index set I which may be infinite (even uncountable), the Cartesian product set $\prod_{i \in I} S_i$ of them can be defined similarly:

$$\prod_{i \in I} S_i = \Big\{ (s_i)_{i \in I} \,\Big|\, s_i \in S_i \text{ for every } i \in I \Big\},$$

i.e. any element $(s_i)_{i \in I}$ is a collection of s_i's indexed by I such that $s_i \in S_i$; if there is no confusion, we can suppress the subscript to write it as (s_i); s_j is also said to be the jth-component of the element (s_i). A more mathematical formulation for $\prod_{i \in I} S_i$ will be stated in §1.4.

Exercises 1.1

1. Let S, T be sets. Prove:
   1) $S \cup T = (S - T) \cup (T - S) \cup (T \cap S)$ and that the right-hand side is a disjoint union.
   2) $S \subset T$ if and only if $S \cup T = T$ and if and only if $S \cap T = S$.
   3) $S = T$ if and only if $S \cup T = S \cap T$.
2. Prove the formulas in Prop.1.1.1.
3. Let T and S_i, $i \in I$, be sets. Prove:
   1) $T \cup (\cap_{i \in I} S_i) = \cap_{i \in I}(T \cup S_i)$;
   2) $T \cap (\cup_{i \in I} S_i) = \cup_{i \in I}(T \cap S_i)$.
4. Let $S_1, \cdots, S_n$ be finite sets, Prove that $|S_1 \times \cdots \times S_n| = |S_1| \cdots |S_n|$.

1.2   Logic

We just sketch the necessary logical notations used in this book.

There are several ways to combine two statements into a new statement. For example, let P stand for the statement "$s \in S$" and Q stand for the statement "$t \in T$"; then "P or Q" is a new statement which stands for "$s \in S$ or $t \in T$", and it is usually denoted by $P \vee Q$ in logic. Thus $\vee$ is an "operation" for statements.

1.2.1 Notation. Let P and Q be statements. The following are some "operations" for statements:

$P \lor Q$	reads	"P or Q";
$P \land Q$	reads	"P and Q"
$P \implies Q$	reads	"P implies Q";
$P \impliedby Q$	reads	"P is implied by Q";
$P \iff Q$	reads	"P if and only if Q", or "P iff Q" in short;
$\neg P$	reads	"not P" (i.e. the negation of P).

Remark. Here the symbols $\land$, $\implies$ etc. are regarded as composition operators to combine two statements into another statement; they are different from the argument processing for statements. For example, we have an argument like this: "since $a \in S$ and $S \subset T$, we have $a \in T$"; then, for brevity, sometimes we write it as "$a \in S$ and $S \subset T \implies a \in T$". In mathematical logic, this is involved in the question of so-called "language" and "meta-language". But here we just make a remark. The two usages of these symbols in this book can usually be distinguished from the context. For example, the conclusion (1) of the Prop.1.2.2 below is stated as:

"$P \implies Q$" is equivalent to "$(\neg Q) \implies (\neg P)$";

not as:

"$P \implies Q$" $\iff$ "$(\neg Q) \implies (\neg P)$".

In mathematical logic, a statement P is usually called a proposition; but note that this does not mean that the statement is true. So, instead of "proposition", we prefer to use "statement"; and reserve the word "proposition" to express a true statement as in usual mathematics. On the other hand, A statement P, e.g. the statement "$s \in S$", is not necessarily true. We say that P gets value 1 if it is true; otherwise P gets value 0. This value is called the *truth value* of P.

From a general knowledge, we know that, for every truth values of P and Q, a unique truth value of a composed statement is determined. For example, the truth value of $P \lor Q$ is as follows:

P	0	0	1	1
Q	0	1	0	1
$P \lor Q$	0	1	1	1

Such a table is called the *truth value table* of $P \lor Q$.

If two composed statements always take the same truth values, then we say that the two statements are equivalent to each other.

1.2.2 Proposition.   *Let P and Q be statements. Then*

    1) *"$P \implies Q$" is equivalent to "$(\neg Q) \implies (\neg P)$".*

    2) *"$\neg(P \vee Q)$" is equivalent to "$(\neg Q) \wedge (\neg P)$".*

    3) *"$\neg(P \wedge Q)$" is equivalent to "$(\neg Q) \vee (\neg P)$".*

One could find their explanations in the common knowledge. Here we prove the first one by the following truth value table:

P	Q	$\neg P$	$\neg Q$	$P \implies Q$	$(\neg Q) \implies (\neg P)$
0	0	1	1	1	1
0	1	1	0	1	1
1	0	0	1	0	0
1	1	0	0	1	1

The other two conclusions can be proved in the same way.   □

Example. Consider the statement "$s \in S$ and $t \in T$"; by Prop. 1.2.2(3), its negation is "$s \notin S$ or $t \notin T$". Therefore, if we want to prove a proposition like this "If $w \in W$ then $s \in S$ and $t \in T$", by Prop.1.2.2(1), it is enough to prove "If $s \notin S$ or $t \notin T$ then $w \notin W$". In practice, such arguments appear usually in the version of proving by contradiction.

In some statements the subjects are allowed to vary within some ranges. For example, in the statement "In the Cartesian plane there is a triangle Δ such that Δ is regular", its subject "triangle Δ" is a variable within all the triangles in Cartesian plane. Let $P(\Delta)$ stand for the statement "triangle Δ is regular", where Δ is a variable subject and P stands for the predicate "is regular"; let T be the set of all triangles in the Cartesian plane. Then the above statement can be restated in logical notation as follows: "$(\exists \Delta \in T)P(\Delta)$", where the symbol "$\exists$" is called the *existential quantifier*. We know that this is a true statement. Another quantifier is the *universal quantifier* "$\forall$", with which we can construct another statement "$(\forall \Delta \in T)P(\Delta)$", which is read as "for all Δ in T, Δ is regular"; of course, this is a false statement.

1.2.3 Notation. A statement $P(x)$ with a variable subject x is called a *predicate*; with the *quantifiers* $\forall$ and $\exists$, we can construct two statements:

    $(\forall x)P(x)$ reads: "for all x we have $P(x)$";

    $(\exists x)P(x)$ reads: "there exists an x such that $P(x)$".

Sometimes we use the symbol "∃1" to stand for "there exists a unique".

From the general knowledges the following hold:

1.2.4 Proposition.

1) *"¬((∃x)P(x))" is equivalent to "(∀x)(¬P(x))";*

2) *"¬((∀x)P(x))" is equivalent to "(∃x)(¬P(x))".* □

Example. Let S, T be sets. Consider the statement "If $S \cap T = \emptyset$, then for all x, $x \notin S$ or $x \notin T$"; by Prop.1.2.1(1) and the above (2) we know that it is equivalent to the statement "If there exists x such that $x \in T$ and $x \in S$, then $S \cap T \neq \emptyset$"; the latter is clearly true, so we have that the former statement is true.

Exercises 1.2

1. Tabulate the truth value tables of "$\neg P$" and "$P \wedge Q$".
2. Tabulate the truth value tables of "$P \implies Q$" and "$P \iff Q$".
3. Prove Prop.1.2.2.
4. Let P and Q be statements. Prove:
   1) $P \implies Q$ is equivalent to $(\neg P) \vee Q$.
   2) $\neg(P \implies Q)$ is equivalent to $P \wedge (\neg Q)$.

1.3   Relations

In our life, a relation in a set means that for every two members a and b (not necessarily distinct) in the set we can determine whether or not a is related to b.

In mathematics, a *relation* in a set S means a subset R of the Cartesian product $S \times S$; for $a, b \in S$, we say that a has the relation R to b if $(a, b) \in R$, and write aRb in this case. Since relations in a set S are the subsets of $S \times S$, we can speak of the inclusion and equality for relations; in fact, for relations R_1 and R_2 in S, we say that R_1 is finer than R_2 if $R_1 \subset R_2$. Note that, instead of the capital letters, the symbol $\sim$, $<$ etc. are sometimes used for relations, and write $a \sim b$ instead of $(a, b) \in \sim$.

Example. The usual relation "<" ("less than") in the set $\mathbf{R}$ of all real numbers is just the subset of $\mathbf{R} \times \mathbf{R}$ formed by the upper open half plane above the line $y = x$.

Example. In arithmetic, given an integer m, the *congruence relation* modulo m "$\equiv$" in the set $\mathbf{Z}$ of integers is defined by that, for $a, b \in \mathbf{Z}$, $a \equiv b$ iff $m|a - b$; as we known, such congreunce relation is usually written as $a \equiv b \pmod{m}$ to indicate the modulo number m.

1.3.1 Definition. Let "$\prec$" be a relation on a set S. We say that "$\prec$" is a *partial order relation* if the following three conditions hold:

(R1) *reflexity*: $a \prec a$, $\forall a \in S$;

(R2) *transitivity*: $(a \prec b$ and $b \prec c) \implies (a \prec c)$, $\forall\, a, b, c \in S$;

(R3a) *anti-symmetry*: $(a \prec b$ and $b \prec a) \implies (a = c)$, $\forall\, a, b \in S$.

Note that the relation "<" in $\mathbf{R}$ mentioned above does not satisfy the reflexity. But the usual relation "$\leq$" ("not greater than") in $\mathbf{R}$ is a partial order relation.

1.3.2 Definition. Let "$\sim$" be a relation in a set S. We say that $\sim$ is an *equivalence relation* if the above (R1) and (R2) and the following (R3) hold:

(R3) *symmetry*: $a \sim b \implies b \sim a$, $\forall\, a, b \in S$.

Further, let "$\sim$" be an equivalence relation in a set S. For any $a \in S$, the following subset is called the *equivalence class* determined by a with respect to the equivalence relation "$\sim$":

$$\bar{a}_\sim = \{x \mid x \in S \text{ and } x \sim a\} \subset S,$$

and a is said to be a *representative* of $\bar{a}_\sim$ (cf. Prop.1.3.3 below). We write $\bar{a} = \bar{a}_\sim$ and refer to it as the equivalence class of a if there is no confusion. All the equivalence classes $\bar{a}$ form a subset of the power set $\mathcal{P}(S)$ of S, called the *quotient set* of S by the equivalence relation "$\sim$", and denoted by $S/\sim$, or by $\overline{S}$ for short:

$$S/\sim\, = \overline{S} = \{\bar{a} \mid a \in S\} \subset \mathcal{P}(S).$$

Example (see Exer.1). It is easy to check that the congruence relation modulo m in $\mathbf{Z}$ shown above is an equivalence relation. More concretely, consider $m = 2$ and the congruence relation modulo 2 in $\mathbf{Z}$. The equivalence class $\bar{1} = 1 + 2\mathbf{Z} = \{1 + 2n \mid n \in \mathbf{Z}\}$, which is called the *congruence class* of 1 (modulo 2). Though for every $a \in \mathbf{Z}$ we have a congruence class $\bar{a}$, there are only two congruence classes modulo 2 altogether: $\bar{0} = 2\mathbf{Z} = \{2n \mid n \in \mathbf{Z}\}$ and $\bar{1} = 1 + 2\mathbf{Z}$; in fact, for any $1 + 2n \in 1 + 2\mathbf{Z}$ it is easy to check that $\overline{1 + 2n} = \bar{1}$; and it is similar for $\bar{0}$. Further, it is an obvious fact that $\bar{0} \cup \bar{1} = \mathbf{Z}$ and $\bar{0} \cap \bar{1} = \emptyset$.

Let "$\sim$" be an equivalence relation in a set S. Since $a \sim a$ by the reflexity (R1), the subset $\bar{a}$ contains a itself; hence the union of all equivalence classes covers the whole S, i.e. $\bigcup_{\bar{a} \in \overline{S}} \bar{a} = S$. Of course, $\bar{a}$ may contain more than one element. Let $a' \in \bar{a}$; we have $a' \sim a$ by Def.1.3.2, hence $a \sim a'$ by the symmetry (R3); so $a \in \bar{a'}$ by Def.1.3.2 again. For any $x \in \bar{a}$ we have $x \sim a$ still by Def.1.3.2, thus $x \sim a'$ by transitivity (R3), hence $x \in \bar{a'}$; this shows that $\bar{a} \subset \bar{a'}$. And $\bar{a'} \subset \bar{a}$ in the same way since $a \in \bar{a'}$ as proved above. Thus we get the following fact which suggests the reasonability of calling any element of $\bar{a}$ a *representative* of the equivalence class, see Def.1.3.2 above.

1.3.3 Proposition. *Let "$\sim$" be an equivalence relation in a set S and $a \in S$. Then $a \in \bar{a}$, and $\bar{a} = \bar{a'}$ for any $a' \in \bar{a}$, or equivalently, for any $a' \sim a$.* $\qquad\square$

As an immediate consequence, any two equivalence classes $\bar{a}$ and $\bar{b}$ either coincide with each other or are disjoint (in fact, if $\bar{a} \cap \bar{b} \neq \emptyset$, then there is c contained in both $\bar{a}$ and $\bar{b}$, hence by Prop.1.3.3 we have $\bar{a} = \bar{c} = \bar{b}$); in particular, $S = \bigcup_{\bar{a} \in \overline{S}} \bar{a}$ is a disjoint union.

1.3.4 Definition. Let S be a set and $\mathcal{T} \subset \mathcal{P}(S)$ be a subset of the power set of S. We say that $\mathcal{T}$ is a *partition* of the set S if $\emptyset \notin \mathcal{T}$ and all the members in $\mathcal{T}$ are disjoint and the union $\bigcup_{T \in \mathcal{T}} T = S$. Further, in this case, we have a relation $\sim_{\mathcal{T}}$ determined by $\mathcal{T}$ as follows ($\forall x, y \in S$): $x \sim_{\mathcal{T}} y$ iff $\exists T \in \mathcal{T}$ such that both $x \in T$ and $y \in T$ (and then such T must be unique since $\mathcal{T}$ is a partition).

In fact, the partitions of a set correspond one-to-one to the equivalence relations in the set; any quotient set of a set is just a partition of the set.

1.3.5 Theorem. *Let S be a set.*

1) *If "$\sim$" is an equivalence relation in S, then the quotient set $\overline{S}$ of S by $\sim$ is a partition of S, and the relation $(\sim_{\overline{S}}) = (\sim)$.*

2) *If $\mathcal{T}$ is a partition of S, then the relation "$\sim_{\mathcal{T}}$" determined by $\mathcal{T}$ is an equivalence relation, and the quotient set $\overline{S} = \mathcal{T}$.*

Proof. For (1) we have shown that $\overline{S}$ is a partition of S in the above. Let $a, b \in S$. Assume that $a \sim b$; then both a and b belong to $\overline{b}$ by Def.1.3.2, thus $a \sim_{\overline{S}} b$ by Def.1.3.4. Conversely, assume that $a \sim_{\overline{S}} b$; by Def.1.3.4 there is a $\overline{c} \in \overline{S}$ such that $a \in \overline{c}$ and $b \in \overline{c}$, hence $a \sim c$ and $b \sim c$ by Def.1.3.2; then, by symmetry (R3) and transitivity (R2) of the relation $\sim$, we have that $a \sim b$. In short, $(\sim_{\overline{S}}) = (\sim)$.

For (2), the reflexity and the symmetry for $\sim_{\mathcal{T}}$ are obvious. If $a \sim_{\mathcal{T}} b$ and $b \sim_{\mathcal{T}} c$, then by Def.1.3.4 there are $T_1, T_2 \in \mathcal{T}$ such that $a, b \in T_1$ and $b, c \in T_2$; so $b \in T_1 \cap T_2$ hence $T_1 \cap T_2 \neq \emptyset$; thus $T_1 = T_2$, since $\mathcal{T}$ is a partition (hence its distinct two members must be disjoint); thus a and c belong to the same T_1, i.e. $a \sim_{\mathcal{T}} c$; the transitivity for $\sim_{\mathcal{T}}$ holds. Therefore $\sim_{\mathcal{T}}$ is an equivalence relation. Now, let $\overline{a} \in \overline{S}$; since $\bigcup_{T \in \mathcal{T}} T = S$, there is a $T \in \mathcal{T}$ such that $a \in T$; by Def.1.3.2 and Def.1.3.4, any $b \in \overline{a}$ must belong to T; i.e. $\overline{a} \subset T$; on the other hand, $t \sim_{\mathcal{T}} a$ for any $t \in T$ by Def.1.3.4, hence $t \in \overline{a}$; thus $\overline{a} = T$. Conversely, let $T \in \mathcal{T}$; then there is an element $a \in T$ since $T \neq \emptyset$ by Def.1.3.4; and the above argument has shown that $a \in T$ implies $\overline{a} = T$. Summarizing the above, we see that $\overline{S} = \mathcal{T}$. $\qquad \square$

Exercises 1.3

+1. Let $0 \neq m \in \mathbf{Z}$. For any $a, b \in \mathbf{Z}$, we write $a \equiv b \pmod{m}$ if $m | a - b$. Then "$\equiv$" is an equivalence relation in $\mathbf{Z}$ and the equivalence class $[a]_m = a + m\mathbf{Z} = \{a + mk \mid k \in \mathbf{Z}\}$; and there are m equivalence classes altogether. (*Remark*: Instead of $\overline{a}$, later on we denote the congruence class of a modulo m by $[a]_m$ to emphasize "modulo m" and call it a *residue class* modulo m; we can further shorten it as $[a]$ if there is no confusion. And, instead of $\mathbf{Z}/\equiv$ and $\overline{\mathbf{Z}}$, we denote the quotient set by this congruence relation by $\mathbf{Z}/\langle m \rangle$, or $\mathbf{Z}_m$ for brevity. These notations will appear more frequently later.)

2. Let S be a set and $\mathcal{W} \subset \mathcal{P}(S)$. If $\emptyset \notin \mathcal{W}$ and for any $x \in S$ there is a unique $W \in \mathcal{W}$ such that $x \in W$, then $\mathcal{W}$ is a partition of S.

3. Let S be a set and $\mathcal{W} \subset \mathcal{P}(S)$. Define a relation $R \subset S \times S$ as follows: $(a, b) \in R$ iff $(\exists W \in \mathcal{W})(a \in W \wedge b \in W)$. Prove:

   1) The relation R is symmetric.

   2) $\bigcup_{W \in \mathcal{W}} W = S$ iff the relation R is reflexive.

   3) If all the members in $\mathcal{W}$ are disjoint, then the relation R is transitive. Give an example to show that the inverse is not necessarily true.

4. Let S be a set and $R \subset S \times S$ be a relation in S. For every $a \in S$ define a subset of S as $W_a = \{x \mid x \in S \text{ and } (x,a) \in R\}$; let $\mathcal{W} = \{W_a \mid a \in S\}$. Prove:

   1) $a \in W_a$ for every $a \in S$ iff the relation R is reflexive.

   2) If the relation R is both symmetric and transitive, then all members of $\mathcal{W}$ are disjoint.

+5. Let S be a set. A partition $\mathcal{T}_1$ is said to be *finer* than a partition $\mathcal{T}_2$ if $\forall T_1 \in \mathcal{T}_1$ there is a $T_2 \in \mathcal{T}_2$ such that $T_1 \subset T_2$. Prove that $\mathcal{T}_1$ is finer than $\mathcal{T}_2$ iff $(\sim_{\mathcal{T}_1}) \subset (\sim_{\mathcal{T}_2})$.

+6. Let S be a set and $\mathcal{T}_1, \mathcal{T}_2$ be two partitions of S. Assume that $\mathcal{T}_1$ is finer than $\mathcal{T}_2$. In the quotient set $\overline{S}_{\mathcal{T}_1}$ (whose members are the subsets $\overline{a}_{\mathcal{T}_1}$) define: $\overline{a}_{\mathcal{T}_1} \sim_2 \overline{b}_{\mathcal{T}_1}$ iff a and b belong to the same member of $\mathcal{T}_2$. Then "$\sim_2$" is an equivalence relation in $\overline{S}_{\mathcal{T}_1}$ and the quotient set $(\overline{S}_{\mathcal{T}_1})/\sim_2 = \overline{S}_{\mathcal{T}_2}$.

1.4 Maps

Let S and T be sets (not necessarily distinct). A subset f of $S \times T$ is said to be a *map* from S to T if for every element $s \in S$ there exists a unique element $t \in T$ such that $(s,t) \in f$. In other words, a map f from S to T means a way in which to every $s \in S$ there corresponds a unique $t \in T$. A map is also called a "mapping", or a "function".

Let f be a map from a set S to a set T. S is called the *domain* of f while T is called the *co-domain* of f. We usually describe it as $f : S \to T$, or $S \xrightarrow{f} T$ in short, or more precisely as follows:

$$f : \ S \longrightarrow T, \qquad s \longmapsto t \ (= f(s))$$

where $s \mapsto t$ is read as "element s is mapped to t (by the map)"; and, since t is the unique one corresponding to s, we denote it by $t = f(s)$ and call it the *image* of s under f; further, the subset $\mathrm{Im}(f) = \{f(s) \mid s \in S\} \subset T$ is called the image (or range) of f. For any subset $S' \subset S$ we denote $f(S') = \{f(x) \mid x \in S'\}$; in particular, $\mathrm{Im}(f) = f(S)$. On the other hand, every $t \in T$ determines a subset $f^{-1}(t) = \{s \mid s \in S \text{ and } f(s) = t\}$ of S which is called the *full inverse image* of t under f, where the attribute "full" is to distinguish it from any single $s \in f^{-1}(t)$, such an s is called an *inverse image* of t under f. And for any subset $T' \subset T$ we denote $f^{-1}(T') = \bigcup_{y \in T'} f^{-1}(y)$ and call it the full inverse image of T'. Note that $f^{-1}(T')$ may be empty even if T' is not empty. However, it is always true that $f^{-1}(T) = S$, since $s \in f^{-1}(f(s))$ $\forall s \in S$.

Two maps f and g from a set S to a set T are said to be equal and written as $f = g : S \to T$ if $f(s) = g(s)$ $\forall s \in S$. Note that this means $f = g$ as subsets of $S \times T$. Of course, we cannot compare two maps which have different domains or different co-domains.

We list several common maps.

For any set S, $\text{id}_S : S \to S, s \mapsto s$ is a map from S to S; i.e. $\text{id}_S(s) = s$ $\forall s \in S$. So we call id_S the *identity map* of S; sometimes, we denote it by 1_S, or even by 1 if there is no confusion.

For Cartesian product set $S_1 \times \cdots \times S_n$ of sets $S_1, \cdots, S_n$, given $1 \le i \le n$, $\rho_i : S_1 \times \cdots \times S_n \to S_i, (s_1, \cdots, s_n) \mapsto s_i$ is a map, which is called the *projection* from the Cartesian product to its ith-component S_i.

1.4.1 Example. Consider the set $\mathbf{Z}$ of integers and the set $\mathbf{Z}_m$ of congruence classes $[a]_m$ modulo m, see Exer.1.3.1. Then

$$\sigma : \ \mathbf{Z} \longrightarrow \mathbf{Z}_m, \qquad a \longmapsto [a]_m$$

is clearly a map; and $\sigma^{-1}([a]_m) = \{a + km \mid k \in \mathbf{Z}\}$. Remember that $[a]_m = \{a + km \mid k \in \mathbf{Z}\}$; but note that here $[a]_m$ is an element of the co-domain $\mathbf{Z}_m$ while $\{a + km \mid k \in \mathbf{Z}\}$ is regarded as a subset of the domain $\mathbf{Z}$.

But, we should define a map carefully. For example, the following is not a map

$$f : \ \mathbf{Z}/\langle 3 \rangle \longrightarrow \mathbf{Z}/\langle 2 \rangle, \qquad [a]_3 \longmapsto [a]_2 \ ;$$

for, e.g. $[1]_3 = [4]_3$ in the domain but $[1]_2 \ne [4]_2$ in the co-domain; so, in the above expression, two different elements $[1]_2$ and $[4]_2$ in the co-domain correspond to the same element $[1]_3$ of the domain; this contradicts the definition of maps. In mathematics, we say that this map *is not well-defined*. In Exer.1, we can see when such a map is indeed well-defined.

Let S, T and W be sets, let $f : S \to T$ and $g : T \to W$ be maps. To any $s \in S$, there corresponds a unique $t = f(s) \in T$ under f; further, to $t = f(s)$ there corresponds a unique $w = g(t) = g(f(s)) \in W$. In this way, we get a map, denoted by $g \circ f$, as follows:

$$g \circ f : \ S \longrightarrow W, \qquad s \longmapsto g(f(s)),$$

which is called the *composition map* of f and g. Sometimes, suppressing the symbol "∘", we denote fg for $f \circ g$ in short. Note that $(gf)(s) = g(f(s))$ for $s \in S$ by the above definition. If $d : S \to W$ is another map, we can illustrate the situation by a diagram:

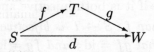

We say that this diagram is *commutative* if $gf = d$.

Moreover, let $h : W \to X$ be a map too, it is clear that we can define the composition of the three maps as follows:

$$h \circ g \circ f : \ S \longrightarrow X, \qquad s \longmapsto h(g(f(s))).$$

Since $\forall s \in S$ we have

$$
\begin{aligned}
((h \circ g) \circ f)(s) &= (h \circ g)(f(s)) = h(g(f(s))) \\
&= h((g \circ f)(s)) = (h \circ (g \circ f))(s),
\end{aligned}
$$

the following holds:

(1.4.2) $\qquad (h \circ g) \circ f = h \circ (g \circ f)$ $\qquad$ (associative law).

It is easy to see that the composition $f_n \circ \cdots \circ f_2 \circ f_1$ can be defined for finitely many maps $f_1, f_2, \cdots, f_n$ provided the co-domain of any f_i of them coincides with the domain of the next one f_{i+1} (in particular, this is the case if all f_i are maps from S to S itself); i.e.

(1.4.3) $\qquad (f_n \circ \cdots \circ f_2 \circ f_1)(s) = f_n(\cdots(f_2(f_1(s)))\cdots), \qquad \forall s \in S.$

And, similarly to the above for three maps, we can check that, inserting brackets to $f_n \circ \cdots \circ f_2 \circ f_1$ in any way provided they make sense, the result map is always equal to the whole composition $f_n \circ \cdots \circ f_2 \circ f_1$ defined in (1.4.3); i.e. the result is independent of the way of inserting brackets. Such a property is said to be the so-called *generalized associative law*. In fact, it is true in general that the generalized associative law holds provided the associative law (for three members) holds. We will formulate this fact precisely in §2.1.

We remark an easy fact. For any map $f : S \to T$ it is clear that

(1.4.4) $\qquad \mathrm{id}_T \circ f = f = f \circ \mathrm{id}_S$ $\qquad$ (identity law)

1.4.5 Definition. Let $f : S \to T$ be a map. Then

f is said to be *injective* (or *one-to-one*) if $f(s) = f(s') \implies s = s'$ for any $s, s' \in S$;

f is said to be *surjective* (or *onto*) if $\forall t \in T \; \exists s \in S$ such that $f(s) = t$ (obviously, this is equivalent to $f(S) = T$);

f is said to be *bijective* if it is both injective and surjective.

1.4.6 Proposition.   *A map $f : S \to T$ is bijective if and only if there is a map $g : T \to S$ such that $gf = \mathrm{id}_S$ and $fg = \mathrm{id}_T$; and such a g is unique if this is the case.*

Proof. Assume that f is bijective. For any $t \in T$, by the surjectivity there is an $s \in S$ such that $f(s) = t$, and by the injectivity such s is unique; so we can define a map $g : T \to S$ by assigning the unique s to $t \in T$; i.e. $g(t)$ is the unique element of S such that $f(g(t)) = t$. In particular, $fg = \mathrm{id}_T$. Then, for any $s \in S$, by the associative law we have that $f(gf(s)) = (fg)(f(s)) = \mathrm{id}_T(f(s)) = f(s)$; but f is injective, so $gf(s) = s$; i.e. $gf = \mathrm{id}_S$.

Conversely, assume that the map $g : T \to S$ with $gf = \mathrm{id}_S$ and $fg = \mathrm{id}_T$ exists. For any $s, s' \in S$, if $f(s) = f(s')$, then, mapping both sides by g, we have $s = \mathrm{id}_S = g(f(s)) = g(f(s')) = \mathrm{id}_S(s') = s'$; thus f is injective. On the other hand, for any $t \in T$, we have $t = \mathrm{id}_T(t) = f(g(t))$ and $g(t) \in S$; in other words, f is surjective. So f is bijective.

At last, in the first paragraph of the proof we have seen that the map $g : T \to S$ with $gf = \mathrm{id}_S$ and $fg = \mathrm{id}_T$ must be unique. However, we mention another proof for the uniqueness of g. Suppose $g' : T \to S$ is a map satisfying both $g'f = \mathrm{id}_S$ and $fg' = \mathrm{id}_T$; then

$$g' = g' \circ \mathrm{id}_T = g'(fg) = (g'f)g = \mathrm{id}_S \circ g = g. \qquad \square$$

1.4.7 Definition. Let $f : S \to T$ be a map. If there exists a map $g : T \to S$ such that $gf = \mathrm{id}_S$ and $fg = \mathrm{id}_T$, then f is said to be *invertible* and g is called the *inverse map* of f; and such g is unique by Proposition 1.4.6, hence we denote it by f^{-1}. It is clear that, if f is invertible, then so is f^{-1} and $(f^{-1})^{-1} = f$.

From Prop.1.4.6 we see that the bijectivity and invertibility for a map are equivalent.

Recall that we denote the inverse image of a subset $T' \subset T$ by $f^{-1}(T')$ even if f is not invertible; but note that "f^{-1}" in the notation

$f^{-1}(T')$ is only a formal symbol and does not mean its invertibility. Once f is really invertible, then the subset $f^{-1}(T')$ coincides indeed with the image of T' under the inverse map f^{-1}.

Return to Example 1.4.1, which is in fact of general interest.

Consider any set S and any equivalence relation "$\sim$" in S, and let $\overline{S}$ be the quotient set of S by $\sim$ as in Def.1.3.2. Then the following is a well-defined map (easy to check):

$$\sigma: \ S \longrightarrow \overline{S}, \qquad s \longmapsto \overline{s}$$

which is called the *natural* or *canonical* map (associated with the equivalence relation "$\sim$"); and it is also easy to check that $\sigma^{-1}(\overline{a}) = \{x \mid x \in S \text{ and } x \sim a\}$. These show us that all the full inverse images of the elements of the co-domain $\overline{S}$ form a partition of the domain S. This is not an accidental phenomena.

Let S and T be sets and $f : S \to T$ be a map. As we shown above, may be $f^{-1}(t) = \emptyset$ for $t \in T$; so the subsets $f^{-1}(t)$, $\forall t \in T$, may not form a partition of S. But this is the only obstruction; it is easy to see that, deleting the empty ones, we can collect the $f^{-1}(t)$'s to form a partition of S. By Theorem 1.3.5, an alternative way to do this is to consider the relation in the domain determined by the map.

1.4.8 Theorem (Fundamental Decomposition of Maps). *Let* $f : S \to T$ *be a map from a set* S *to a set* T. *For* $s, s' \in S$ *we write* $s \sim_f s'$ *if* $f(s) = f(s')$. *Then*

1) "$\sim_f$" *is an equivalence relation in* S*; let* $\sigma : S \to \overline{S}$ *denote the natural map from* S *to the quotient set* $\overline{S}$ *by "*$\sim_f$*";*

2) *there exists a unique map* $\overline{f} : \overline{S} \to T$ *such that* $f = \overline{f} \circ \sigma$, *and* $\overline{f}$ *is injective; moreover,* $\overline{f}$ *is bijective provided* f *is surjective.*

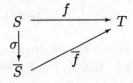

Proof. (1). This is obvious by the definition of "$\sim_f$"; we just remark that the equivalence class $\overline{s}$ for $s \in S$ is just the full inverse image $f^{-1}(f(s))$ of $f(s) \in T$.

(2). For any $\bar{s} \in \overline{S}$, choosing a representative $s \in \bar{s}$ (cf. Prop.1.3.3), we assign $f(s) \in T$ to $\bar{s}$. Suppose $s' \in \bar{s}$ is another representative of the equivalence class, then $s' \sim_f s$, i.e. $f(s') = f(s)$; in other words, $f(s)$ is independent of the choice of representatives. Thus we have a map:

(1.4.9) $\overline{f}: \overline{S} \longrightarrow T,$ $\qquad \bar{s} \longmapsto f(s)$ $\qquad$ (i.e. $\overline{f}(\bar{s}) = f(s)\ \forall \bar{s} \in \overline{S}$).
Then it is clear that

$$(\overline{f} \circ \sigma)(s) = \overline{f}(\sigma(s)) = \overline{f}(\bar{s}) = f(s);$$

i.e. $f = \overline{f} \circ \sigma$. Now suppose $\widetilde{f}: \overline{S} \to T$ is also a map such that $f = \widetilde{f} \circ \sigma$; then for any $\bar{s} \in \overline{S}$, we have

$$\widetilde{f}(\bar{s}) = \widetilde{f}(\sigma(s)) = (\widetilde{f} \circ \sigma)(s) = f(s) = \overline{f}(\bar{s}) .$$

Thus $\widetilde{f} = \overline{f}$; the uniqueness of $\overline{f}$ is proved.

Let $\bar{a}, \bar{b} \in \overline{S}$ such that $\overline{f}(\bar{a}) = \overline{f}(\bar{b})$; then, by the definition (1.4.9), we see that $f(a) = f(b)$ hence $a \sim_f b$ by the definition of "$\sim_f$"; so $\bar{a} = \bar{b}$ by Prop.1.3.3; i.e. $\overline{f}$ is injective. At last, assume f is surjective; then for any $t \in T$ there is an $s \in S$ such that $f(s) = t$; by (1.4.9) again, $\overline{f}(\bar{s}) = f(s) = t$; thus $\overline{f}$ is surjective hence bijective. $\qquad\qquad\square$

1.4.10 Remark. We usually refer to the above $\overline{f}$ as the map *induced by f*. If f is not surjective, recall that $\mathrm{Im}(f) \subset T$ and let $\iota : \mathrm{Im}(f) \to T$ denote the inclusion map; then the induced map $\overline{f} : \overline{S} \to T$ can be written as a composition $\overline{S} \overset{\widetilde{f}}{\to} \mathrm{Im}(f) \overset{\iota}{\to} T$, where $\widetilde{f} : \overline{S} \to \mathrm{Im}(f)$ is always bijective, and the theorem can be described as another commutative diagram:

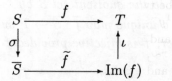

At last, we give a mathematical description of the Cartesian products of sets introduced in §1.1. Let S_i, $i \in I$, be sets indexed by an index set I. The Cartesian product $\prod_{i \in I} S_i$ of the sets S_i, $i \in I$, is the set of all the maps $f : I \to \cup_{i \in I} S_i$ such that $f(i) \in S_i$. Rewrite the image $f(i)$ as $f_i \in S_i$, the collection $(f_i)_{i \in I}$ is just the member of $\prod_{i \in I} S_i$ described in §1.1.

Exercises 1.4

1. The following is a well-defined map iff $n|m$:
$$\mathbf{Z}_m \longrightarrow \mathbf{Z}_n, \qquad [a]_m \longmapsto [a]_n.$$

2. Let S and T be sets. By T^S we denote the set of all maps from S to T. Assume $|S| = m$ and $|T| = n$. Calculate $|T^S|$.

3. Let $f : S \to T$ be a map. Prove:
   1) f is injective iff $|f^{-1}(t)| \leq 1 \ \ \forall t \in T$.
   2) f is surjective iff $|f^{-1}(t)| \geq 1 \ \ \forall t \in T$.
   3) f is bijective iff $|f^{-1}(t)| = 1 \ \ \forall t \in T$.

4. Let $S \xrightarrow{f} T \xrightarrow{g} W$ be maps. Prove:
   1) $(gf)^{-1}(w) = f^{-1}(g^{-1}(w))$ for $w \in W$.
   2) gf is injective (surjective, bijective resp.) if both f and g are injective (surjective, bijective resp.).
   3) f is injective if gf is injective.
   4) g is surjective if gf is surjective.

+5. Let $f : S \to T$ be a map. Assume $|S| = |T| < \infty$; prove that f is bijective if f is injective or f is surjective. Give examples to show that the conclusion is not true if $|S| = |T| = \infty$.

6. Let $f : S \to T$ be a map. Prove:
   1) f is injective iff there is a map $g : T \to S$ such that $gf = \mathrm{id}_S$.
   2) f is surjective iff there is a map $g : T \to S$ such that $fg = \mathrm{id}_T$.

7. Assume that $f : S \to T$ is a bijective map, and $g : T \to S$ is a map. Prove:
   1) if $gf = \mathrm{id}_S$, then $g = f^{-1}$.
   2) if $fg = \mathrm{id}_T$, then $g = f^{-1}$.

+8. Let $S \xrightarrow{f} T \xrightarrow{g} W$ be two invertible maps. Prove that gf is also invertible and $(gf)^{-1} = f^{-1}g^{-1}$.

+9. Let $f : S \to T$ and $g : S \to W$ be maps of sets S, T and W, and g be surjective; let "$\sim_f$" and "$\sim_g$" resp. be the equivalence relations in S determined by f and g resp. Then there exists a map $h : W \to T$ such that $f = h \circ g$ if and only if $(\sim_g) \subset (\sim_f)$; Further, if this is the case, such h is unique, and it is bijective provided f is surjective; what about the relation "$\sim_h$" determined by h?

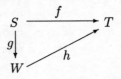

(Hint: let $[a]_f$ denote the equivalence class in S determined by f, etc.; for $w \in W$, similar to (1.4.9), $h(w) = f(a)$ for $a \in g^{-1}(w)$; $w_1 \sim_h w_2$ iff $g^{-1}(w_1)$ and $g^{-1}(w_2)$ belong to one and the same $[a]_f$, compare with Exer.1.3.6.)

1.5   Zorn's Lemma

1.5.1 Definition. If S is a set with a partial order relation $\prec$, then $(S, \prec)$ is called a *partial order set* (or *poset* for short); we also say that S is a poset, if the partial order relation $\prec$ is specified by context. Further $(S, \prec)$ is called a *total order set* if it is a poset and any two elements $a, b \in S$ are *comparable*, i.e. either $a \prec b$ or $b \prec a$.

Assume $(S, \prec)$ is a poset. An element $s_0 \in S$ is said to be *maximal* if for any $s \in S$ such that $s_0 \prec s$ we have $s_0 = s$. The *minimal elements* are defined similarly. A subset $T \subset S$ is said to be a *chain* if $(T, \prec)$ is a total set. An element $u \in S$ is called an *upper bound* of a subset T if $t \prec u$ for all $t \in T$

In this book the following statement is adopted as an axiom.

1.5.2 Zorn's Lemma. *Let $(S, \prec)$ be a poset. If any chain of S has an upper bound, then S has a maximal element.* $\qquad\square$

Chapter 2

Groups

2.1 Transformations and Permutations

We are familiar with mathematical objects such as $\mathbf{Z}$ of integers, $\mathbf{R}$ of real numbers etc. For example, $\mathbf{Z}$ is a set, and there are several arithmetic operations on it as well, e.g. addition, multiplication etc., some laws such as associativity, commutativity and distributivity hold.

In fact, there are some objects which we are also familiar with, but they do not consist of numbers, and they do have operations but not "arithmetic operations".

Let X be a set. Any map from X to itself is called a *transformation* of X. Let $\text{Tran}(X)$ denote the set of all transformations of X. For any $\alpha, \beta \in \text{Tran}(X)$, from §1.4 we know that the composition $\beta \circ \alpha$ is a map from X to itself; i.e. $\beta \circ \alpha \in \text{Tran}(X)$. So the composition "$\circ$" is an operation on the set $\text{Tran}(X)$. Before going on, we state what is an algebraic operation on a set in general.

2.1.1 Definition. An *algebraic operation* "$\circ$" on a set T means a map $\circ : T \times T \to T$; for $(a, b) \in T \times T$ the image $\circ(a, b) \in T$ is called the result of the operation of elements a and b; and, according to our convention, instead of $\circ(a, b)$ the result is usually denoted by $a \circ b$. An algebraic operation is also called an operation for brevity. Moreover, for an operation $\circ$, we say that

— "$\circ$" is *associative* if $(a \circ b) \circ c = a \circ (b \circ c)$, $\forall a, b, c \in T$;

19

—— "∘" is *commutative* if $a \circ b = b \circ a$, $\forall a, b \in T$;

and, if two elements a and b satisfy $a \circ b = b \circ a$, then we just say that a and b *commute* (with respect to "∘"). The "*distributive law*" is stated for two operations; let "$\oplus$" be another operation on T, we say that

—— "∘" is left distributive for "$\oplus$" if $a \circ (b \oplus c) = (a \oplus b) \circ (a \oplus c)$, $\forall a, b, c \in T$;

—— "∘" is right distributive for "$\oplus$" if $(b \oplus c) \circ a = (b \oplus a) \circ (c \oplus a)$, $\forall a, b, c \in T$.

The symbols representing operations are varied, e.g. "+", "×", "·" etc.; in addition, as in §1.4, the composition of two transformations f and g can be denoted by gf with "∘" omitted. We remark that, in a more general formulation (cf. Chapter 7), the above is in fact called a *binary operation* on a set.

2.1.2 Remark. Return to the set $\mathrm{Tran}(X)$ of all transformations of a set X. We have seen that:

(G1) There is an operation "∘" on $\mathrm{Tran}(X)$;

for convenience we also call the operation "∘" a *multiplication on* the set $\mathrm{Tran}(X)$ and call $ab = a \circ b$ the product of a and b. Further, by (1.4.2) we have:

(G2) $(a \circ b) \circ c = a \circ (b \circ c)$, $\forall a, b, c \in \mathrm{Tran}(X)$;

and, by (1.4.4), there is a distinguished element $\mathrm{id}_X \in \mathrm{Tran}(X)$ such that:

(G3) $a \circ \mathrm{id}_X = a = \mathrm{id}_X \circ a$, $\forall a \in \mathrm{Tran}(X)$.

But, note that multiplication on $\mathrm{Tran}(X)$ is not commutative in general, see Exer.1.

Now, we consider another set $\mathrm{Sym}(X)$ which consists of all invertible transformations of X. Though $\mathrm{Sym}(X)$ is a proper subset of $\mathrm{Tran}(X)$ in general, the composition "∘" is still an operation on $\mathrm{Sym}(X)$, since, by Exer.1.4.8, $\beta \circ \alpha \in \mathrm{Sym}(X)$ for all $\alpha, \beta \in \mathrm{Sym}(X)$.

2.1.3 Definition. If "∘" is an operation on a set T, and if $G \subset T$ such that $a \circ b \in G$ $\forall a, b \in G$, then we say that G is *closed* under the operation "∘"; and restricting the map $\circ : T \times T \to T$ to $G \times G$, we can get a map $G \times G \to G$, $(f, g) \mapsto f \circ g$ which is called the restricted operation of "∘" to G; or for convenience, we just say that "∘" is also an operation on G.

2.1.4 Remark. Thus, for a set X we have another object $\mathrm{Sym}(X)$ consisting of all invertible transformations of X, and (G1) holds; it is easy to see that (G2) and (G3) also hold for $\mathrm{Sym}(X)$, since $\mathrm{id}_X \in \mathrm{Sym}(X)$ clearly. Moreover, by Prop.1.4.6, there is one more property for $\mathrm{Sym}(X)$:

(G4) for any $a \in \mathrm{Sym}(X)$ there is an $a' \in \mathrm{Sym}(X)$ such that $aa' = \mathrm{id}_X = a'a$ (such a' is in fact unique by Prop.1.4.6).

In the following, assume the set $X = \{s_1, s_2, \cdots, s_n\}$ is finite; then we can describe elements in $\mathrm{Tran}(X)$ and $\mathrm{Sym}(X)$ just by listing the images of all elements. To save spaces, for $\alpha \in \mathrm{Tran}(X)$ we list all elements of X in a row and write the image of every element right under the element:

$$\alpha = \begin{pmatrix} s_1 & s_2 & \cdots & s_n \\ \alpha(s_1) & \alpha(s_2) & \cdots & \alpha(s_n) \end{pmatrix}.$$

We must note that the order of the elements in the brackets is not essential, since $s_1, \cdots, s_n$ form a set *not* a sequence. However, for some purposes, we can fix an order of them, say $s_1, \cdots, s_n$. For example, in this way we can easily calculate $|\mathrm{Sym}(X)|$ as follows: the above α belongs to $\mathrm{Sym}(X)$ iff the second row is a sequence such that every $s \in X$ occurs exactly once in the sequence; such a sequence is usually called a *permutation* of the elements $s_1, \cdots, s_n$. So we also take the following terminology:

Definition. An element of $\mathrm{Sym}(X)$ is called a *permutation* of X.

It is well-known that, since X has n elements, the total number of such permutations is $n!$. Thus we get

2.1.5 Proposition. *If $|X| = n$, then $|\mathrm{Sym}(X)| = n!$.* □

We show another way of expressing the permutations of a finite set; this is not valid for elements of $\mathrm{Tran}(X)$ even if X is finite.

In order to simplify the symbols, we take the following notation.

2.1.6 Notation. Let $X = \{1, 2, \cdots, n\}$ and denote $S_n = \mathrm{Sym}(X)$; the elements of X are called letters; and the elements of S_n are called permutations of *degree n*.

We will see in §2.3 that this assumption is enough for discussing permutations of any finite set.

Consider such a permutation $\left(\begin{smallmatrix}12345\cdots n\\23145\cdots n\end{smallmatrix}\right)$, it fixes every letter $t \geq 4$, and changes $1, 2, 3$ as a "cycle":

This picture suggests to us to denote such a permutation by (123) and to call it a *cycle*, or more precisely, a cycle of *length* 3, or a 3-cycle.

Similarly, the cycle (2368) means a permutation which maps 2 to 3, 3 to 6, 6 to 8, 8 to 2, and fixes the other letters.

2.1.7 Definition. By $(a_1 a_2 \cdots a_k)$, where the letters are different from each other, we denote a permutation which maps a_1 to a_2, a_2 to a_3, $\cdots$, a_{k-1} to a_k, a_k to a_1, and fixes the other letters; and we call it a *cycle* of length k (or, a *k-cycle*). A 2-cycle is also called a *transposition*. A 1-cycle is said to be trivial. We say that two cycles $(a_1 a_2 \cdots a_k)$ and $(b_1 b_2 \cdots b_h)$ are *disjoint* if $a_i \neq b_j$ for every $1 \leq i \leq k$ and $1 \leq j \leq h$.

2.1.8 Remark. By the definition, a 1-cycle (a) for any $a \in X$ is clearly the identity permutation id_X, we call it a *trivial cycle*. With the exception of this occurring, it is clear that the letters occurring in a cycle are uniquely determined by the cycle; but the expressions may be different; e.g. $(a_1 a_2 \cdots a_k) = (a_2 \cdots a_k a_1) = \cdots$; in fact, every a_i could lead $(a_i a_{i+1} \cdots)$, and then the expression of the cycle is determined. In particular, a k-cycle has exactly k expressions.

By Def.2.1.7 we immediately have the following fact.

2.1.9 Lemma.   *Let $\gamma_1, \cdots, \gamma_k$ be disjoint cycles and $\alpha = \gamma_1 \cdots \gamma_k$, let $t \in X$. If t occurs in γ_i for some $1 \leq i \leq k$, then $\alpha(t) = \gamma_i(t)$; otherwise $\alpha(t) = t$.*   □

The order of the factors of $\gamma_1 \cdots \gamma_k$ in the above lemma is not essential, since we have the following corollary.

2.1.10 Corollary.   *Two disjoint cycles commute.*

Proof. If cycles $\alpha = (a_1 a_2 \cdots a_k)$ and $\beta = (b_1 b_2 \cdots b_h)$ are disjoint, then for any $t \in X$ we have $\alpha\beta(t) = \beta\alpha(t)$ by the above lemma.   □

2.1.11 Example. Let $\alpha = \left(\begin{smallmatrix}1234567\\6217534\end{smallmatrix}\right)$ for $n = 7$. We start from the letter 1; then 1 is mapped to 6, and 6 is mapped to 3, then 3 is mapped to 1; so a cycle (163) is formed. Deleting 1, 6 and 3 and considering the permutation of α restricted to the remaining letters $\{2,4,5,7\}$, we start from 2, get a 1-cycle (2). And so on. At last we get

$$\alpha = \begin{pmatrix} 1 & 2 & 3 & 4 & 5 & 6 & 7 \\ 6 & 2 & 1 & 7 & 5 & 3 & 4 \end{pmatrix} = (163)(2)(47)(5) \,;$$

i.e. α is decomposed into a product of disjoint cycles where every letter appears exactly once. We illustrate this as follows:

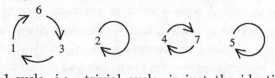

Since any 1-cycle, i.e. trivial cycle, is just the identity permutation id_X, by deleting 1-cycles we have $\alpha = \binom{1234567}{6217534} = (163)(47)$ which is a product of disjoint non-trivial cycles; by Cor.2.1.10 this product is independent of the order of the disjoint cycles.

This example suggests the following fact, the proof is just a precise general formulation of the above idea.

2.1.12 Proposition (Cyclic Decomposition of Permutations).   *Any non-identity permutation $\alpha \in S_n$ can be written as a product of disjoint non-trivial cycles in a unique way up to the order of factors.*

Proof. Recall that $S_n = \mathrm{Sym}(X)$ with $X = \{1, 2, \cdots, n\}$. We prove the existence by induction on n. It is trivial if $n = 1$ (note that S_1 contains no non-identity permutation). Assume $n \geq 2$. Starting a letter a_1, we get a sequence in X:

$$a_1, \quad a_2 = \alpha(a_1), \quad a_3 = \alpha(a_2), \quad \cdots .$$

Since X is finite, there must have repeated letters. Let a_{h+1}, $h \geq 1$, be the first repeated letter, i.e. $a_1, \cdots a_h$ are different from each other, but $a_{h+1} = a_i$ for some $1 \leq i \leq h$; suppose $i > 1$, then the two different letters a_{i-1} and a_h are mapped by α to the same letter a_i, this contradicts the bijectivity of α. So $a_{h+1} = a_1$; hence, restricting α to the subset $X' = \{a_1, a_2, \cdots, a_h\} \subset X$, we have $\alpha|_{X'} = \gamma_1 = (a_1 a_2 \cdots a_h)$. Further, α maps the letters of $X_1 = X - X'$ to X_1, i.e. $\alpha|_{X_1}$ is a permutation on X_1; however, $|X_1| < n$; so by induction we get $\alpha|_{X_1} = \gamma_2 \cdots \gamma_k$ with disjoint cycles. By Lemma 2.1.9, $\alpha = \gamma_1 \gamma_2 \cdots \gamma_k$ is a product of disjoint cycles. Deleting the trivial cycles, we write α as a product of disjoint non-trivial cycles.

Suppose we have two such expressions:

$$\alpha = \gamma_1\gamma_2 \cdots \gamma_k = \gamma_1'\gamma_2' \cdots \gamma_{k'}' \ ;$$

our aim is to show that $k = k'$ and, relabelling them suitably (recall that the orders are not crucial), we have $\gamma_i = \gamma_i'$ for $1 \leq i \leq k$.

We prove this by induction on the number of factors k. First, we show

(2.1.13). *if γ_i and γ_j' have a letter in common, then $\gamma_i = \gamma_j'$.*
In fact, letting a_1 occurs in both γ_i and γ_j', we can write $\gamma_i = (a_1 a_2 \cdots)$ and $\gamma_j' = (a_1 a_2' a_3' \cdots)$; then by Lemma 2.1.9, $a_2' = \gamma_j'(a_1) = \alpha(a_1) = \gamma_i(a_1) = a_2$, and $a_3' = \gamma_j'(a_2) = \gamma_i(a_2) = a_3$, and so on; thus $\gamma_i = \gamma_j'$.

Assume $k = 1$, i.e. $\gamma_1 = \gamma_1'\gamma_2' \cdots \gamma_{k'}'$. A letter of γ_1 occurs in some γ_i', say in γ_1'; so $\gamma_1 = \gamma_1'$ by (2.1.13); hence $\alpha = \gamma_1' \cdots \gamma_{k'}'$ which fixes every letter out of γ_1'. Then, since $\gamma_1', \cdots, \gamma_{k'}'$ are disjoint, every γ_i', $i > 1$, must be trivial; hence $k' = 1$ since every γ_i' is non-trivial. We are done for $k = 1$.

Next assume $k > 1$, similar to the above we can assume that $\gamma_1 = \gamma_1'$. Since the two expressions are products of disjoint cycles, by Lemma 2.1.9 we have $\gamma_2 \cdots \gamma_k = \gamma_2' \cdots \gamma_{k'}'$ as permutations on the letters out of $\gamma_1 = \gamma_1'$; then, by induction on the number of factors, we obtain that $k - 1 = k' - 1$, hence $k = k'$, and $\gamma_i = \gamma_i'$ for $2 \leq i \leq k$ up to relabelling. □

Some more properties on permutations are stated in Exer.4–7.

Return to $\mathrm{Sym}(X)$ or S_n, they are sets with four conditions (G1)—(G4).

Now we explain how such a structure which consists of a set and the four conditions (G1)—(G4) happens naturally.

Consider a square in a plane with four vertices labelled by $1, 2, 3, 4$:

and move it as a "rigid body". Some moves make it coincide with its original position, e.g. rotate it by $\pi/2$ radian through its center. We want to find the set D_4 of all such moves. Note that we are concerned only with the result of the move, not with the process of the move; so it is clear that every such move can be expressed by a permutation of $\{1, 2, 3, 4\}$; e.g. the rotation of $\pi/2$ radian is just $\binom{1234}{2341} = (1234)$. Then one can find without difficulty that

(**2.1.14**)
$$D_4 = \{(1), (1234), (13)(24), (1432), (12)(34), (14)(23), (13), (24)\}.$$

Even without calculation and just from intuition, we can convince ourselves that $\beta\alpha \in D_4$ for any $\alpha, \beta \in D_4$, since the composition of two moves which make the square coincide with itself still makes the square coincide with itself. By Def.2.1.3, the composition of permutations is an operation on D_4, i.e. conditions (G1) holds for D_4. Furthermore, (G2), (G3) and ($G4$) all hold for D_4. Thus we get a set D_4 with the four conditions (G1)—(G4); and such a structure describes the symmetries of the square. For example, if the above picture is a rectangle, not a square, then we get

(**2.1.15**) $$K_4 = \{(1), (13)(24), (12)(34), (14)(23)\},$$

which also satisfies the conditions (G1)—(G4) and its operation is the same as that of the above D_4. But K_4 is a proper subset of D_4; this is reasonable from the viewpoint of intuitive, since the symmetries of a rectangle are really a part of the symmetries of a square.

2.1.16 Example. An algebraic example similar to the above geometric ones. Consider a polynomial $g(X) = x_1^2 + x_2^2 + x_3^2 - 3x_1x_2x_3$ of variables $X = \{x_1, x_2, x_3\}$. For any permutation $\alpha \in S_3$, let α permute the variables as follows: x_1 is changed to $x_{\alpha(1)}$, x_2 is changed to $x_{\alpha(2)}$ etc. Then α changes a polynomial $f(X)$ to a polynomial denoted by $\alpha f(X)$; e.g. if $\gamma = (123)$ and $h(X) = x_1 - x_2 - x_3$, then $\gamma h(X) = x_2 - x_3 - x_1$. If $\alpha f(X) = f(X)$, we say that α fixes $f(X)$. We want to find the set G_f of the permutations which fix a given polynomial $f(X)$. For the above $h(X)$ this is easy:

$$G_h = \{(1), (23)\}.$$

On the other hand, for the above $g(X)$ one can find:

$$G_g = \{(1), (123), (132), (12), (13), (23)\} = S_3.$$

Similar to the above geometric observation, we can easily convince ourselves that G_f is a set with conditions (G1)—(G4), and such a structure describes the symmetries of the polynimial $f(X)$. In fact, the above polynomial $g(X)$ is really a "symmetric polynomial" in the usual sense, while the above $h(X)$ is symmetric only for x_2 and x_3, much less than $g(X)$.

Exercises 2.1

+1. 1) If $n \geq 3$, then there are $\alpha, \beta \in S_n$ which do not commute.

    2) The operation "$\circ$" on the set $\mathrm{Tran}(X)$ is not commutative if $|X| \geq 2$.

2. Prove that $\mathrm{Sym}(X)$ is a proper subset of $\mathrm{Tran}(X)$ if $|X| > 1$.

3. Assume $|X| = n$. Calculate $|\mathrm{Tran}(X)|$.

4. Show that any permutation $\alpha \in S_n$ can be expressed as a product of disjoint cycles (including 1-cycles) such that every letter $i \in \{1, \cdots, n\}$ appears exactly once, and such expression is unique up to order.

    Definition. The above expression is said to be the *cycle decomposition* of α. If for $1 \leq i \leq n$ there are t_i cycles of length i in the above expression of α, then we say that the *type* of α is $(t_1, t_2, \cdots, t_n)$; it is clear that every $t_i \geq 0$ and $t_1 + \cdots + t_n = n$.

5. Let $a_1, \cdots, a_h, b_1, \cdots, b_k$ be pairwise distinct letters (may be $h, k = 1$). Prove:

    1) $(a_1 b_1)(a_1 \cdots a_h)(b_1 \cdots b_k) = (a_1 \cdots a_h b_1 \cdots b_k)$.

    2) $(a_1 b_1)(a_1 \cdots a_h b_1 \cdots b_k) = (a_1 \cdots a_h)(b_1 \cdots b_k)$.

+6. *(Transposition Decomposition).* Let $\alpha \in S_n$ be of type $(t_1, \cdots, t_n)$. Prove:

    1) α can be written as a product of transpositions; more precisely, a product of $T(\alpha)$ transpositions, where $T(\alpha) = \sum_{i=1}^{n}(t_i - 1)$.

    Definition. The above expression for α is called a *transposition decompositions* of α.

    (Hint: Any cycle $(a_1 a_2 \cdots a_k) = (a_1 a_k) \cdots (a_1 a_3)(a_1 a_2)$.)

    2) The above expressions for α are not unique.

    (Hint: e.g. (12)(23)(13)=(23))

    3) If w is the number of transpositions in a transposition decomposition of α, then $w \equiv T(\alpha) \pmod 2$.

    Definition. α is called an *even* permutation if w is even, otherwise it is called an *odd* permutation.

    (Hint: assume $\alpha = \tau_w \tau_{w-1} \cdots \tau_1$; then $\tau_w \alpha = \tau_{w-1} \cdots \tau_1$, and, by induction on w, we have $w - 1 \equiv T(\tau_w \alpha) \pmod 2$; on the other hand, $T(\tau_w \alpha) \equiv T(\alpha) - 1 \pmod 2$ by Exer.5.)

7. 1) The product of two even permutations (two odd permutations) is an even permutation.

    2) The product of an even permutation and an odd permutation is an odd permutation.

2.2 Groups

With the concrete models of the last section in mind, we begin an abstract formulation of the mathematical structure so-called "group".

2.2.1 Definition. A non-empty set G with the following four conditions is called a *group*:

(G1) there is an operation "$\cdot$" on G;

(G2) (associativity): $(a \cdot b) \cdot c = a \cdot (b \cdot c)$, $\forall a, b, c \in G$;

(G3) there is an element $e \in G$ such that $e \cdot a = a = a \cdot e$, $\forall a \in G$;

(G4) for any $a \in G$ there is an $a' \in G$ such that $a \cdot a' = e = a' \cdot a$,

where e is the element specified in (G3).

Thus, $\mathrm{Sym}(X)$ in (2.1.4), S_n in (2.1.6), D_4 in (2.1.14), and K_4 in (2.1.15) are all groups. After some necessary remarks, we will show some more examples.

2.2.2 Remark. 1) By the definition, a group is an object consisting of a non-empty set G and four structures (G1)—(G4) which are called the group structure on the set G; it is far from a single set. Thus, sometimes a group is denoted by $(G, \cdot, e)$ to indicate the group structure. But, we usually say "Let G be a group" or "a group G ..." for brevity. If we disregard the group structure on a group G and only consider its set, then we call the set G the *underlying set* of the group.

2) The element e in (G3) is in fact unique, though the uniqueness is not specified by the definition; for, if e' also satisfies (G3), then $e' = e' \cdot e = e$. So we can denote it by a special symbol 1_G, or just 1, if there is no confusion; and call it the *identity element* of the group G, or the *unity* of the group G.

3) The element a' for a in (G4) is unique; for, if a'' also satisfies (G4), then $a'' = a'' \cdot 1 = a'' \cdot (a \cdot a') = (a'' \cdot a) \cdot a' = 1 \cdot a' = a'$. So we denote it by a^{-1} specially, and call it the *inverse element* of a, or the *inverse* of a. It is clear that $(a^{-1})^{-1} = a$, since $a^{-1} \cdot a = 1 = a \cdot a^{-1}$ by definition. These are consistent with those for $\mathrm{Sym}(X)$ in §2.1.

4) Though the symbol for the group operation may be varied as shown in Def.2.1.1, however, as we saw in §2.1 and will see below, a group usually appears in a concrete question; so the operation symbol comes from the question. For example, the operation symbol "∘" for Sym(X) is the common symbol representing the composition of maps. For an abstract group, as stated in (G1), "·" is the convenient symbol for the operation, and it is usually called "multiplication", and its result is called "product" and denoted $a \cdot b = ab$ with "·" omitted. But note that we will continue this remark in (2.2.16) below.

2.2.3 Definition. **1)** A group G is said to be *commutative* or *Abelian*, if one more condition holds:

(Commutativity): $a \cdot b = b \cdot a$, $\forall a, b \in G$.

2) The cardinality $|G|$ of a group is called the *order* of the group. A group G is said to be finite if its order is finite.

2.2.4 Example. (Sym(X), ∘, id$_X$) shown in (2.1.4) is a group, it is called the *symmetric group of the set X*.

(S_n, ∘, id) shown in (2.1.6) is a group, it is called the *symmetric group of degree n*. S_n is a finite group of order $n!$ by Prop. 2.1.5.

Moreover, some sets of S_n with the operation "∘" also form groups e.g. D_4 in (2.1.14) and K_4 in (2.1.15); such groups are called *permutation groups of degree n*.

Note that for the above groups we also call $a \circ b = ab$ with "∘" omitted the product, as mentioned above.

2.2.5 Example. ($\mathbf{Z}$, +, 0) is an Abelian group, where $\mathbf{Z}$ is the set of the integers, and the operation "+" is the usual addition of numbers, the identity element is the number zero 0, and the inverse element of n is the negation $-n$ of n. This is called the *additive group of integers*. The other similar examples are shown in Exer.2.

2.2.6 Example. Let $M_n(\mathbf{C})$ be the set of all complex matrices of dimension n. Then $(M_n(\mathbf{C}), +, 0)$ is an Abelian group, where "+" is the matrix addition and 0 is the zero matrix of dimension n. In a similar way, we have the Abelian groups: $(M_n(\mathbf{R}), +, 0)$, $(M_n(\mathbf{Z}), +, 0)$, and so on.

2.2.7 Example. Let $\mathrm{GL}_n(\mathbf{C})$ be the set of all invertible complex matrices of dimension n. Then $(\mathrm{GL}_n(\mathbf{C}), \cdot, 1)$ is a group, where "·" is the matrix multiplication and 1 is the identity matrix of dimension n.

This is called the *general linear group* of degree n over $\mathbf{C}$. It is a non-Abelian group provided $n > 1$. In the similar way, we have the general linear groups over $\mathbf{R}$, $\mathbf{Q}$ etc.: $(\mathrm{GL}_n(\mathbf{R}), \cdot, 1)$, $(\mathrm{GL}_n(\mathbf{Q}), \cdot, 1)$, and so on.

For the following two examples, we show in more details.

2.2.8 Example. Let $\mathbf{Z}_m$ be the set defined in Exer.1.3.1. We define an operation "+" on $\mathbf{Z}_m$ as follows. For $[a], [b] \in \mathbf{Z}_m$, let $[a]+[b] = [a + b]$. We have to show that this is independent of the choice of representatives: suppose $[a'] = [a]$ and $[b'] = [b]$, then $m|a' - a$ and $m|b' - b$, hence $m|((a' + b') - (a - b))$; i.e. $[a' + b'] = [a + b]$. Thus "+" on $\mathbf{Z}_m$ is a well-defined operation which is called the addition on $\mathbf{Z}_m$. Next, we have

$$([a] + [b]) + [c] = [a + b] + [c] = [a + b + c] = [a] + [b + c] = [a] + ([b] + [c]);$$

the associative law holds for the operation "+". And

$$[a] + [0] = [a + 0] = [a] = [0 + a] = [0] + [a];$$

i.e. $[0]$ is the identity element for the operation "+". At last,

$$[a] + [-a] = [a + (-a)] = [0] = [(-a) + a] = [-a] + [a];$$

every element $[a]$ has its inverse element $[-a]$ for the operation "+". So, $(\mathbf{Z}_m, +, [0])$ is a group. Further, $[a] + [b] = [a + b] = [b + a] = [b] + [a]$; and $|\mathbf{Z}_m| = m$ by Exer.1.3.1. Therefore $\mathbf{Z}_m$ is a finite Abelian group. We call this group the *additive group of the residue classes modulo m*.

2.2.9 Example. Consider the set $\mathbf{Z}_m$ again. We can define another operation so-called multiplication "$\cdot$" on $\mathbf{Z}_m$ as follows (the products are written down with "$\cdot$" omitted as convention). For $[a], [b] \in \mathbf{Z}_m$, let $[a][b] = [ab]$; and the associativity also holds for "$\cdot$"; and $[1]$ is the identity element for the operation "$\cdot$"; and the operation "$\cdot$" is commutative. All of these can be checked similarly to the above. But, $(\mathbf{Z}_m, \cdot, [1])$ is *not* a group; for, $[0][a] = [0]$ hence $[0]$ cannot have its inverse element for the operation "$\cdot$". However, there exist some elements $[a], [b]$ such that $[a][b] = [1]$; e.g. $[1][1] = [1]$; we call such elements the invertible elements of $\mathbf{Z}_m$; and by $\mathbf{Z}_m^*$ we denote the set of all invertible elements in $\mathbf{Z}_m$. Then it is easy to check that $\mathbf{Z}_m^*$ is closed under "$\cdot$" hence "$\cdot$" is an operation on $\mathbf{Z}_m^*$, cf. (2.1.3); and $(\mathbf{Z}_m^*, \cdot, [1])$ is a finite Abelian group; see Exer.3. This is called the *multiplicative group of* *invertible residue classes modulo m.*

One can find the similarity of the present situation for $\mathbf{Z}_m$ and $\mathbf{Z}_m^*$ from that for $M_n(\mathbf{C})$ and $\mathrm{GL}_n(\mathbf{C})$ above, or from that for $\mathrm{Tran}(X)$ and $\mathrm{Sym}(X)$ in §2.1.

Let G be a group. Note that, for a sequence $a_1, \cdots, a_n$ of elements in G and $n > 2$, the product $a_1 a_2 \cdots a_n$ is *not defined*; since the multiplication on G can be done only for two elements, see (G1) and (2.1.1). This is different from that for transformations on which any finite product can be defined by (1.4.3). However, (1.4.3) suggests to us to extend the associativity to the generalized associativity. In order to state it generally, let us introduce:

2.2.10 Definition. Let S be a non-empty set.

1) If (G1) and (G2) in (2.2.1) hold for S, then we say that $(S, \cdot)$ is a *semigroup*.

2) If (G1) and (G2) and (G3) in (2.2.1) hold for S, then we say that $(S, \cdot, e)$ is a *monoid*.

A semigroup (a monoid) is said to be *commutative* or *Abelian* if the operation is commutative.

Note that here we still adopt the convention in Remark 2.2.2(4).

2.2.11 Proposition (Generalized Associativity).   *Let $(S, \cdot)$ be a semigroup and $a_n, \cdots, a_2, a_1$ be a sequence of elements in S. If an element a is calculated from this sequence in this order by iterating the operation "$\cdot$" through bracketing suitably, then*

$$a = a_n(\cdots (a_3(a_2 a_1)) \cdots);$$

so, a is independent of the bracketing and we can write $a = a_n \cdots a_2 a_1$.

Proof. The case $n = 1, 2$ is trivial. The case $n = 3$ is just the associativity of "$\cdot$". For $n > 3$, we consider the last step of the iterating, then there is an m with $n > m \geq 1$ such that $a = bc$ where b is calculated from $a_n, \cdots, a_{m+1}$ and c is calculated from $a_m, \cdots, a_1$. Thus, by induction on n, we can write

$$b = a_n \cdots a_{m+1} \qquad \text{and} \qquad c = a_m \cdots a_1$$

and we can bracket a sequence of length$< n$ in any way provided it makes sense; so

$$a = bc = (a_n \cdots a_{m+1})(a_m \cdots a_1) = (a_n(a_{n-1} \cdots a_{m+1}))(a_m \cdots a_1)$$

$$= a_n((a_{n-1} \cdots a_{m+1})(a_m \cdots a_1)) = a_n(a_{n-1} \cdots a_{m+1}a_m \cdots a_1)$$
$$= a_n(a_{n-1}(\cdots (a_3(a_2 a_1)) \cdots)) \qquad \square$$

2.2.12 Proposition. *Let $(S, \cdot)$ be a semigroup. If $a_1, \cdots, a_n \in S$ commute pairwise, then for any permutation $(i_1, \cdots, i_n)$ of $1, \cdots, n$ we have*
$$a_{i_1} \cdots a_{i_n} = a_1 \cdots a_n \,;$$
hence the product is independent of the order.

Remark. *In particular, this is always true in Abelian semigroups; and the notation $\prod_{i=1}^n a_i$ makes sense.*

Proof. Transposing a_1 in $a_{i_1} \cdots a_1 \cdots a_{i_n}$ with its left one repeatedly, we can get $a_{i_1} \cdots a_1 \cdots a_{i_n} = a_1(a_{i_1} \cdots a_n)$; then apply induction on the number of factors. $\qquad \square$

Return to a group G. For any $a \in G$ and $n \in \mathbf{Z}$, by the above proposition, we can define the power a^n as follows:

$$(2.2.13) \quad a^n = \begin{cases} a \cdots a & \text{with } a \text{ repeated } n \text{ times} & \text{if } n > 0 \\ 1 & & \text{if } n = 0 \\ a^{-1} \cdots a^{-1} & \text{with } a^{-1} \text{ repeated } n \text{ times} & \text{if } n < 0 \end{cases}$$

Now we state several facts for groups.

2.2.14 Proposition. *Let G be a group. For any $a, b, c \in G$ and $m, n \in \mathbf{Z}$ we have:*

1) $a^m a^n = a^{m+n}$.

2) $(a^m)^n = a^{mn}$.

3) *if $ab = ba$ then $a^n b = b a^n$ and $(ab)^n = a^n b^n$.*

4) $(ab)^{-1} = b^{-1} a^{-1}$.

5) *(left and right cancellation) if $ab = ac$ or $ba = ca$, then $b = c$;*

6) *equations $ax = b$ and $xa = b$ with variable x have unique solutions in G.* $\qquad \square$

Proof. For (1), if $m = 0$ or $n = 0$ it is obvious; if both $m, n < 0$ then

$$a^m a^n = (\overbrace{a^{-1} \cdots a^{-1}}^{-m})(\overbrace{a^{-1} \cdots a^{-1}}^{-n}) = \overbrace{a^{-1} \cdots a^{-1}}^{-(m+n)} = a^{m+n};$$

it is the same for both $m, n > 0$; if $m < 0$ and $n > 0$, then, deleting

$a^{-1}a = 1$ n times (if $|m| \geq n$) or $|m|$ times (if $|m| < n$), we have

$$a^m a^n = \overbrace{a^{-1} \cdots a^{-1}}^{-m} \overbrace{a \cdots a}^{n} = \overbrace{a^{-1} \cdots a^{-1}}^{-m+1} \overbrace{a \cdots a}^{n-1} = \cdots$$

$$= \overbrace{a^{-1} \cdots a^{-1}}^{-(m+n)} \text{ (if } |m| \geq n), \text{ or } \overbrace{a \cdots a}^{(m+n)} \text{ (if } |m| < n)$$
$$= a^{m+n};$$

it is the same for $m > 0$ and $n < 0$.

For (2), using (1) repeatedly, we have

$$(a^m)^n = \overbrace{a^m a^m \cdots a^m}^{n} = a^{\overbrace{m + \cdots + m}^{n}} = a^{mn}$$

For (3), it is trivial for $n = 0$; if $n > 0$, using $ab = ba$ for n times, we can transpose b in $a^n b = a \cdots ab$ to the first position to get $a^n b = ba^n$; if $n < 0$, then $a^{-1}b = ba^{-1}$ by Exer.6(3), and by the above (2) we have $a^n = (a^{-1})^{-n}$, hence it is reduced to the positive exponent.

By the condition (G4), (4) is proved by the following calculations:
$$(b^{-1}a^{-1})(ab) = (b^{-1}(a^{-1})a)b = b^{-1}b = 1 \text{ and } (ab)(b^{-1}a^{-1}) = 1.$$

For (5), multiplying a^{-1} to the left of both sides of $ab = ac$, we get $b = c$ at once. It is the same for the other one.

For (6), it is clear that $x = a^{-1}b$ is a solution of $ax = b$; conversely, if $ax = b$ for $x \in G$, then, multiplying a^{-1} to the left, we have $x = a^{-1}b$.
$\square$

2.2.15 Remark. The definition (2.2.13) and Prop.2.2.14(1), (2) and (3) are valid for semigroups and positive exponents; and valid for monoids and non-negative exponents.

2.2.16 Remark. Up to now, for abstract groups and many concrete groups, we use the *multiplicative notation* remarked in (2.2.2). However, so many examples such as $(\mathbf{Z}, +, 0)$, $(M_n(\mathbf{C}), +, 0)$ etc. suggest to us that it is convenient to take the *additive notation* for Abelian groups sometimes. In such notation, for an abstract Abelian group, the operation is denoted by "$+$" and called "addition" and its result $a + b$ is called "sum"; and 0 stands for the "identity element" or "unity" in (G3), but the terminology for it *must be changed* to "zero element"; and so on.   We list the multiplicative notation and the additive notation

for comparison; but note that the additive notation is usually applied
to Abelian groups.

Symbols	reads	Symbols	reads
$(G, \cdot, 1)$	a group	$(G, +, 0)$	an additive group
$\cdot$	multiplication	$+$	addition
1	idetity element or unity	0	zero element or zero
a^{-1}	inverse element or inverse	$-a$	negative element or negation
ab	product of a and b	$a + b$	sum of a and b
a^n	n is the exponent	na	n is the coefficient
$\prod_{i=1}^n a_i$	the product of a_i's	$\sum_{i=1}^n a_i$	the sum of a_i's

Exercises 2.2

1. 1) In S_7, $\alpha = (1357)(246)$, $\beta = (25)(367)$; calculate $\alpha\beta$ and $\beta\alpha$.

   2) In S_7, $\alpha = (1357)(246)$; find α^{-1}.

+2. Check that the following are Abelian groups:

   1) $(\mathbf{Q}, +, 0)$, where $\mathbf{Q}$ is the set of rationals, "+" is the usual addition of numbers, 0 is the zero. (this is called the *additive group of rationals*).

   2) $(\mathbf{Q}^*, \cdot, 1)$, where $\mathbf{Q}^*$ is the set of all non-zero rationals, "$\cdot$" is the usual multiplication of numbers, 1 is the identity. (this is called the *multiplicative group of rationals*).

   3) The additive group of reals: $(\mathbf{R}, +, 0)$, similar to (1).

   4) The multiplicative group of reals: $(\mathbf{R}^*, \cdot, 1)$, similar to (2).

   5) The additive group of complexes: $(\mathbf{C}, +, 0)$, similar to (1).

   6) The multiplicative group of complexes: $(\mathbf{C}^*, \cdot, 1)$, similar to (2).

+3. Notations are as in (2.2.9). Prove:

   1) $(\mathbf{Z}_m^*, \cdot, [1])$ in (2.2.9) is an Abelian group.

   2) $[a] \in \mathbf{Z}_m^*$ iff a is prime to m.

   3) $|\mathbf{Z}_m| = \varphi(m)$, where $\varphi(m)$ is the number-theoretic *Euler function* which stands for the number of positive integers less than m and prime to m.

4. Let W be the set of all non-bijective transformations of a finite set and "∘" stands for the composition of transformations. Show that "∘" is an operation on W. Is W with this operation a semigroup, a monoid, or a group?

5. Define $a * b = a + b - ab$ $\forall a, b \in \mathbf{Q}$; then "$*$" is an operation on $\mathbf{Q}$. Is the operation commutative? Is $(\mathbf{Q}, *)$ a semigroup, a monoid, or a group?

+6. In a group G prove (all the elements come from G):

   1) if $a^2 = a$ then $a = 1$.
   2) $a^2 = 1$ iff $a = a^{-1}$.
   3) if $ab = ba$ then $a^{-1}b = ba^{-1}$.
   4) $(a_1 a_2 \cdots a_n)^{-1} = a_n^{-1} \cdots a_2^{-1} a_1^{-1}$.

7. Prove: a group G is Abelian iff $(ab)^2 = a^2 b^2$, $\forall a, b \in G$.

8. There are two modes to bracket abc such that the multiplications can be done iteratively: $(ab)c$ and $a(bc)$. Write out all such modes to bracket $abcd$; and check the generalized associativity for this case if the associativity holds.

   (As a hint for Exer.9–11, please cf. the proof of Prop.2.4.3 below.)

+9. Let $(G, \cdot)$ be a semigroup. If there is an $e \in G$ such that $(\forall a \in G)(ea = a)$ and $(\forall a \in G)((\exists a' \in G)(a'a = e))$, then G with the operation "$\cdot$" is a group.

10. Let $(G, \cdot)$ be a semigroup. If for any $a, b \in G$ the equations $ax = b$ and $xa = b$ with variable x have unique solutions in G, then G with the operation "$\cdot$" is a group.

11. Let $(G, \cdot)$ be a finite semigroup. If both left cancellation and right cancellation hold for G, then G is a group.

12. For a finite set $S = \{a_1, \cdots, a_n\}$, we describe an operation "$*$" on S by listing all the results of the operation in a so-called operation table:

$*$	a_1	$\cdots$	a_n
a_1	$a_1 * a_1$	$\cdots$	$a_1 * a_n$
$\cdots$	$\cdots$	$\cdots$	$\cdots$
a_n	$a_n * a_1$	$\cdots$	$a_n * a_n$

   1) Let $S = \{e\}$, then there is only one operation on S; and, with the operation, S is a group.

   2) Let $S = \{e, a, b, c\}$, check that both the following two tables define group structures on S, so that there are at least two different groups of order 4.

$*$	e	a	b	c
e	e	a	b	c
a	a	b	c	e
b	b	c	e	a
c	c	e	a	b

$\bullet$	e	a	b	c
e	e	a	b	c
a	a	e	c	b
b	b	c	e	a
c	c	b	a	e

2.3 Subgroups

2.3.1 Definition. Let G be a group. A non-empty subset H of G is said to be a *subgroup* of G, denoted by $H < G$ or $H \leq G$, if H is closed under the operation of G and, with the same operation, H is also a group.

2.3.2 Example. For any group G, it is clear that $G \leq G$ and $\{1\} \leq G$; i.e. G has at least these two different subgroups if $G \neq \{1\}$, they are called the *trivial subgroups* of G. And H is said to be a *proper subgroup* of G, denoted by $H \lneq G$, if $H \leq G$ and $H \neq G$.

2.3.3 Example. Let G be a group and let

$$Z(G) = \{a \mid a \in G \text{ and } ax = xa \ \forall x \in G\};$$

then $Z(G)$ is a subgroup of G, called the *center* of G.

2.3.4 Example. Notations as in (2.2.4). Any subgroup of the symmetric group $\text{Sym}(X)$ on the set X is called a *transformation group on the set* X; and it is called a *permutation group* on X if $|X| < \infty$; and any subgroup of S_n is called a *permutation group of degree n*. So, both K_4 in (2.1.15) and D_4 in (2.1.14) are permutation groups of degree 4, since they are subgroups of S_4.

2.3.5 Example. Notations as in (2.2.7). Any subgroup of the general linear group $\text{GL}_n(\mathbf{C})$ is called a *linear group* of dimension n over $\mathbf{C}$. Similarly, we have the linear groups over $\mathbf{R}$, $\mathbf{Q}$ etc. For example, it is easy to check that

$$K = \left\{ \begin{pmatrix} 1 & 0 \\ 0 & 1 \end{pmatrix}, \begin{pmatrix} 1 & 0 \\ 0 & -1 \end{pmatrix}, \begin{pmatrix} -1 & 0 \\ 0 & 1 \end{pmatrix}, \begin{pmatrix} -1 & 0 \\ 0 & -1 \end{pmatrix} \right\}$$

with the matrix multiplication is a group, hence it is a linear group of dimension 2 over $\mathbf{Q}$.

Before more examples, we show some criterions for subgroups.

2.3.6 Proposition. *Let G be group and $H \subset G$. The following are equivalent:*

i) $H \leq G$;

ii) $1_G \in H$ *and, for any $a, b \in H$, $ab \in H$ and $a^{-1} \in H$;*

iii) $H \neq \emptyset$ *and, for any $b \in H$, $ab^{-1} \in H$.*

Proof. (i) $\Longrightarrow$ (ii): By (i), $1_H^2 = 1_H$; this is also an equality in G, by Exer.2.2.6(1) we get $1_G = 1_H \in H$. For $a, b \in H$, $ab \in H$ by (i); and

a has an inverse $a' \in H$, i.e. $a'a = 1_H = 1_G = aa'$; hence $a^{-1} = a' \in H$ since the inverse is unique.

(ii) $\Longrightarrow$ (iii):   Clear.

(iii) $\Longrightarrow$ (i):   By (iii), $H \neq \emptyset$ hence there is a $c \in H$, and $1_G = cc^{-1} \in H$. Next, for $a, b \in H$, we have $1_G, b \in H$, hence $b^{-1} = 1_G b^{-1} \in H$; from $a, b^{-1} \in H$ and by (iii) we have $ab = a(b^{-1})^{-1} \in H$; thus H is closed under the operation of G, and this operation is an operation on H which is clearly associative since it is associative on G. $1_G \in H$ is clearly also the identity element of H. At last, for $a \in H$ we have shown that $a^{-1} \in H$ which is the inverse of a in H. Therefore H is a group, consequently, a subgroup of G. $\qquad\qquad\qquad\qquad\qquad\qquad\quad\square$

For finite subgroups, we have an easier criterion.

2.3.7 Proposition.   *A finite non-empty set H of a group G is a subgroup provided H is closed under the operation of G.*

Proof. We remark that this proposition is obvious by Exer.2.2.11. But, without citing the exercises of §2.2, we sketch a proof, which is also a hint for Exer.2.2.9–11; and the technique for λ_a and τ_a below is in fact a very important idea in the group theory which will appear in the next section again. The finiteness of H is applied only for the bijectivity of λ_a and τ_a.

For $a \in H$, by the closedness of the operation for H we have a map
$$\lambda_a : \quad H \longrightarrow H , \qquad h \longmapsto ah$$
which is injective since $\lambda_a(h) = \lambda_a(h') \Longrightarrow ah = a'h \Longrightarrow h = h'$, by right cancellation; consequently, λ_a is bijective since H is finite, cf. Exer.1.4.5. Similarly, we have
$$\tau_a : \quad H \longrightarrow H , \qquad h \longmapsto ha$$
is a bijective map. (Note: these imply that equations $ax = b$ and $xa = b$ with variable x have unique solutions in H.)

Since $H \neq \emptyset$, we can choose $c \in H$. From the above, there is an $e \in H$ such that $ec = c$; and for $a \in H$ arbitrary, there is a $b \in H$ such that $a = cb$; so $ea = ecb = cb = a$. That is
$$(\forall a \in H)\,(ea = a).$$
Next, there is an $a' \in H$ such that $a'a = e$; i.e.
$$(\forall a \in H)(\exists a' \in H)\,(a'a = e).$$

At last, for any $a \in H$, we already have $a' \in H$ such that $a'a = e$, then for a' we have an $a'' \in H$ such that $a''a' = e$, so $aa' = eaa' = a''a'aa' = a''a' = e$; and $ae = aa'a = ea = a$. Thus, (G3) and (G4) in (2.2.1) hold for H. □

Now we give two more examples which can be checked by the criterions.

Example. For any $n \in \mathbf{Z}$, $n\mathbf{Z} = \{nk \mid k \in \mathbf{Z}\}$ is a subgroup of the additive group $\mathbf{Z}$.

Example. $(\mathbf{Q}^*, \cdot, 1)$ is a subgroup of $(\mathbf{R}^*, \cdot, 1)$; but, $(\mathbf{Q}^*, \cdot, 1)$ is not a subgroup of $(\mathbf{R}, +, 0)$, because the operations are different.

We consider how to construct subgroups of a group.

2.3.8 Proposition. *Let G be a group and H_i, $i \in I$, be subgroups of G. Then $\bigcap_{i \in I} H_i$ is a subgroup of G.*

Proof. $1_G \in H_i \ \forall i \in I$; so $1_G \in \bigcap_{i \in I} H_i$. For any $a, b \in \bigcap_{i \in I} H_i$, $a, b \in H_i$ hence $ab \in H_i \ \forall i \in I$, consequently, $ab \in \bigcap_{i \in I} H_i$; and $a^{-1} \in H_i \ \forall i \in I$, hence $a^{-1} \in \bigcap_{i \in I} H_i$. □

Remark. The union of subgroups is not a subgroup in general. For example, consider the additive group $\mathbf{Z}$, both $4\mathbf{Z} \leq \mathbf{Z}$ and $6\mathbf{Z} \leq \mathbf{Z}$, but $4\mathbf{Z} \cup 6\mathbf{Z} \not\leq \mathbf{Z}$; since, e.g. $4 + 6 = 10 \notin 4\mathbf{Z} \cup 6\mathbf{Z}$. But one can see that, if we define $(4\mathbf{Z}) + (6\mathbf{Z}) = \{4m + 6n \mid m, n \in \mathbf{Z}\}$, then $4\mathbf{Z} + 6\mathbf{Z} \leq \mathbf{Z}$. Thus we can construct a subgroup suitably from two subgroups.

We show how to construct a subgroup from a subset suitably.

2.3.9 Definition. Let G be a group and $S \subset G$. Define

$$\langle S \rangle = \bigcap_{S \subset H \leq G} H$$

where the subscript means that H runs over the subgroups of G which contains S. By Prop.2.3.8, $\langle S \rangle \leq G$; we call it the subgroup of G *generated by S*. If $S = \{a, b, \cdots\}$, then we denote $\langle S \rangle = \langle a, b, \cdots \rangle$ in short.

Thus $\langle S \rangle$ is in fact the smallest subgroup of G which contains S.

2.3.10 Proposition. *Notations as above. Then*

$$\langle S \rangle = \{a_1^{n_1} \cdots a_k^{n_k} \mid a_i \in S, n_i \in \mathbf{Z} \text{ for } 1 \leq i \leq k, \text{ with } k \in \mathbf{Z}, k \geq 0\}$$

(where $k \in \mathbf{Z}$ and $k \geq 0$ mean that the set (on the right) consists of all such finite products $a_1^{n_1} \cdots a_k^{n_k}$ with the usual agreement that the product is 1_G if $k = 0$. Another expression of $\langle S \rangle$ is stated in Exer.2.5.1(2) below).

Proof. By W we denote the set on the right-hand side. Then, for any $a_1^{n_1} \cdots a_k^{n_k} \in W$, it is clear that $a_1^{n_1} \cdots a_k^{n_k} \in H$ for all H satisfying $S \subset H \leq G$. That is: $W \subset \langle S \rangle$. On the other hand, $1_G \in W$; and for $a_1^{n_1} \cdots a_k^{n_k} \in W$ and $b_1^{m_1} \cdots b_h^{m_h} \in W$, $(a_1^{n_1} \cdots a_k^{n_k})(b_1^{m_1} \cdots b_h^{m_h})^{-1} = a_1^{n_1} \cdots a_k^{n_k} b_h^{-m_h} \cdots b_1^{-m_1} \in W$, so that $W \leq G$ by Prop. 2.3.6; and clearly, $a \in W$ for all $a \in S$, i.e. $S \subset W$. Thus $\langle S \rangle \subset W$. Combining the above, we get $\langle S \rangle = W$. □

Example. $S_3 = \{(1), (123), (132), (12), (23), (13)\} = \langle (12), (13) \rangle$, since $(123) = (13)(12)$, $(132) = (12)(13)$ and $(23) = (12)(13)(12)$.

Example. The additive group $\mathbf{Z} = \langle 1 \rangle = \langle -1 \rangle$, since for $n \in \mathbf{Z}$ arbitrary, $n = n1 = (-n)(-1)$.

2.3.11 Remark. We remark that in the above the n in $n1$, and $-n$ in $(-n)(-1)$, are the coefficients. Recall from Remark 2.2.16 that for additive group the Prop.3.3.10 appears as

$$\langle S \rangle = \Big\{ \sum_{i=1}^{k} n_i a_i \mid a_i \in S, n_i \in \mathbf{Z} \text{ for } 1 \leq i \leq k, \text{ with } k \in \mathbf{Z}, k \geq 0 \Big\}.$$

2.3.12 Definition. Let G be a group and $H \leq G$.

1) If $H = \langle a_1, \cdots, a_n \rangle$ for finitely many $a_1, \cdots, a_n \in H$, then H is said to be a *finitely generated* subgroup, and $\{a_1, \cdots, a_n\}$ is called a system of generators for H.

2) If $H = \langle a \rangle$ for some $a \in H$, then H is called a *cyclic subgroup* generated by a, and a is called a *generator*. G is called a *cyclic group* generated by a if $G = \langle a \rangle$.

3) For $a \in G$, the order $|\langle a \rangle|$ is also called the order of the element a, and denoted by $|a|$.

Thus, the additive group $\mathbf{Z}$ is a cyclic group, and either 1 (or -1) is a generator.

In fact, it is not difficult to classify all the cyclic groups. But first we should know how to compare groups and when we can say that two groups have the same structures; these are considered in the next section.

Exercises 2.3

1. In any group show that $\langle \emptyset \rangle = \{1\}$ and $\langle 1 \rangle = \{1\}$.

2. Prove: for any group G the center $Z(G)$ is an Abelian subgroup.

3. Assume that H_i, $i = 1, 2, \cdots$, are subgroups of G and $H_1 \subset H_2 \subset \cdots$. Prove that the union $\bigcup_{i=1}^{\infty} H_i \leq G$.

4. Prove that a group cannot be a union of two proper subgroups. Can a group be a union of three proper subgroups? (Hint: for proper subgroups H_1 and H_2 consider $a \in H_1 - H_2$ and $b \in H_2 - H_1$; and consider K_4 in (2.1.15).)

5. Let G be a group and $H, K, L \leq G$. If $L \subset H \cup K$, then either $L \subset H$ or $L \subset K$. (cf. the above hint for Exer.4.)

$^{+}$6. Recall that a complex number ω is said to be an *nth root of unity* if $\omega^n = 1$; moreover, an nth root of unity is said to be *primitive* if $\omega^k \neq 1$ for any $0 < k < n$. In the multiplicative group $\mathbf{C}^*$ prove:

   1) The subset R consisting of all roots of unity is a subgroup of $\mathbf{C}^*$.

   2) The set R_n consisting of all nth roots of unity is a cyclic subgroup of order n; and $R_n = \langle \omega \rangle$ iff ω is a primitive nth root of unity.

7. Prove: if the additive group $\mathbf{Z} = \langle a \rangle$, then either $a = 1$ or $a = -1$.

$^{+}$8. A non-empty subset N of a semigroup $(M, \cdot)$ is said to be a *subsemigroup* of M if $(\forall a, b \in N)(ab \in N)$.

   A non-empty subset N of a monoid $(M, \cdot, 1)$ is said to be a *submonoid* of M if $1 \in N$ and $(\forall a, b \in N)(ab \in N)$.

   1) Prove: $(M_2(\mathbf{C}), \cdot, 1)$ (matrix multiplication as the operation) is a monoid.

   2) Let $N = \left\{ \begin{pmatrix} a & 0 \\ 0 & 0 \end{pmatrix} \Big| a \in \mathbf{C} \right\}$; then $\emptyset \neq N \subset M_2(\mathbf{C})$ and, with the matrix multiplication, N is a monoid but not a submonoid of the above monoid.

2.4 Homomorphisms, Isomorphisms

Recall that groups are sets with certain mathematical structures; it is reasonable to consider the maps between groups which preserve the mathematical structures.

2.4.1 Definition. Let G and H be groups. A map $f : G \to H$ is called a *homomorphism* if

$$f(ab) = f(a)f(b) \qquad\qquad \forall a, b \in G;$$

and, f is said to be a group injection (group surjection resp.) if it is an injective (surjective resp.) homomorphism; and f is said to be a group *isomorphism* if it is both injective and surjective homomorphism.

If there is an isomorphism $f : G \to H$ between groups G and H, then we say that group G *is isomorphic* to group H and denote $G \overset{f}{\cong} H$, or $f : G \overset{\cong}{\to} H$, or simply $G \cong H$.

Further, if $G = H$, then a homomorphism $f : G \to G$ is said to be an *endomorphism* of G, while an isomorphism $f : G \overset{\cong}{\to} G$ is said to be an *automorphism* of G.

2.4.2 Remark. We emphasize that ab on the left of $f(ab) = f(a)f(b)$ is the operation on the domian G, while the one on the right $f(a)f(b)$ is the operation on the co-domain H; thus, in practice, the two operations may be expressed in different symbols.

Though the definition of homomorphisms only says that "$f(ab) = f(a)f(b) \; \forall a, b \in G$" which means the operations are preserved, a homomorphism really preserves all the group structures, since we have the following proposition which implies that $f(1_G) = 1_H$ and $f(a^{-1}) = f(a)^{-1}$.

2.4.3 Proposition.   *Let $f : G \to H$ be a homomorphism from a group G to a group H. Then $f(a^n) = f(a)^n$ for all $a \in G$ and all $n \in \mathbf{Z}$.*

Proof. If $n > 0$, then

$$f(a^n) = f(\overbrace{a \cdots a}^{n}) = \overbrace{f(a) \cdots f(a)}^{n} = f(a)^n.$$

Taking $a = 1_G$ and $n = 2$, we have $f(1_G) = f(1_G^2) = f(1_G)^2$ in H; by Exer.2.2.6(1) we get

$$f(1_G) = 1_H;$$

which is the case of $n = 0$ for the proposition. For $n < 0$, noting that $-n > 0$, we have (cf. Prop.2.2.14):

$$1_H = f(1_G) = f(a^n a^{-n}) = f(a^n)f(a^{-n}) = f(a^n)f(a)^{-n}$$

consequently,

$$f(a^n) = (f(a)^{-n})^{-1} = f(a)^n . \qquad\qquad \square$$

2.4.4 Corollary. *Notations as above, and define the* kernel *of* f *as follows:*
$$\text{Ker}(f) = \{a \mid a \in G \text{ and } f(a) = 1_H\}.$$
Then the image $\text{Im}(f) \le H$ *while the kernel* $\text{Ker}(f) \le G$.

Proof. For $a', b' \in \text{Im}(f)$ there are $a, b \in G$ such that $a' = f(a)$ and $b' = f(b)$; by the above proposition we have $f(b)^{-1} = f(b^{-1})$, hence $a'(b')^{-1} = f(a)f(b)^{-1} = f(a)f(b^{-1}) = f(ab^{-1}) \in \text{Im}(f)$. So $\text{Im}(f) \le H$ by Prop.2.3.6.

For $a, b \in \text{Ker}(f)$, we have $f(b^{-1}) = f(b)^{-1} = 1_H^{-1} = 1_H$; hence $f(ab^{-1}) = 1_H 1_H = 1_H$; i.e. $ab^{-1} \in \text{Ker}(f)$. Thus $\text{Ker}(f) \le G$ by Prop.2.3.6 again. □

Remark. In fact, we have a futher property for $K = \text{Ker}(f) \le G$:
$$aya^{-1} \in K, \qquad \forall a \in G, \ y \in K.$$
This motivates a further discussion in §2.6 below.

2.4.5 Proposition. 1) *If* $G \xrightarrow{f} H$ *and* $H \xrightarrow{g} K$ *are homomorphisms between groups, then so is their composition* $gf : G \to K$.

2) *If* $f : G \to H$ *is an isomorphism between groups, then so is its inverse map* $f^{-1} : H \to G$.

Proof. 1). For any $a, b \in G$, we have
$$\begin{aligned} gf(ab) &= g(f(a \cdot b)) = g(f(a) \cdot f(b)) \\ &= g(f(a)) \cdot g(f(b)) = gf(a) \cdot gf(b). \end{aligned}$$

2). For any $x, y \in H$, we have
$$\begin{aligned} f^{-1}(x) \cdot f^{-1}(y) &= (f^{-1}f)\Big(f^{-1}(x) \cdot f^{-1}(y)\Big) \\ &= f^{-1}\Big(f(f^{-1}(x) \cdot f^{-1}(y))\Big) \\ &= f^{-1}\Big(ff^{-1}(x) \cdot ff^{-1}(y)\Big) = f^{-1}(x \cdot y). \quad \square \end{aligned}$$

We give some examples.

2.4.6 Example. 1) For any groups G and H we have a homomorphism $G \to H$ which maps every element of G to 1_H; this is called the *zero homomorphism*. On the other hand, for any group G the identity map id_G is obviously an endomorphism (automorphism in fact), this is called the identity endomorphism.

2) (The Example 1.4.1). Let $\mathbf{Z}$ and $\mathbf{Z}_m$ resp. be the groups as in (2.2.5) and (2.2.8) resp. Then

$$f: \ \mathbf{Z} \longrightarrow \mathbf{Z}_m \ , \qquad n \longmapsto [n]$$

is a surjective homomorphism, but not injective.

3) Let $\mathbf{Z}_m$ and $\mathbf{C}^*$ be as in Example.2.2.8 and Exer.2.2.2(6) resp. and let ω be an mth root of unity (i.e. $\omega^m = 1$). Then

$$\chi_\omega: \ \mathbf{Z}_m \longrightarrow \mathbf{C}^* \ , \qquad [n] \longmapsto \omega^n$$

is a homomorphism, it is certainly not surjective, and it is injective if ω is a primitive mth root of unity (i.e. $\omega^k \neq 1$ for $0 < k < m$). Usually, a homomorphism from an Abelian group to the group $\mathbf{C}^*$ is also called a *character of Abelian group* of the Abelian group.

4) Let $\mathbf{R}$ and $\mathbf{R}^*$ resp. be the groups as in Exer.2.2.2(3) and (4) resp. Let $a \neq 1$ be a positive real number. Then

$$f: \ \mathbf{R} \longrightarrow \mathbf{R}^* \ , \qquad x \longmapsto a^x$$

is an injective homomorphism. In this example the condition in Def.2.4.1 appears as $f(x+y) = f(x)f(y), \forall x, y \in \mathbf{R}$.

Now we give an example of non-Abelian groups. Recall that we said after (2.1.6) that, to consider $\text{Sym}(Y)$ where $Y = \{a_1, \cdots, a_n\}$, it is enough to consider $S_n = \text{Sym}(X)$ with $X = \{1, \cdots, n\}$. In fact, from intuition we can imagine that a_1 corresponds to 1, a_2 to 2, etc.; any permutation α of Y can easily corresponds to a permutation α' of X such that α' maps any $x \in X$ just like the α maps the correspondent of x:

$$\begin{array}{ccc} y & \to & x \\ \alpha \downarrow & & \downarrow \alpha' \\ \alpha(y) & \to & \alpha'(x) \end{array}$$

(for example $\begin{pmatrix} a_1 a_2 a_3 \\ a_2 a_3 a_1 \end{pmatrix} \mapsto \begin{pmatrix} 123 \\ 231 \end{pmatrix}$); and such correspondence should preserves the composition of permutations; in other words, $\text{Sym}(Y)$ has the same group structure as S_n.

We give a precise formulation for this intuition.

2.4.7 Proposition.   *Let X and Y be sets and $\rho: Y \to X$ be a bijective map and $\rho^{-1}: X \to Y$ be the inverse map. Then*

$$f: \ \text{Sym}(Y) \longrightarrow \text{Sym}(X) \ , \qquad \alpha \longmapsto \rho\alpha\rho^{-1}$$

is an isomorphism of groups.

Proof. For any $\alpha, \beta \in \mathrm{Sym}(Y)$,
$$f(\alpha\beta) = \rho(\alpha\beta)\rho^{-1} = \rho\alpha \cdot \mathrm{id}_Y \cdot \beta\rho^{-1}$$
$$= \rho\alpha \cdot (\rho^{-1}\rho) \cdot \beta\rho^{-1} = (\rho\alpha\rho^{-1})(\rho\beta\rho^{-1}) = f(\alpha)f(\beta).$$

So f is a homomorphism. Let
$$f' : \ \mathrm{Sym}(X) \longrightarrow \mathrm{Sym}(Y) \,, \qquad \alpha' \longmapsto \rho^{-1}\alpha'\rho.$$
Then for any $\alpha \in \mathrm{Sym}(Y)$
$$f'f(\alpha) = f'(\rho\alpha\rho^{-1}) = \rho^{-1}(\rho\alpha\rho^{-1})\rho = \alpha\,.$$
That is, $f'f = \mathrm{id}_{\mathrm{Sym}(Y)}$. Similarly, $ff' = \mathrm{id}_{\mathrm{Sym}(X)}$. Thus f is a bijective map hence an isomorphism. □

We show that any group can be represented as a transformation group.

Let G be an arbitrary group. To any $a \in G$ we can assign a transformation on the underlying set G:

(2.4.8) $\qquad\qquad \lambda_a : \ G \longrightarrow G \,, \qquad x \longmapsto ax\,,$

which is called the *left translation* by a. First, observe λ_1 by $1 = 1_G$; for any $x \in G$, $\lambda_1(x) = 1x = x$; hence

$\lambda_1 = \mathrm{id}_G$ (the identity transformation of the underlying set G).

Let $a, b \in G$, we compare λ_{ab} and $\lambda_a\lambda_b$; for this, letting $x \in G$, we calculate:
$$\lambda_{ab}(x) = (ab)x = a(bx) = a \cdot \lambda_b(x) = \lambda_a(\lambda_b(x)) = \lambda_a\lambda_b(x)\,,$$
which shows that

(2.4.9) $\qquad\qquad \lambda_{ab} = \lambda_a\lambda_b \qquad\qquad \forall a, b \in G.$

In particular, $\lambda_{a^{-1}}\lambda_a = \lambda_{a^{-1}a} = \lambda_1 = \mathrm{id}_G$ and $\lambda_a\lambda_{a^{-1}} = \mathrm{id}_G$. That is
$$\lambda_{a^{-1}} = (\lambda_a)^{-1} \qquad\qquad \forall a \in G.$$

Thus λ_a must be bijective, cf. Prop. 1.4.6; hence we get a map

(2.4.10) $\qquad\qquad \lambda : \ G \longrightarrow \mathrm{Sym}(G) \,, \qquad a \longmapsto \lambda_a.$

By (2.4.9) this is a group homomorphism. Moreover, if $\lambda_a = \lambda_b$, then $a = a1_G = \lambda_a(1_G) = \lambda_b(1_G) = b1_G = b$; i.e. (2.4.10) is injective.

Definition. In general, a homomorphism of a group to a transformation (permutation) group on a set (finite set) is said to be a *transformation (permutation) representation* of the group on the set; further, the representation is said to be *faithful* if the homomorphism is injective.

Summarizing the above, we reach our aim:

2.4.11 Theorem (Cayley's Left Regular Representation). *For any group G, mapping $a \in G$ to the left translation by a is a faithful transformation (permutation if $|G| < \infty$) representation of G (on the underlying set G).*   □

Just like Remark 1.4.10, it is easy to reform the above to get an isomorphism. Let

$$L(G) = \mathrm{Im}(\lambda) = \{\lambda_a \mid a \in G\} \leq \mathrm{Sym}(G);$$

then we get an isomorphism $\lambda : G \overset{\sim}{\to} L(G)$.

2.4.12 Corollary. *Any group is isomorphic to the transformation group consisting of the left translations by the elements of the group.* □

In order to understand homomorphisms well, we should consider what about the relation in the domain determined by the homomorphism, see (1.4.8). In the next section, we will discuss such relations determined by general subgroups.

Exercises 2.4

1. 1) Let $\mathrm{GL}_n(\mathbf{C})$ and $\mathbf{C}^*$ resp. be the groups as in (2.2.7) and Exer.2.2.2(6) resp. Show that

   $$\mathrm{GL}_n(\mathbf{C}) \longrightarrow \mathbf{C}^* , \qquad A \longmapsto \det(A)$$

   is a surjective homomorphism of groups, where $\det(A)$ is the determinant of the matrix A.

   2) Let $M_n(\mathbf{C})$ and $\mathbf{C}$ resp. be the groups as in (2.2.6) and Exer.2.2.2(5) resp. Show that

   $$M_n(\mathbf{C}) \longrightarrow \mathbf{C} , \qquad A \longmapsto \mathrm{tr}(A)$$

   is a surjective homomorphism of groups, where $\mathrm{tr}(A)$ is the trace of the matrix A (i.e. the sum of the diagonal entries of A).

2. Let A and B be groups. If there is a surjective homomorphism $A \to B$ and A is Abelian, then B is also Abelian.

+3. Let G be a group, let H be a set with an operation "$*$". If $f : G \to H$ is a surjective map such that $f(ab) = f(a) * f(b)$, $\forall a, b \in G$, then H with the operation "$*$" forms a group, hence f is a group homomorphism.

+4. Let G be a group. For $a \in G$ define $\tau_a : G \to G$, $x \mapsto xa$. Then:

   1) $\tau_a \in \mathrm{Sym}(G)$ (τ_a is called the *right translation* by a);

2) $\tau : G \to \mathrm{Sym}(G)$, $a \mapsto \tau_a$ is an injective homomorphism (called the *right regular representation* of G);

3) let $R(G) = \{\tau_a \mid a \in G\}$, then $R(G)$ with the composition of transformations is a group and $G \cong R(G)$.

5. Let G be a group. For $a \in G$, the map $\sigma_a : G \to G$, $x \mapsto axa^{-1}$ is an automorphism of G.

$^+$6. Let $f : G \to \overline{G}$ be a homomorphism between groups G and $\overline{G}$. Prove:

1) If $H \leq G$, then $f(H) \leq \overline{G}$.

2) If $\overline{H} \leq \overline{G}$, then $f^{-1}(\overline{H}) \leq G$.

$^+$7. Let G be a group.

1) The set $\mathrm{End}(G)$ of all endomorphisms of G and the composition of endomorphisms form a monoid.

2) The set $\mathrm{Aut}(G)$ of all automorphisms of G and the composition of automorphisms form a group (called the automorphism group of G).

8. Prove that the group K_4 in (2.1.15) is isomorphic to the group K in (2.3.5).

$^+$9. Let $f, f' : G \to H$ be two homomorphisms of groups G and H; assume $G = \langle S \rangle$ is generated by a subset S. If the restricted maps $f|_S = f'|_S$, then $f = f'$.

2.5 Cosets

We have looked at the following example several times (1.4.1 & 2.4.6):

$$f : \mathbf{Z} \longrightarrow \mathbf{Z}_m , \qquad n \longmapsto [a] ;$$

which is a group homomorphism. We are concerned with the relation "$\sim_f$" determined by f and the associated partition of $\mathbf{Z}$. From the observation in (1.4.1) we have

$$a \sim_f b \iff a \equiv b (\mathrm{mod}\ m) \iff a - b \in m\mathbf{Z} = \{mk \mid k \in \mathbf{Z}\},$$

and the full inverse image in $\mathbf{Z}$ of $[a]$ is

$$a + m\mathbf{Z} = \{a + mk \mid k \in \mathbf{Z}\};$$

and, in our present notation, $m\mathbf{Z} \leq \mathbf{Z}$.

These suggest the important concept of cosets for any subgroup H of any group G. Note that $a - b = a + (-b)$ in an additive group; but an arbitrary group G need not be commutative, so that we can have two distinct "differences": right difference ab^{-1}, and left difference $b^{-1}a$.

Just like the familiar expression $a + m\mathbf{Z}$, we first introduce:

2.5.1 Notation. Let G be a group and $S \subset G$ and $T \subset G$. Denote
$$ST = \{st \mid s \in S, t \in T\}.$$
If $S = \{a\}$, then we write ST (TS resp.) as aT (Ta resp.) in short.

It is easy to verify that associativity holds for the above "multiplication" of subsets (see Exer.1(1)); thus, such notation $HaHb$, xSx^{-1} etc. make sense in a group.

2.5.2 Proposition. *Let G be a group and $H \leq G$. For all $a, b \in G$ define: $a \equiv b(\mathrm{mod}_r H)$ iff $ab^{-1} \in H$. Then:*

1) *"$\equiv (\mathrm{mod}_r H)$" is an equivalence relation in G;*

2) *the equivalence class of a is the subset $Ha = \{ha \mid h \in h\}$, which is called the* right *coset of H in G containing a;*

3) *$|Ha| = |H|$ for any right coset Ha.*

Proof. (1). Reflexity: $aa^{-1} = 1 \in H$. Transitivity: if $ab^{-1} \in H$ and $bc^{-1} \in H$ then $ac^{-1} = ab^{-1}bc^{-1} \in H$. Symmetry: if $ab^{-1} \in H$ then $ba^{-1} = (ab^{-1})^{-1} \in H$.

(2). For $a \in G$, $b \equiv a(\mathrm{mod}_r H)$ iff $ba^{-1} \in H$ iff there is an $h \in H$ such that $ba^{-1} = h$, or equivalently $b = ha$ iff $b \in Ha$.

(3). It is clear that $H \to Ha$, $h \mapsto ha$ is a bijective.   □

In the similar way, we have:

2.5.2$'$ Proposition. *Let G be a group and $H \leq G$. For all $a, b \in G$ define: $a \equiv b(\mathrm{mod}_l H)$ iff $b^{-1}a \in H$. Then:*

1) *"$\equiv (\mathrm{mod}_l H)$" is an equivalence relation in G;*

2) *the equivalence class of a is the subset $aH = \{ah \mid h \in h\}$, which is called the* left *coset of H in G containing a;*

3) *$|aH| = |H|$ for any left coset aH.*   □

Thus we have two partitions on G: G is the disjoint union of all the left cosets of H in G; and it is also the disjoint union of all the right cosets of H in G. Further, we have the following notation.

2.5.3 Proposition. *Let G be a group and $H \leq G$. Then the set of all left cosets of H in G and the set of all right cosets of H in G have the same cardinality, which is called the* index *of H in G and denoted $|G : H|$.*

Proof. Consider the map $\nu : G \to G$, $x \mapsto x^{-1}$ which is clearly bijective since $\nu\nu = \mathrm{id}_G$ (identity transformation). Then ν induces a map (by ν again for brevity)

$$\nu : \quad \mathcal{P}(G) \longrightarrow \mathcal{P}(G) , \qquad S \longmapsto \nu(S) = S^{-1} = \{s^{-1} \mid s \in S\}$$

and it is also true that $\nu\nu = \mathrm{id}_{\mathcal{P}(G)}$. For any left coset $aH \in \mathcal{P}(G)$ it is easy to check (Exer.1(3)(4)) that $\nu(aH) = Ha^{-1}$ which is a right coset of H; it is also true that ν maps a right coset to a left coset. Thus ν induces a bijective map from the set of all left cosets of H to the set of all right cosets of H. □

Combining the above, we get the following at once.

2.5.4 Lagrange's Theorem. *Let G be a group and $H \leq G$. Then $|G| = |H| \cdot |G : H|$. In particular, $|H| \big| |G|$ if G is finite.* □

Exer.7 shows that the converse of this theorem is false.

2.5.5 Corollary. *If G is a finite group, then $|a| \big| |G|$ for any element $a \in G$.*

Proof. By Def.2.3.12 and the above theorem. □

2.5.6 Notation. Let H be a subgroup of a group G. We can label all the right cosets of H in G by an index set I, then by Prop.2.5.3 we also label all the left cosets of H in G by I. For every right coset of H in G we select a representative a_i, i.e. it is the right coset Ha_i; and $\{a_i \mid i \in I\}$ is a *system of right representatives* of H in G, which is also called a *right transversal* of H in G. Similarly, we have *systems of left representatives* of H in G, or *left transversals* of H in G. From a right transversal $\{a_i \mid i \in I\}$ of H in G, by Prop.2.5.3 and its proof, $a_i^{-1}H$ is the left coset of H labelled by $i \in I$, and $\{a_i^{-1} \mid i \in I\}$ is a left transversal of H in G; by Props. 2.5.2 and 2.5.2′ we have the following two disjoint unions of G:

$$G = \bigcup_{i \in I} Ha_i = \bigcup_{i \in I} a_i^{-1}H .$$

Example. Consider $G = S_3$, the symmetric group of degree 3, and $H = \{(1), (12)\}$ a subgroup of order 2. Then $|G : H| = 3$ by Theorem 2.5.4; and $a_1 = (1)$, $a_2 = (123)$, $a_2 = (132)$ form a right transversal of H in G, and the following is the disjoint union of the corresponding right cosets:

$$G = \{(1)(1), (12)(1)\} \bigcup \{(1)(123), (12)(123)\} \bigcup \{(1)(132), (12)(132)\}.$$

Exercises 2.5

+1. Let G be a group and S, T, W be subsets of G, and $S^{-1} = \{s^{-1} \mid s \in S\}$. Prove:

1) $(ST)W = S(TW)$; hence we can write it as STW;

2) in this notation, (2.3.10) can be rewritten as (where $\overline{S} = S \cup S^{-1}$)
$$\langle S \rangle = \{1\} \cup \overline{S} \cup \overline{SS} \cup \overline{SSS} \cup \cdots;$$

3) $(ST)^{-1} = T^{-1}S^{-1}$;

4) if $W \leq G$ then $W^{-1} = W$ and $WW = W$;

5) if $W \leq G$ and $1 \in S \subset W$, then $WS = SW = W$.

2. Complete the proof of Proposition 2.5.2′.

3. Write down the corresponding left transversal and disjoint union of the example following Remark 2.5.6.

4. Let G be a group, $H \leq G$ and $a \in G$. Then $aH = H \iff a \in H$; in particular, $1_G H = H = H 1_G$ is both a left and right coset.

5. Let G be a group and $H \leq G$ and $a, b \in G$. The following are equivalent:
   i). $aH \bigcap bH \neq \emptyset$;   ii). $aH = bH$;   iii). $a^{-1}b \in H$;
   iv). $b^{-1}a \in H$;   v). $a \in bH$;   vi). $b \in aH$.

+6. Let $K \leq H \leq G$ be groups. Prove: $|G : K| = |G : H| \cdot |H : K|$.
   (Hint: for a left transversal a_i, $i \in I$, of H in G, and a left transversal h_j, $j \in J$, of K in H, prove that $a_i h_j$, $i \in I$ and $j \in J$, form a left transversal of K in G.)

7. Prove: the following is a permutation group of degree 4
   $$A_4 = \{ \ (1), (12)(34), (13)(24), (14)(23),$$
   $$(123), (132), (124), (142), (134), (143), (234), (243) \ \}$$
   of order 12 but has no subgroup of order 6.
   (Hint: A_4 is closed under the composition. Suppose $H \leq A_4$ and $|H| = 6$, then $a = (123) \in H$; for, if $a \notin H$, then $aH \neq a^2 H \neq H$, hence $|G : H| \geq 3$, which is impossible; so every 3-cycle of A_4 belongs to H, hence H contains at least 8 elements, impossible again.)

8. Let G be a group and $H, K \leq G$. Then $HK = KH$ iff $HK \leq G$. In particular, if G is an additive group and $H, K \leq G$, then $H + K \leq G$.

+9. In the additive group $\mathbf{Z}$, prove:

1) $m\mathbf{Z} + n\mathbf{Z} = (m, n)\mathbf{Z}$, where (m, n) denotes the largest common divisor of m and n.

 2) $m\mathbf{Z} \cap n\mathbf{Z} = [m, n]\mathbf{Z}$, where $[m, n]$ denotes the least common multiple of m and n.

*10. Let G be a group and $H, K \leq G$. If $\{h_i \mid i \in I\}$ is a left transversal of $H \cap K$ in H, show that $HK = \bigcup_{i \in I} h_i K$ which is a disjoint union; deduce that

$$|HK| = (|H| \cdot |K|)/|H \cap K| \qquad \text{and} \qquad |H : H \cap K| \leq |G : K|.$$

*11. Notations as above Exer.10. Prove:

 1) $|G : H \cap K| \leq |G : H| \cdot |G : K|$.

 2) If both $|G : H|$ and $|G : K|$ are finite and they are coprime, then the equality in the above (1) holds.

 (Hint: for (1), $|G : H \cap K| \leq |HK : H \cap K| = |HK : H| \cdot |H : H \cap K| \leq |G : H| \cdot |G : K|$; for (2), $|G : H| \big| |G : H \cap K|$ and $|G : K| \big| |G : H \cap K|$.)

2.6 Normal Subgroups, Quotient Groups

Let $f : G \to H$ be a homomorphism between groups G and H, we want to make the Decomposition of Maps (1.4.8) to have group-theoretical sense. Let $K = \text{Ker}(f)$ be the kernel of f, we know that $K \leq G$, see Cor.2.4.4. Further, for any $a, b \in G$, $f(a) = f(b)$ iff $f(a)f(b)^{-1} = 1_H$ iff $f(ab^{-1}) = 1_H$ iff $ab^{-1} \in K$; so, by Prop.2.5.2, the equivalence relation "$\sim_f$" determined by f is just the relation "$\equiv (\text{mod}_r K)$", hence the quotient set determined by f is just the partition of all right cosets of K in G. However, one can see at once that, the same argument shows that the quotient set determined by f is also the partition of all left cosets of K in G. Thus, the left cosets of K in G must coincide with the right cosets of K in G. On the other hand, we have remarked after Cor.2.4.4 that $aya^{-1} \in K$ for all $a \in G$ and $y \in K$.

2.6.1 Definition. A subgroup K of a group G is said to be *normal* in G, denoted by $K \lhd G$ or $K \unlhd G$, if $aKa^{-1} = K$ for all $a \in G$.

It is clear that the trivial subgroups $\{1\}$ and G are normal subgroups of G, which are called the trivial normal subgroups.

2.6.2 Proposition. *Let G be a group and $K \leq G$. The following four are equivalent:*

 i) $K \unlhd G$;

ii) $aKa^{-1} \subset K$ for all $a \in G$;

iii) $axa^{-1} \in K$ for all $a \in G$ and $x \in K$;

iv) $aK = Ka$ for all $a \in G$.

Proof. (i) $\Longrightarrow$ (ii) $\Longrightarrow$ (iii):   Trivial.

(iii) $\Longrightarrow$ (iv):   For any $ax \in aK$, $y = axa^{-1} \in K$ by (iii), then $ax = ya \in Ka$; so $aK \subset Ka$ for any $a \in G$. Conversely, for any $xa \in Ka$, $z = a^{-1}xa \subset K$ by (iii) again, then $xa = az \in aK$; so $Ka \subset aK$.

(iv) $\Longrightarrow$ (i): $aKa^{-1} = (aK)a^{-1} = (Ka)a^{-1} = K(aa^{-1}) = K$. $\square$

By (iv) of the proposition, every left coset of a normal subgroup is a right coset of the normal subgroup, and *vice versa*; so we can speak of "cosets" without the attributives "left" or "right"; and, collecting them to form a set, we get a partition, or a quotient set (see Theorem 1.3.5), of the underlying set of the group; and the observation of the beginning of this section can be stated as:

2.6.3 Proposition. *If $f : G \to H$ is a homomorphism of groups and $K = \mathrm{Ker}(f)$ is its kernel, then $K \trianglelefteq G$, and $f(a) = f(b)$ for $a, b \in G$ iff $ab^{-1} \in K$ iff $b^{-1}a \in K$, i.e. the quotient set determined by f is just the set of all cosets of K in G.* $\square$

2.6.4 Proposition.   *Let G be a group and $K \trianglelefteq G$, let G/K denote the set consisting of all the cosets of K in G. Then,*

$$(aK)(bK) = (ab)K \qquad\qquad \text{for any cosets } aK, \ bK$$

is a well-defined operation on G/K, and, with this operation, G/K is a group; moreover, the map

$$\sigma: \ G \longrightarrow G/K \ , \qquad a \longmapsto aK$$

is a surjective group homomorphism with kernel K.

Proof. If $a'K = aK$ and $b'K = bK$, then

$$\begin{aligned} a'b'K &= a'(b'K) = a'(b'Kb'^{-1})b' = a'Kb' \\ &= (a'K)b' = aKb' = a(Kb') = a(b'K) = abK; \end{aligned}$$

thus the definition of the multiplication on G/K is independent of the choice of the representatives of the cosets. This operation is associative by Exer.2.5.1(1). It is easy to check that the coset K is the identity element for this operation, and $a^{-1}K$ is the inverse element of the element

$aK \in G/K$. It is obvious that σ is surjective and $\sigma^{-1}(K) = K$. At last, for $a, b \in G$ we have (cf. Exer.2.5.1(5)):

$$\sigma(ab) = abK = a(bKK) = a(KbK) = (aK)(bK) = \sigma(a)\sigma(b). \qquad \square$$

2.6.5 Definition. Notations as above proposition. The group G/K is called the *quotient group* of G with respect to K, and the group homomorphism σ is called the *natural or canonical homomorphism*.

2.6.6 Remark. It is clear that $|G/K| = |G : K|$. For the trivial normal subgroups we have: $G/\{1\} \cong G$ and $G/G \cong \{1\}$, see Exer.10; in other words, any non-trivial group has at least two quotient groups.

2.6.7 Definition. A non-trivial group is called a *simple group* if it has no normal subgroup other than the trivial normal subgroups.

It is easy to determine all Abelian simple groups, see Exer.2.8.2; however, it is very hard to determine all non-Abelian simple groups. In fact, all *finite* non-Abelian simple groups are determined in 1980's, it was one of the great achievements of Mathematics in the 20th century.

Exercises 2.6

$^{+}$1. Every subgroup of an Abelian group is a normal subgroup.

$^{+}$2. The center $Z(G)$ of a group G (see (2.3.3)) is a normal subgroup of G . Moreover, every subgroup of the center $Z(G)$ is a normal subgroup of G.

3. Let G be a group and $N \leq G$. The following are equivalent:

   i) $N \trianglelefteq G$;

   ii) every left coset of N in G is a right coset of N in G;

   iii) The product of every two left cosets of N in G is a left coset of N in G again.

$^{+}$4. The intersection of two normal subgroups is again a normal subgroup.

$^{+}$5. Let G be a group and $H, N \leq G$.

   1) If $N \trianglelefteq G$, then $HN = NH$ and it is a subgroup of G.

   2) If both $H, N \trianglelefteq G$, then $HN \trianglelefteq G$.

6. Let H be a subgroup of G with index 2. Prove that $H \trianglelefteq G$ (a generalization is stated in Exer.11 and Exer.12 below).

7. If $N \trianglelefteq G$ and $|N| = 2$, then $N \subset Z(G)$.

8. Consider the permutation group A_4 in Exer.2.5.7. Prove:

   1) $K_4 = \{(1), (12)(34), (13)(24), (14)(23)\}$ is a normal subgroup of A_4.

   2) $C_2 = \{(1), (12)(34)\}$ is a normal subgroup of K_4.

3) C_2 is not a normal subgroup of A_4.

9. 1) If $f : G \to \overline{G}$ is a homomorphism of groups and $\overline{N} \trianglelefteq \overline{G}$ then $f^{-1}(\overline{N}) \trianglelefteq G$.

2) If $f : G \to \overline{G}$ is a surjective homomorphism of groups and $N \trianglelefteq G$ then $f(N) \trianglelefteq \overline{G}$. Give an example to show that the condition "surjective" is necessary.

(Hint: Consider $G = \{(1), (12)\}$ and $\overline{G} = S_3$ and the inclusion homomorphism $G \to \overline{G}$. Compare with Exer.2.4.6.)

+10. Let G be a group. Prove: 1). $G/\{1\} \cong G$;   2). $G/G \cong \{1\}$.

*11. Let G be a finite group and $H \leq G$. If $|G : H|$ equals the least prime factor of $|G|$, then $H \trianglelefteq G$.

(Hint: for $x \in G - H$, let $K = x^{-1}Hx$; then $|HK| = |H| \cdot (|K|/|H \cap K|)$ by Exer.2.5.10, and $(|K|/|H \cap K|)||G|$ by Theorem 2.5.4; suppose $K \neq H$, by the hypothesis we have $|HK| = |G|$, i.e. $G = HK$; so there are $h, h' \in H$ such that $x = hx^{-1}h'x$, which implies $x \in H$.)

*12. Let G be a group and $H \leq G$.

1) $H_G = \bigcap_{x \in G} x^{-1}Hx \trianglelefteq G$ (H_G is called the *core* of H in G).

2) If $\{x_i \mid i \in I\}$ is a right transversal of H in G, then $\bigcap_{i \in I} x_i^{-1} H x_i = H_G \trianglelefteq G$.

3) If $|G : H|$ is finite, then $|G : H_G|$ is finite.

4) Show that the hypothesis that G is finite in Exer.11 is not necessary; how to restate and prove Exer.11 without this hypothesis?

2.7   Homomorphism Theorems

2.7.1 Fundamental Theorem (for Group Homomorphisms).   *Let G and H be groups and $f : G \to H$ be a homomorphism with kernel K. Then $K \trianglelefteq G$ and there exists a unique group homomorphism $\overline{f} : G/K \to H$ such that $f = \overline{f} \circ \sigma$, where $\sigma : G \to G/K$ is the natural homomorphism; and such $\overline{f}$ is injective; moreover, $\overline{f}$ is an isomorphism provided f is surjective.*

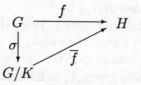

Proof. From Prop.2.6.3, the quotient set of G determined by f is just G/K. By the Fundamental Decomposition of Maps 1.4.8, as a map such $\overline{f}$ exists and unique and it must be defined by (1.4.9); thus it remains only to prove that the map $\overline{f}$ in (1.4.9) is indeed a group homomorphism in the present case. For any $aK, bK \in G/K$, by the definition (1.4.9), $\overline{f}(aK) = f(a)$, etc.; then the following completes the proof:

$$\overline{f}\big((aK)(bK)\big) = \overline{f}(abK) = f(ab) = f(a)f(b) = \overline{f}(aK)\overline{f}(bK). \qquad \square$$

2.7.2 Remark. Similar to Remark 1.4.10, we usually refer the above $\overline{f}$ as the homomorphism *induced by f*. And, if $f : G \to H$ is not a surjective homomorphism, we can consider the inclusion homomorphism $\iota : \mathrm{Im}(f) \to H$; then the induced homomorphism $\overline{f} : G/K \to H$ can be written as a composition $G/K \overset{\tilde{f}}{\to} \mathrm{Im}(f) \overset{\iota}{\to} H$, where $\tilde{f} : G/K \to \mathrm{Im}(f)$ is always an isomorphism.

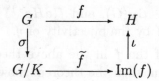

In particular, we have

2.7.3 Corollary (First Isomorphism Theorem). *If $f : G \to H$ is a group homomorphism, then $G/\mathrm{Ker}(f) \cong \mathrm{Im}(f)$; in particular, $G/\mathrm{Ker}(f) \cong H$ provided f is surjective.* $\square$

2.7.4 Theorem (Correspondence of Subgroups). *Let $f : G \to \overline{G}$ be a group surjective homomorphism and $K = \mathrm{Ker}(f)$ be its kernel; let $\mathcal{L}(\overline{G}) = \{\overline{H} \mid \overline{H} \leq \overline{G}\}$ while $\mathcal{L}_K(G) = \{H \mid H \leq G \text{ and } H \supset K\}$. Then*

1) $\eta : \mathcal{L}_K(G) \to \mathcal{L}(\overline{G})$, $H \mapsto f(H)$ is a bijective map;

2) for $H_1, H_2 \in \mathcal{L}_K(G)$, $H_1 \subset H_2$ iff $f(H_1) \subset f(H_2)$, and $|H_2 : H_1| = |f(H_2) : f(H_1)|$ in this case;

3) for $H \in \mathcal{L}_K(G)$, $H \unlhd G$ iff $f(H) \unlhd \overline{G}$.

Proof. (1). $f(H) \in \mathcal{L}(\overline{G})$ by Exer.2.4.6; thus η is well-defined. Since $K = f^{-1}(\{1_{\overline{G}}\}) \subset f^{-1}(\overline{H})$ for any $\overline{H} \in \mathcal{L}(\overline{G})$, the following is again a well-defined map (cf. Exer.2.4.6):

(2.7.5) $\qquad \xi : \mathcal{L}(\overline{G}) \longrightarrow \mathcal{L}_K(G) , \qquad \overline{H} \longmapsto f^{-1}(\overline{H}).$

For $\overline{H} \in \mathcal{L}(\overline{H})$, $\eta\xi(\overline{H}) = \overline{H}$ by the surjectivity of f. For $H \in \mathcal{L}_K(G)$, it is clear that $H \subset \xi\eta(H)$; on the other hand, for $a \in \xi\eta(H)$, there is an $h \in H$ such that $f(h) = f(a)$, hence $ah^{-1} \in K$ by Prop.2.6.3; consequently, $a \in Kh \subset KH = H$ by Exer.2.5.1(5) because $K \subset H$; so $\xi\eta(H) \subset H$. Therefore η and ξ are invertible, and both are bijective.

(2). "only if" is obvious. Conversely, assume that $f(H_1) \subset f(H_2)$, then, by the above (1), $H_1 = f^{-1}(f(H_1)) \subset f^{-1}(f(H_2)) = H_2$. Let b_i, $i \in I$ be a left transversal of H_1 in H_2, then $H_2 = \bigcup_{i \in I} b_i H_1$ and $b_i^{-1} b_{i'} \in H_1$ iff $i = i'$; thus $f(H_2) = \bigcup_{i \in I} f(b_i) f(H_1)$, and $f(b_i) = f(b_{i'})$ iff $b_i^{-1} b_{i'} \in K \subset H_1$ iff $i = i'$; so $f(b_i)$, $i \in I$ form a left transversal of $f(H_1)$ in $f(H_2)$. Hence $|H_2 : H_1| = |I| = |f(H_2) : f(H_1)|$.

(3). If $H \trianglelefteq G$, for any $\overline{a} \in \overline{G}$, there is an $a \in G$ such that $f(a) = \overline{a}$; then $\overline{a} f(H) \overline{a}^{-1} = f(a) f(H) f(a^{-1}) = f(aHa^{-1}) = f(H)$; i.e. $f(H) \trianglelefteq \overline{G}$. Conversely, assume that $f(H) \trianglelefteq \overline{G}$, then, for any $a \in G$, $K = aKa^{-1} \subset aHa^{-1}$ hence $aHa^{-1} \in \mathcal{L}_K(G)$, and $f(aHa^{-1}) = f(a) f(H) f(a)^{-1} = f(H)$, thus $aHa^{-1} = H$ by the bijectivity of η. $\square$

2.7.6 Remark. If the f in the above theorem is also injective, i.e. f is an isomorphism, then the theorem shows that two isomorphic groups have all the same properties on subgroups. As we saw in §2.4, two isomorphic groups are really the same provided we map one of them to the other by an isomorphism.

2.7.7 Corollary (Second Isomorphism Theorem).   *Let G be a group and $H, K \leq G$. If $K \trianglelefteq G$, then $HK \leq G$ and $K \trianglelefteq HK$ and $H \cap K \trianglelefteq H$, and we have a group isomorphism:*

$$H/(H \cap K) \xrightarrow{\cong} (HK)/K , \qquad h(H \cap K) \longmapsto hK .$$

Proof. $HK \leq G$ by Exer.2.6.5(1). Of course, $aKa^{-1} = K$ for any $a \in HK$ since $K \trianglelefteq G$ and $a \in G$; i.e. $K \trianglelefteq HK$. For any $h \in H$, it is trivial that $h(H \cap K)h^{-1} \subset H$; and $h(H \cap K)h^{-1} \subset K$ since $K \trianglelefteq G$ and cf. Prop.2.6.2(iii); thus $h(H \cap K)h^{-1} \subset H \cap K$, hence $(H \cap K) \trianglelefteq H$ by Prop.2.6.2(ii). Therefore, both sides of the desired isomorphism make sense. To prove the isomorphism, consider the composition of the inclusion homomorphism $H \to HK$ and the natural homomorphism $HK \longrightarrow (HK)/K$:

$$H \longrightarrow HK \longrightarrow (HK)/K , \qquad h \longmapsto hK .$$

Since the kernel of $HK \to (HK)/K$ is K (see Prop.2.6.4), it is clear that only the elements in $H \cap K$ are mapped by the above composition map to the identity element of $(HK)/K$; i.e. the kernel of the above composition homomorphism is $H \cap K$; on the other hand, for any $(hk)K \in (HK)/K$, it is clear that $(hk)K = hK$ which is the image of $h \in H$, i.e. the above composition is surjective; by Theorem 2.7.1 and cf. Cor.2.7.3, this composition homomorphism induces the desired isomorphism. □

The following result is a corollary of Theorem 2.7.1, and also a complement for Theorem 2.7.4(3).

2.7.8 Corollary (Third Isomorphism Theorem). *Let G be a group and both $N \subset K$ normal subgroups of G. Then $K/N \trianglelefteq G/N$ and we have a group isomorphism:*

$$G/K \xrightarrow{\cong} (G/N)\big/(K/N) \ , \qquad aK \longmapsto (aN)(K/N) .$$

Proof. Consider the natural homomorphism $G \to G/N$, then $K/N \trianglelefteq G/N$ by Theorem 2.7.4(3); hence we have natural homomorphism $G/N \to (G/N)/(K/N)$. Thus we have the composition homomorphism:

$$G \longrightarrow G/N \longrightarrow (G/N)\big/(K/N) \ , \qquad a \longmapsto (aN)(K/N),$$

which is a surjective homomorphism since both natural homomorphisms are surjective; on the other hand, the inverse image in G/N of the identity element of $(G/N)\big/(K/N)$ is K/N by Prop.2.6.4, and the inverse image in G of $K/N \subset G/N$ is K by Theorem 2.7.4(1); thus the kernel of the above composition homomorphism is K. By Theorem 2.7.1, the composition homomorphism induces the wanted isomorphism. □

We conclude this section by an example.

2.7.9 Example. Let S_n be the symmetric group of degree n, see Examples 2.2.4 and 2.1.6. On the other hand, $\{\pm 1\}$ is clearly a group with respect to the number multiplication. For $\alpha \in S_n$ define :
$$\text{sign}(\alpha) = \begin{cases} 1 & \text{if } \alpha \text{ is even} \\ -1 & \text{if } \alpha \text{ is odd} \end{cases} \quad : \quad \text{cf. Exer.2.1.6. Then it is easy}$$
(cf Exer.2.1.7) to see that $\alpha \mapsto \text{sign}(\alpha)$ is a surjective homomorphism from S_n onto $\{\pm 1\}$, and the kernel of this homomorphism is $A_n =$

{even permutations of degree n}. Hence, by the Fundamental Theorem 2.7.1, $A_n \lhd S_n$ and $S_n/A_n \cong \{\pm1\}$. The details are left as Exer.3.

Exercises 2.7

1. Let $f : G \to H$ be a group homomorphism. Prove:
   1) f is injective if and only if $\text{Ker}(f) = \{1\}$.
   2) f is surjective if and only if $\text{Im}(f) = H$.
2. Consider the homomorphism
$$\rho_m :\ \mathbf{Z} \longrightarrow \mathbf{Z}_m\ , \qquad n \longmapsto [n]_m\ .$$
   Prove that $\text{Ker}(\rho_m) = \langle m \rangle = m\mathbf{Z}$, and $\mathbf{Z}/\text{Ker}(\rho_m) = \mathbf{Z}_m$; hence ρ_m is just the natural homomorphism.
3. Complete the details of Example 2.7.9.
+4. (Factorization Theorem for Group Homomorphisms) Let $f : G \to X$ and $g : G \to H$ be group homomorphisms and g be surjective. Then there is a homomorphism $h : H \to X$ such that $f = hg$ if and only if $\text{Ker}(g) \subset \text{Ker}(f)$; and, if this is the case, such h is unique and $\text{Ker}(h) = g(\text{Ker}(f))$.

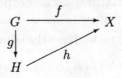

(Hint: cf. Exer.1.4.9. The unique homomorphism h is referred to as the *homomorphism induced by f*.)

2.8   Cyclic Groups, Orders of Elements

As an application of the homomorphism theorems, we classify the cyclic groups up to isomorphism, and describe them precisely.

Recall from Def.2.3.12 that a subgroup C of a group G is said to be cyclic if $C = \langle a \rangle = \{a^n \mid n \in \mathbf{Z}\}$ for some $a \in C$ which is called a generator of C; and the order of $\langle a \rangle$ is also called the order of a, and denoted by $|a|$; and G is said to be cyclic if $G = \langle a \rangle$ for some $a \in G$.

2.8.1 Lemma.   *Every subgroup of the additive group $\mathbf{Z}$ is a cyclic subgroup $\langle m \rangle = m\mathbf{Z} = \{mk \mid k \in \mathbf{Z}\}$, and:*

1) *if* $m = 0$*, then* $\mathbf{Z}/\langle m \rangle \cong \mathbf{Z}$*; and* $\{0\}$ *is the unique subgroup of* $\mathbf{Z}$ *of infinite index;*

2) *if* $m \neq 0$*,* $\mathbf{Z}/\langle m \rangle = \mathbf{Z}_m$ *as in (2.2.8); and* $\langle m \rangle$ *is the unique subgroup of* $\mathbf{Z}$ *of index* $|m|$*.*

3) $\langle m_1 \rangle \subset \langle m_2 \rangle$ *iff* $m_2 | m_1$*; and* $|\langle m_2 \rangle : \langle m_1 \rangle| = m_1/m_2$ *in this case.*

Proof. Consider $H \leq \mathbf{Z}$ arbitrary, then $H \neq \emptyset$. If $H = \{0\}$, it is clear that $H = \langle 0 \rangle$ is cyclic and every coset $n + \{0\} = \{n\}$ consists of only one element n; and $\mathbf{Z} \to \mathbf{Z}/\langle 0 \rangle$, $n \mapsto \{n\}$ is clearly an isomorphism. This is the case of (1). Next, assume $H \neq 0$; then there is $0 \neq k \in H$; if $k < 0$, then $0 < -k \in H$. Thus H contains positive integers; let m be the least positive integer in H. For any $h \in H$, by Euclid Division we have $h = qm + r$ for $q, r \in \mathbf{Z}$ and $0 \leq r < m$; hence $r = h - qm \in H$; it forces that $r = 0$, i.e. $h = qm \in \langle m \rangle$. Thus $H = \langle m \rangle$ is cyclic. Further, since $\langle m \rangle = m\mathbf{Z}$, any coset of $\langle m \rangle$ in $\mathbf{Z}$ is as $a + \langle m \rangle = a + m\mathbf{Z}$; then, it is easy to see that the operation on the quotient group $\mathbf{Z}/\langle m \rangle$ defined in (2.6.4) is just the addition on $\mathbf{Z}_m$ defined in (2.2.8). So $\mathbf{Z}/\langle m \rangle = \mathbf{Z}_m$; consequently, $|\mathbf{Z} : \langle m \rangle| = |m|$. At last, noting that all the subgroups H of $\mathbf{Z}$ have been considered, we get the uniqueness of both (1) and (2).

The first conclusion of (3) is clear, and the second one follows from $|\mathbf{Z} : \langle m \rangle| = m$ and Exer.2.5.6. $\qquad\square$

2.8.2 Theorem. *Assume that* $G = \langle a \rangle$ *is a cyclic group.*

1) *If* $|G| = \infty$*, then* $\mathbf{Z} \overset{\cong}{\to} G$*,* $n \mapsto a^n$*.*

2) *If* $|G| = m < \infty$*, then* $\mathbf{Z}_m \overset{\cong}{\to} G$*,* $[n] \mapsto a^n$*.*

Proof. The following is clearly a surjective homomorphism:

(2.8.3) $\qquad\qquad \rho : \mathbf{Z} \longrightarrow G , \qquad n \longmapsto a^n .$

By Lemma 2.8.1 we can assume that the kernel of ρ is $\langle m \rangle$; and by Theorem 2.7.1, ρ induces a group isomorphism:

$$\bar{\rho} : \mathbf{Z}/\langle m \rangle \overset{\cong}{\longrightarrow} G , \qquad n + \langle m \rangle \longmapsto a^n .$$

Therefore, the theorem is proved by Lemma 2.8.1. $\qquad\square$

2.8.4 Corollary. *Let* G *be a group and* $a \in G$*, let* $n_1.n_2 \in \mathbf{Z}$*.*

1). *If* $|a| = \infty$*, then* $a^{n_1} = a^{n_2}$ *iff* $n_1 = n_2$*.*

2). *If* $|a| = m < \infty$*, then* $a^{n_1} = a^{n_2}$ *iff* $n_1 \equiv n_2 \pmod{m}$*; in* *particular, $a^k = 1$ iff $k \equiv 0 \pmod{m}$; hence $a^{|G|} = 1$, $\forall a \in G$, provided $|G| < \infty$.*

Proof. Consider the cyclic subgroup $\langle a \rangle$ of G.

1). By (1) of the above theorem, $\mathbf{Z} \to \langle a \rangle$, $n \mapsto a^n$, is injective; this is just the conclusion of (1).

2). By (2) of the above theorem, $\mathbf{Z}_m \to \langle a \rangle$, $n \mapsto a^n$ is injective. The last conclusion is a restatement of Cor.2.5.5.   □

2.8.5 Remark. Therefore, there are in fact only two kinds of cyclic groups up to isomorphism: infinite cyclic group $\mathbf{Z}$, and cyclic group $\mathbf{Z}_m$ of finite order m; and, in multiplicative notation, we can list the elements of a cyclic group *without repetition* as follows:

$$\langle a \rangle = \{\cdots, a^{-2}, a^{-1}, 1 = a^0, a = a^1, a^2, \cdots\} \quad \text{if it is infinite;}$$
$$\langle a \rangle = \{1 = a^0, a = a^1, a^2, \cdots, a^{m-1}\} \quad \text{if it is of finite order } m.$$

In fact, one can prove the above (even more, all the results of this section) directly without citing the homomorphism theorems, see Exer.5, which looks more elementary. However, here the idea to treat these objects, e.g. (2.8.3), is common and effective in algebra, and it will appear repeatedly later in this book.

2.8.6 Corollary. *The order of an element a in a group is either the least positive integer m such that $a^m = 1$, or ∞ if such positive integer does not exist.*

Proof. If there is a positive integer n such that $a^n = 1$, then it must be the case (2) of Cor.2.8.4, by which the least one of them is clearly the order of a. If such positive integer does not exist, then the case (2) of Cor.2.8.4 is impossible because it means that $a^m = a^0 = 1$; thus $|a| = \infty$.   □

For a finite group of order m, by Theorem 2.5.4, any subgroup has index dividing m, and its order and its index determine each other; in other words, only the divisors of m are possible to be indexes of some subgroups. But, for an infinite group, any natural number and ∞ are both possible to be an index of some subgroups, e.g. see Lemma 2.8.1.

With the technique of (2.8.3) and Theorem 2.7.4, we extend the result Lemma 2.8.1 to all cyclic groups.

2.8.7 Theorem. *Let $G = \langle a \rangle$ be a cyclic group. Let $k \big| |G|$ if $|G| < \infty$, or, let k be a natural number or ∞ if $|G| = \infty$. Then G has exactly* *one subgroup of index k, and it is cyclic and generated by a^k (with an agreement that $a^\infty = 1$). In particular, any subgroup of a cyclic group is also cyclic.*

Proof. If $|G| = \infty$, then $G \cong \mathbf{Z}$ by Theorem 2.8.2, and the conclusion follows by Lemma 2.8.1 and Theorem 2.7.4 (cf. Remark 2.7.6). Next, assume $|G| = m < \infty$. Consider the surjective homomorphism ρ in (2.8.3), by Theorem 2.7.4 the subgroups of index k in G bijectively correspond to the subgroups of index k in $\mathbf{Z}$, hence the existence and uniqueness are proved by Lemma 2.8.1; more precisely, the subgroup of G of index k is a homomorphism image of the subgroup $\langle k \rangle$ of index k in $\mathbf{Z}$, hence it is just $\langle a^k \rangle$ generated by a^k. $\qquad\square$

2.8.8 Remark. For a finite group, by Theorem 2.5.4, the index and the order of a subgroup determined each other. Thus, the above theorem can be restated as:

"If G is a cyclic group of order $m < \infty$ and $d|m$, then G has a unique subgroup of order d, and it is a cyclic group generated by $a^{m/d}$."

How to judge which element generates the unique subgroup of order d? Of course, only the elements of order d generate it.

2.8.9 Proposition. *If a is an element of finite order m in a group G and n is an integer, then $|a^n| = \frac{m}{(m,n)}$, where (m,n) stands for the greatest common divisor of m and n.*

Proof. By Prop.2.8.4, $(a^n)^k = a^{nk} = 1 = a^0$ iff $nk \equiv 0 \pmod{m}$ iff $\frac{n}{(m,n)} \cdot k \equiv 0 \pmod{\frac{m}{(m,n)}}$; but $\left(\frac{n}{(m,n)}, \frac{m}{(m,n)}\right) = 1$, hence, $(a^n)^k = 1$ iff $k \equiv 0 \pmod{\frac{m}{(m,n)}}$; the least one of such positive integers k is clearly $\frac{m}{(m,n)}$. $\qquad\square$

2.8.10 Corollary. *There are $\varphi(m)$ elements of order m in a cyclic group of order m ($\varphi(m)$ is the number-theoretic Euler function as in Exer.2.2.3).*

Proof. By Remark 2.8.5 the group is

$$\langle a \rangle = \{a^n \mid n = 0, 1, 2, \cdots, m-1\}.$$

Then this follows from the above, since $\frac{m}{(m,n)} = m$ iff $(m,n) = 1$. $\qquad\square$

As an application, for a positive integer m we have the following number-theoretic result (where $d|m$ means d runs over the positive divisors of m):

2.8.11 Corollary. $m = \sum_{d|m} \varphi(d)$.

Proof. First, we remark a general fact. For *any* group G of order m, its elements are classified by orders; let $\psi(d)$ denote the number of elements of order d, then $\psi(d) = 0$ if $d \nmid m$ by Cor.2.5.5; thus

(2.8.12) $m = \sum_{d|m} \psi(d)$;

Note that an element of order d generates a cyclic subgroup of order d.

Now consider a cyclic group G of order m. For any $d|m$, G has exactly one subgroup of order d by Remark 2.8.8; hence all the elements of order d belong to this unique subgroup of order d; thus $\psi(d) = \varphi(d)$ by Cor.2.8.10; and the corollary follows from (2.8.12). $\quad\square$

2.8.13 Theorem. *Let G be a group of order $m < \infty$. G is cyclic if and only if for any positive $d|m$ there is at most one subgroup of order d in G.*

Proof. The necessity follows from Theorem 2.8.7.

Conversely, assume the hypothesis of the theorem holds. Consider the formula (2.8.12) which holds for any groups of order m; an element of order d generates a subgroup of order d; but, from the hypothesis, such subgroup is unique if it exists; hence any elements of order d belong to the unique subgroup of order d, which contains exactly $\varphi(d)$ elements of order d by Cor.2.8.10. Therefore, $\psi(d) \leq \varphi(d)$ for all $d|m$. Further, comparing (2.8.11) and (2.8.12), we have $\psi(d) = \varphi(d)$ for all $d|m$. In particular, $\psi(m) = \varphi(m) > 0$; which implies that there is an element a of order m in G, hence $G = \langle a \rangle$ is a cyclic group. $\quad\square$

A similar argument can be applied to prove the following

2.8.14 Proposition. *Let G be a group of order $m < \infty$. G is cyclic if and only if for any $d|m$ the equation $x^d = 1$ with variable x has at most d solutions in G.*

Proof. Note that, by Cor.2.8.4(2), a solution x_0 of $x^d = 1$ must be of order $k|d$, hence generates a subgroup $\langle x_0 \rangle$ of order $k|d$. If G is cyclic, then G has a unique subgroup D of order d and a unique subgroup of order k by Remark 2.8.8, and D also has a unique subgroup of order k, so $\langle x_0 \rangle \subset D$. In other words, only the d elements of D are the solutions of the equation $x^d = 1$. The necessity is proved.

Conversely, assume the condition of the lemma holds. Note that all the d elements of a subgroup of order d are solutions of the equation

$x^d = 1$ by Cor.2.8.4(2); thus, by the condition, G has at most one subgroup of order d; hence G is cyclic by Theorem 2.8.13. □

Exercises 2.8

1. A group of prime order p is cyclic.

$^+$2. Prove that only the groups of prime order are the Abelian simple groups.

3. The homomorphism image of any cyclic group is also cyclic.

4. Let G and H be cyclic groups. There is a surjective homomorphism from G onto H if and only if either $|G| = \infty$, or $|G|, |H| < \infty$ and $|H| \,||G|$. (Remark: compare with Exer.1.4.1.)

5. Prove Cor.2.8.4 directly without citing homomorphism theorems. (Hint: consider sequence: $\cdots, a^{-2}, a^{-1}, 1 = a^0, a, a^2, \cdots$; if there is no repetition, it is the case (1); otherwise, there are positive integers n such that $a^n = 1$, and the least one m satisfies (2).)

6. Let $G = \langle a \rangle$ be a cyclic group of order $m < \infty$ and $d|m$. Prove that the set of the solutions of the equation $x^d = 1$ with variable x is the subgroup of order d.

7. Prove that any finitely generated subgroup of the additive group $\mathbf{Q}$ (cf. Exer.2.2.2) is a cyclic subgroup; but $\mathbf{Q}$ is not a cyclic group.

8. Let G be a group and $a, b \in G$. Prove:

   1). $|a| = |a^{-1}|$; 2). $|a| = |bab^{-1}|$; 3). $|ab| = |ba|$.

9. If every element of a group has order 2, then it is an Abelian group.

10. Prove:

    1) The number of the elements of order > 2 in a finite group is even.

    2) A group of even order has an element of order 2.

11. In a group, if a and b are two elements of coprime order and $ab = ba$, then $|ab| = |a| \cdot |b|$. Give an example to show that the hypothesis $ab = ba$ is necessary.

    (Hint: let $|a| = m$, $|b| = n$, $|ab| = r$; $(ab)^{mn} = a^{mn}b^{mn} = 1$, so $r|mn$; $1 = (ab)^{mr} = a^{mr}b^{mr} = b^{mr}$, so $n|mr$ hence $n|r$; similarly, $m|r$; thus $mn|r$.)

12. For a group G, the least positive integer e such that $a^e = 1$, $\forall a \in G$, is called the *exponent* of G; if such positive integer does not exist, then the exponent of G is defined to be ∞.

    Prove that, if the exponent of G is finite, then it is the least common multiples of the orders of all elements of G.

    (Hint: Cor.2.8.4(2).)

13. If G is an Abelian group of finite exponent, then there is an $a \in G$ such that the order of a is equal to the exponent of G.

    (Hint: Choose an $a \in G$ such that a has largest order in G; if there is a $b \in G$ such that $|b| \nmid |a|$, then there is a prime integer p such that $|a| = p^s m$ and $|b| = p^t n$ with $(m, p) = 1$ and $(n, p) = 1$ and $s < t$; hence, by Exer.11, $|a^{p^s} b^n| = p^t m > |a|$.)

14. If the exponent of a finite Abelian group G is equal to the order of G, then G is cyclic. Show that the hypothesis "Abelian" is necessary

15. Let $G = \langle a \rangle$ be a cyclic group of order m. Prove that there are exactly m characters of G.

    (Hint: A character of an Abelian group means a homomorphism from the group to the multiplicative group $\mathbf{C}^*$, see Example 2.4.6. A character of the group G of the exercise must map a to an mth root of unity.)

2.9 Direct Products

We begin with an ancient Chinese question. There are several objects such that one remains if you count them three by three, one remains if you count them five by five, and one remains if you count them seven by seven; how many objects are there?

It is known now that the answer is unique modulo $3 \cdot 5 \cdot 7 = 105$; in fact, we have the following basic number-theoretic fact.

2.9.1 Chinese Remainder Theorem (Number-Theoretic Version). *If integers $m_1, \cdots, m_n$ are pairwise coprime, then for any integers $r_1, \cdots, r_n$ the system of equations $x \equiv r_i \pmod{m_i}$, $i = 1, \cdots n$, has an integer solution, and the solution is unique up to modulo the product $m_1 \cdots m_n$.*

It is interesting that the idea contained in the fact appears in various mathematical aspects. In this section we are attempting to understand it from the point of group-theoretic view; later, we will show another version in Chapter 3.

2.9.2 Proposition. *Let $G_1, \cdots, G_n$ be groups. Define an operation on the Cartesian product set $G_1 \times \cdots \times G_n$ as follows:*

$$(a_1, \cdots, a_n) \cdot (a'_1, \cdots, a'_n) = (a_1 a'_1, \cdots, a_n a'_n),$$

for $a_i, a_i' \in G_i$ with $i = 1, \cdots, n$. Then $G_1 \times \cdots \times G_n$ is a group; and, for a fixed $1 \leq i \leq n$,

$$\rho_i : \quad G_1 \times \cdots \times G_n \longrightarrow G_i \ , \qquad (a_1, \cdots, a_n) \longmapsto a_i$$

is a surjective group homomorphism, while

$$\iota_i : \quad G_i \longrightarrow G_1 \times \cdots \times G_n \ , \qquad a_i \longmapsto (1_{G_1}, \cdots, 1_{G_{i-1}}, a_i, 1_{G_{i+1}}, \cdots 1_{G_n})$$

is an injective group homomorphism.

Proof. All the results can be directly checked; we just note that $1_{G_1 \times \cdots \times G_n} = (1_{G_1}, \cdots, 1_{G_n})$ and $(a_1, \cdots, a_n)^{-1} = (a_1^{-1}, \cdots, a_n^{-1})$. □

2.9.3 Definition. The group $G_1 \times \cdots \times G_n$ in the above is called the *direct product* of the groups $G_1, \cdots, G_n$; and the ρ_i is called the *projection* from the direct product to its *ith-component* G_i, while the ι_i is called the *injection* from the ith-component G_i to the direct product.

We remark that, just like the Cartesian product set, the order of the factors of the direct product $G_1 \times \cdots \times G_n$ is not crucial, so that we can denote it also by $\prod_{i=1}^n G_i$.

Note that, if the groups are additive, then we use the additive notation "$G_1 \oplus \cdots \oplus G_n$" or "$\bigoplus_{i=1}^n G_i$" and call it "direct sum"; cf. Remark 2.2.16.

For example, we can construct the direct sum $\mathbf{Z}_3 \oplus \mathbf{Z}_5 \oplus \mathbf{Z}_7$; and the injection $\mathbf{Z}_3 \to \mathbf{Z}_3 \oplus \mathbf{Z}_5 \oplus \mathbf{Z}_7$, etc.

2.9.4 Remark. Note that, by definition, the component G_i is *not* a subgroup of the direct product $G_1 \times \cdots \times G_n$; however, the image of the injection ι_i is indeed isomorphic to G_i since ι_i is injective; hence we can *identify* the image of ι_i with G_i; in this way, G_i is regarded as a subgroup of $G_1 \times \cdots \times G_n$. Thus, in group theory, it is sometimes written as $G_i \leq G_1 \times \cdots \times G_n$. But we should keep in mind that this is just a convenient identification: $a_i \in G_i$ as an element of $G_1 \times \cdots \times G_n$ is in fact the element $(1_{G_1}, \cdots, 1_{G_{i-1}}, a_i, 1_{G_{i+1}}, \cdots 1_{G_n})$.

For example, $\mathbf{Z}_3 \leq \mathbf{Z}_3 \oplus \mathbf{Z}_5 \oplus \mathbf{Z}_7$ means: $[2]_3 \in \mathbf{Z}_3$ is identified with $([2]_3, [0]_5, [0]_7) \in \mathbf{Z}_3 \oplus \mathbf{Z}_5 \oplus \mathbf{Z}_7$.

On the other hand, an integer solution of $x \equiv r_i \pmod{m_i}$ means that x is in fact regarded as the element $[x]_{m_i} \in \mathbf{Z}_{m_i}$. Generally, the map $\mathbf{Z} \to \mathbf{Z}_m, a \mapsto [a]_m$ is a group surjective homomorphism, see Example 2.4.6.

2.9.5 Proposition. *Let G and $G_1, \cdots, G_n$ be groups.*

1) *If $f_i : G \to G_i$, $i = 1, \cdots, n$, are group homomorphisms, then*

$$(f_1, \cdots, f_n) : \quad G \longrightarrow G_1 \times \cdots \times G_n , \qquad a \longmapsto \big(f_1(a), \cdots, f_n(a)\big)$$

is a group homomorphism with kernel $\bigcap_{i=1}^{n} \mathrm{Ker}(f_i)$.

2) *If $f : G \to G_1 \times \cdots \times G_n$ is a group homomorphism, letting $f_i = \rho_i \circ f : G \to G_i$, then $f = (f_1, \cdots, f_n)$.*

Proof. Check them directly.   □

However, it is not so easy to determine the image of the homomorphism $(f_1, \cdots, f_n)$. For the purpose of this section, we are concerned with when $(f_1, \cdots, f_n)$ is surjective. Obviously, a necessary condition is that every f_i is surjective, see Exer.3. But this condition is not sufficient; for example,

$$(\mathrm{id}_G, \mathrm{id}_G) : \quad G \longrightarrow G \times G , \qquad a \longmapsto (a, a) ,$$

is not a surjective homomorphism provided $G \neq \{1\}$ (the image of this homomorphism is usually called the *diagonal subgroup* of $G \times G$).

First, we look at the direct product $G_1 \times G_2$ of two factors, then turn to general case. Assume $f = (f_1, f_2) : G \to G_1 \times G_2$ is surjective. For any $a_1 \in G_1$ there is an $a \in G$ such that $f(a) = (a_1, 1_{G_2})$, i.e. $f_1(a) = a_1$ and $f_2(a) = 1_{G_2}$. Thus, for any $a_1 \in G_1$ there is an $a \in \mathrm{Ker}(f_2)$ such that $f_1(a) = a_1$. Then, for any $x \in G$, letting $f_1(x) = a_1$, we can choose an $a \in \mathrm{Ker}(f_2)$ such that $f_1(a) = f_1(x)$; hence $xa^{-1} \in \mathrm{Ker}(f_1)$ by Prop.2.6.3, so $x \in \mathrm{Ker}(f_1) \cdot a \subset \mathrm{Ker}(f_1) \cdot \mathrm{Ker}(f_2)$. Thus, we have proved the necessity of the following conclusion:

2.9.6. *Let G, G_1, G_2 be groups and both $f_1 : G \to G_1$, $f_2 : G \to G_2$ be surjective homomorphism. Then $f = (f_1, f_2) : G \to G_1 \times G_2$ is surjective if and only if $\mathrm{Ker}(f_1) \cdot \mathrm{Ker}(f_2) = G$.*

Proof. Denote $K_1 = \mathrm{Ker}(f_1)$ and $K_2 = \mathrm{Ker}(f_2)$. It remains to prove the sufficiency. Assume $K_1 K_2 = G$. Let $(a_1, a_2) \in G_1 \times G_2$. Since f_1 is surjective, there is an $a \in G$ such that $f_1(a) = a_1$; but, from the condition, there are $x_1 \in K_1$ and $y_1 \in K_2$ such that $a = x_1 y_1$; then $f_1(y_1) = 1 \cdot f_1(y_1) = f_1(x_1) f_1(y_1) = f_1(x_1 y_1) = f(a) = a_1$. Thus, we have a $y_1 \in G$ such that

$$f_1(y_1) = a_1 \qquad \text{and} \qquad f_2(y_1) = 1.$$

Similarly, we have a $y_2 \in G$ such that
$$f_1(y_2) = 1 \qquad \text{and} \qquad f_2(y_2) = a_2.$$
Set $y = y_1 y_2 \in G$; then
$$\begin{aligned}
f(y_1 y_2) &= \Big(f_1(y_1 y_2), f_2(y_1 y_2) \Big) \\
&= \Big(f_1(y_1) f_1(y_2), f_2(y_1) f_2(y_2) \Big) = (a_1, a_2).
\end{aligned}$$
That is, $f = (f_1, f_2)$ is a surjective homomorphism. $\qquad\qquad \square$

Notation as above; if $K_1 \cap K_2 = \{1\}$, then, by Prop.2.9.5, $G \overset{f}{\cong} G_1 \times G_2$. It is quite a satisfied fact that every element a of the direct product $G_1 \times G_2$ is uniquely expressed as its components a_1 in G_1 and a_2 in G_2. These inspire the following result.

2.9.7.   *Let G be a group and $N_1, N_2 \trianglelefteq G$. Then the following two conditions are equivalent:*

i) $N_1 N_2 = G$ *and* $N_1 \cap N_2 = \{1\}$.

ii) *Every $a \in G$ is uniquely expressed as a product $a = a_1 a_2$ with $a_1 \in N_1$ and $a_2 \in N_2$.*
Moreover, $G \cong N_1 \times N_2$ if this is the case.

Proof. First we note that, if $N_1, N_2 \trianglelefteq G$ and $N_1 \cap N_2 = 1$, then $a_1 a_2 = a_2 a_1$ for all $a_1 \in N_1$ and $a_2 \in N_2$, see Exer.4. Thus the expression in 2.9.7(ii) is in fact $a = a_1 a_2 = a_2 a_1$.

(i) $\implies$ (ii).   Since $N_1 N_2 = G$, there are $a_1 \in N_1$ and $a_2 \in N_2$ such that $a = a_1 a_2$. If $a = a'_1 a'_2$ for $a'_1 \in N_1$ and $a'_2 \in N_2$, then $a_1 a_2 = a'_1 a'_2$, i.e. $(a'_1)^{-1} a_1 = a'_2 (a_2)^{-1} \in N_1 \cap N_2 = \{1\}$; so, $(a'_1)^{-1} a_1 = a'_2 (a_2)^{-1} = 1$; hence $a'_1 = a_1$ and $a'_2 = a_2$.

(ii) $\implies$ (i).   Since any $a \in G$ is written as $a = a_1 a_2$ with $a_1 \in N_1$ and $a_2 \in N_2$, we have $N_1 N_2 = G$. For any $c \in N_1 \cap N_2$, since both $c = c \cdot 1 = 1 \cdot c$ are the expressions described as in (ii), $c = 1$ by the uniqueness of such expression; thus $N_1 \cap N_2 = \{1\}$.

Next, assume the conditions hold. Consider the map
$$\sigma : \; N_1 \times N_2 \longrightarrow G , \qquad (a_1, a_2) \longmapsto a_1 a_2.$$
Since N_1 and N_2 commute element-wise by the conditions, we have
$$\begin{aligned}
\sigma\big((a_1, a_2)(a'_1, a'_2) \big) &= \sigma(a_1 a'_1, a_2 a'_2) = (a_1 a'_1)(a_2 a'_2) \\
&= (a_1 a_2)(a'_1 a'_2) = \sigma(a_1, a_2) \sigma(a'_1, a'_2).
\end{aligned}$$

Thus σ is a group homomorphism. Since every $a \in G$ can be written as $a = a_1 a_2$ with $a_i \in N_i$, σ is surjective; and this expression is unique, thus σ is injective. That is, σ is a group isomorphism. □

We introduce the following notation.

Definition. A group G is said to be an *internal direct product* of its normal subgroups N_1 and N_2, and denoted by $G = N_1 \times N_2$, if the conditions in (2.9.7) hold.

Take $G = \mathbf{Z}$, $G_1 = \mathbf{Z}_3$, $G_2 = \mathbf{Z}_5$, $\rho_3 : \mathbf{Z} \to \mathbf{Z}_3$, $\rho_5 : \mathbf{Z} \to \mathbf{Z}_5$. Then

$$\mathrm{Ker}(\rho_3) + \mathrm{Ker}(\rho_5) = 3\mathbf{Z} + 5\mathbf{Z} = \mathbf{Z}$$

by Exer.2.7.2 and Exer.2.5.9; so

$$(\rho_3, \rho_5) : \quad \mathbf{Z} \longrightarrow \mathbf{Z}_3 \oplus \mathbf{Z}_5$$

is a surjective homomorphism by Prop.2.9.6. This is just the special case $n = 2$ of the number-theoretic fact 2.9.1. On the other hand,

$$\mathrm{Ker}(\rho_3) \bigcap \mathrm{Ker}(\rho_5) = 3\mathbf{Z} \cap 5\mathbf{Z} = 15\mathbf{Z}$$

by Exer.2.7.2 and Exer.2.5.9, so $\mathbf{Z}_{15} \cong \mathbf{Z}_3 \oplus \mathbf{Z}_5$ by Theorem 2.7.1 and Prop.2.9.5; further, it is easy to see that the kernel of $\mathbf{Z}_{15} \to \mathbf{Z}_3$ is $\langle [3]_{15} \rangle \cong \mathbf{Z}_5$, and the kernel of $\mathbf{Z}_{15} \to \mathbf{Z}_5$ is $\langle [5]_{15} \rangle \cong \mathbf{Z}_3$; thus, by (2.9.7), we have the internal direct product

$$\mathbf{Z}_{15} = \Big\langle [5]_{15} \Big\rangle \oplus \Big\langle [3]_{15} \Big\rangle.$$

These explain the Chinese Remainder Theorem (of the special case $n = 2$) from the point of view of the internal structure of $\mathbf{Z}_{15}$.

Now we extend the above observations to general n.

2.9.8 Chinese Remainder Theorem (Group-Theoretic Version).
Let G and G_i, $i = 1, \cdots n$, be groups and $f_i : G \to G_i$, $i = 1, \cdots n$, be surjective homomorphism. Then

$$f = (f_1, \cdots, f_n) : \quad G \longrightarrow G_1 \times \cdots \times G_n$$

is surjective if and only if

$$\mathrm{Ker}(f_i) \cdot \Big(\bigcap_{1 \le j \le n,\ j \ne i} \mathrm{Ker}(f_j) \Big) = G \qquad \text{for} \quad i = 1, \cdots, n.$$

Proof. The necessity is clear by (2.9.6), since for every i

$$f : G \longrightarrow G_i \times \Big(\prod_{1 \le j \le n,\ j \ne i} G_j \Big)$$

is surjective and the kernel of

$$G \to \prod_{1 \le j \le n,\ j \ne i} G_j$$

is just $\bigcap_{1 \le j \le n,\ j \ne i} \mathrm{Ker}(f_j)$ by Prop.2.9.5.

We prove the sufficiency by induction on n. It has been proved in (2.9.6) if $n = 2$. Next assume $n > 2$. Since

$$\bigcap_{1 \le j \le n-1, j \ne i} \mathrm{Ker}(f_j) \quad \supseteq \quad \bigcap_{1 \le j \le n,\ j \ne i} \mathrm{Ker}(f_j),$$

for $1 \le i \le n - 1$ we have

$$\mathrm{Ker}(f_i) \cdot \Big(\bigcap_{1 \le j \le n-1,\ j \ne i} \mathrm{Ker}(f_j) \Big) \ = \ G \ ;$$

thus, by induction, the following is a surjective homomorphism

$$G \longrightarrow G_1 \times \cdots \times G_{n-1}$$

with kernel $\bigcap_{1 \le j \le n-1,} \mathrm{Ker}(f_j)$. Then, since

$$\mathrm{Ker}(f_n) \cdot \Big(\bigcap_{1 \le j \le n-1} \mathrm{Ker}(f_j) \Big) \ = \ G \ ,$$

the surjectivity of $G \to (G_1 \times \cdots \times G_{n-1}) \times G_n$ follows from (2.9.6). $\square$

Looking at the homomorphism $\mathbf{Z} \to \mathbf{Z}_3 \oplus \mathbf{Z}_5 \oplus \mathbf{Z}_7$, we see that the kernels of the natural homomorphisms from $\mathbf{Z}$ to $\mathbf{Z}_3$, $\mathbf{Z}_5$ and $\mathbf{Z}_7$ are $3\mathbf{Z}$, $5\mathbf{Z}$ and $7\mathbf{Z}$ respectively; and the condition of Theorem 2.9.8 is indeed satisfied: $3\mathbf{Z} + (5\mathbf{Z} \cap 7\mathbf{Z}) = 3\mathbf{Z} + 35\mathbf{Z} = \mathbf{Z}$; etc. So Theorem 2.9.8 indeed covers the number-theoretic fact 2.9.1.

We remark that the condition in Theorem 2.9.8 cannot be replaced by $\mathrm{Ker}(f_i) \cdot \mathrm{Ker}(f_j) = G$ for $1 \le i \ne j \le n$; see Exer.5.

In order to extend the (2.9.7), we first note that, if a group G has normal subgroups N_i, $i = 1, \cdots, n$, then $N_i N_j = N_j N_i \trianglelefteq G$ by Exer.2.6.5; hence, it is easy to see that $N_1 \cdots N_n \trianglelefteq G$ and the order of the factors is not crucial; thus we can denote it by $\prod_{i=1}^n N_i \trianglelefteq G$.

The following is the extension of the fact 2.9.7; we just remark that here N_i matches with the $\bigcap_{1 \le j \le n, j \ne i} \mathrm{Ker}(f_j)$ in Theorem 2.9.8, while $\prod_{1 \le j \le n, j \ne i} N_i$ matches with $\mathrm{Ker}(f_i)$ in Theorem 2.9.8.

2.9.9 Theorem. *Let G be a group and $N_i \trianglelefteq G$, $i = 1, \cdots, n$. Then the following two conditions are equivalent:*

i) $\prod_{i=1}^{n} N_i = G$ *and* $N_i \cap \left(\prod_{1 \le j \le n, j \ne i} N_j \right) = \{1\}$ *for* $i = 1, \cdots, n$.

ii) *For any* $a \in G$ *there are unique* $a_i \in N_i$, $i = 1, \cdots, n$ *such that* $a = a_1 \cdots a_n$.

Moreover, any N_i *and* N_j *with* $1 \le i \ne j \le n$ *commute element-wise and* $G \cong N_1 \times \cdots \times N_n$ *if this is the case.*

Proof. (i) $\implies$ (ii). The existence of the expression follows from the condition $\prod_{i=1}^{n} N_i = G$. The uniqueness is in fact derived from the condition

$$N_i \cap \left(\prod_{1 \le j \le n, j \ne i} N_j \right) = \{1\} \qquad \text{for } i = 1, \cdots, n \ ;$$

we argue it by induction. Assume $a = a_1 \cdots a_n = a_1' \cdots a_n'$ for $a_i, a_i' \in N_i$, $i = 1, \cdots, n$. Then

$$(a_{n-1}')^{-1} \cdots (a_1')^{-1} a_1 \cdots a_{n-1} = a_n'(a_n)^{-1} \in N_n \cap \left(\prod_{1 \le j \le n-1} N_j \right) = \{1\};$$

thus $a_n' = a_n$ and $a_1' \cdots a_{n-1}' = a_1 \cdots a_{n-1}$. But, since it is clear that

$$\prod_{1 \le j \le n-1, j \ne i} N_j \subset \prod_{1 \le j \le n, j \ne i} N_j \ ,$$

for $n - 1$ the same condition holds:

$$N_i \cap \left(\prod_{1 \le j \le n-1, j \ne i} N_j \right) = \{1\} \qquad \text{for } i = 1, \cdots, n-1 \ .$$

Thus, from the equation $a_1' \cdots a_{n-1}' = a_1 \cdots a_{n-1}$, we have $a_{n-1}' = a_{n-1}$, $\cdots$, $a_1' = a_1$.

(ii) $\implies$ (i). The conclusion that $\prod_{i=1}^{n} N_i = G$ follows directly from that $a = a_1 \cdots a_n$ with $a_i \in N_i$, $i = 1, \cdots, n$, for every $a \in G$. Let $c \in N_i \cap \left(\prod_{1 \le j \le n, j \ne i} N_j \right)$; then there are unique $x_i \in N_i$, $i = 1, \cdots, n$, such that

$$c = x_1 \cdots x_{i-1} x_i x_{i+1} \cdots x_n \ ;$$

but, since $c \in N_i$, $c = 1 \cdots 1 c 1 \cdots 1$ is clearly such an expression; thus $x_j = 1$ for $1 \le j \ne i \le n$; on the other hand, $c \in \prod_{1 \le j \le n, j \ne i} N_j$, i.e. there are $z_j \in N_j$ with $1 \le j \ne i \le n$ such that

$$c = z_1 \cdots z_{i-1} z_{i+1} \cdots z_n = z_1 \cdots z_{i-1} 1 z_{i+1} \cdots z_n;$$

hence $x_i = 1$ by the uniqueness of the expressions of c again; therefore,

$c = 1$; and $N_i \cap \left(\prod_{1 \le j \le n,\, j \ne i} N_j \right) = \{1\}$.

At last, assume the conditions hold. The element-wise commutativity follows from the condition (i) and Exer.4. Consider the map

$$(2.9.10) \qquad \sigma : \; N_1 \times \cdots \times N_n \longrightarrow G \;, \qquad (a_1, \cdots, a_n) \longmapsto a_1 \cdots a_n \;.$$

By the element-wise commutativity of N_i and N_j for $i \ne j$, we have

$$\begin{aligned}
\sigma((a_1, \cdots, a_n)(a_1', \cdots, a_n')) &= \sigma(a_1 a_1', \cdots, a_n a_n') = (a_1 a_1') \cdots (a_n a_n') \\
&= (a_1 \cdots a_n)(a_1' \cdots a_n') \\
&= \sigma(a_1, \cdots, a_n)\sigma(a_1', \cdots, a_n').
\end{aligned}$$

Thus σ is a group homomorphism. Since every $a \in G$ is written as $a = a_1 \cdots a_n$ with $a_i \in N_i$, σ is surjective; and this expression is unique, thus σ is injective. That is, σ is a group isomorphism. $\qquad \square$

2.9.11 Definition. A group G is said to be an *internal direct product* of its normal subgroups $N_1, \cdots, N_n$, and denoted by $G = N_1 \times \cdots \times N_n = \times_{i=1}^{n} N_i$, if the conditions in Theorem 2.9.9 hold.

Exercises 2.9

1. Complete the details of the proof of Prop.2.9.2.
2. Complete the details of the proof of Prop.2.9.5.
3. If the homomorphism $(f_1, \cdots, f_n)$ in Prop.2.9.5 is surjective, then every f_i is surjective.
+4. If G is a group and $N_1, N_2 \trianglelefteq G$ and $N_1 \cap N_2 = 1$, then $a_1 a_2 = a_2 a_1$ for all $a_1 \in N_1$ and $a_2 \in N_2$. (Hint: $a_1(a_2 a_1^{-1} a_2^{-1}) \in N_1$ and $(a_1 a_2 a_1^{-1}) a_2^{-1} \in N_2$; $a_1 a_2 a_1^{-1} a_2^{-1} \in N_1 \cap N_2$.)
5. Take $G = K_4$, the permutation group of degree 4 in (2.1.15) (cf. (2.2.4)). G has three normal subgroups $H_1 = \langle (12)(34) \rangle$, $H_2 = \langle (13)(24) \rangle$, $H_3 = \langle (14)(23) \rangle$. Show that $H_i H_j = G$ for $1 \le i \ne j \le 3$, but $H_1(H_2 \cap H_3) \ne G$.
6. If G_i, $i = 1, \cdots, n$ are groups and $H_i \le G_i$, $i = 1, \cdots, n$. then
$$H_1 \times \cdots \times H_n \; \le \; G_1 \times \cdots \times G_n \;.$$
7. If G_i, $i = 1, \cdots, n$ and H_i, $i = 1, \cdots, n$ are groups and $\varphi_i : H_i \to G_i$, $i = 1, \cdots, n$, are group homomorphisms, then
$$\begin{aligned}
\varphi_1 \times \cdots \times \varphi_n : \quad H_1 \times \cdots \times H_n \quad &\longrightarrow \quad G_1 \times \cdots \times G_n \;, \\
(h_1, \cdots, h_n) \quad &\longmapsto \quad (\varphi_1(h_1), \cdots, \varphi_n(h_n))
\end{aligned}$$
is a group homomorphism with kernel $\mathrm{Ker}(\varphi_1) \times \cdots \times \mathrm{Ker}(\varphi_n)$.

8. 1) The associative law holds for the direct product of groups, i.e. $(G_1 \times G_2) \times G_3 = G_1 \times (G_2 \times G_3)$.

   2) The associative law holds for the internal direct product in a group, i.e. $(N_1 \times N_2) \times N_3 = N_1 \times (N_2 \times N_3)$.

   (Hint: the first one is by definiton; for the second one, consider 2.9.9(ii).)

9. Let p be a prime integer. A group is called a *p-group* if its order is a power of p. Show that a cyclic p-group is not isomorphic to a direct product of two non-trivial groups.

   (Hint: if $G = N_1 \times N_2$ with $N_1 \neq \{1\} \neq N_2$, let $r = \text{Max}(|N_1|, |N_2|)$, then $r < |G|$ and $a^r = 1$, $\forall a \in G$; contradicts to that G is cyclic. Another proof can be obtained by using 2.8.7 and 2.9.7.)

Chapter 3

Rings

3.1 Fundamentals

The objects discussed in Chapter 1 have one operation. Many mathematical systems familiar to us have two operations and the two operations are connected by the distributive law.

For example, the set $\mathbf{Z}$ of integers have the addition and multiplication; moreover, $\mathbf{Z}$ with the addition forms an additive group; while the multiplication is associative, i.e. $\mathbf{Z}$ with the multiplication forms a semigroup (monoid in fact); and, the multiplication is left and right distributive for the addition.

Another familiar example comes from $\mathbf{Z}_m$. As shown in §2.2, the set $\mathbf{Z}_m$ of residue classes modulo m have two operations: the addition defined in (2.2.8) and the multiplication defined in (2.2.9); $\mathbf{Z}_m$ with the addition forms an additive group; while the multiplication is associative, i.e. $\mathbf{Z}_m$ with the multiplication forms a semigroup (monoid in fact); and, the multiplication is left and right distributive for the addition.

There are some systems whose multiplicative semigroups are not commutative. Let $M_n(\mathbf{C})$ be the set of all complex matrices of dimension n. It is known from the elementary linear algebra that there are addition and multiplication in $M_n(\mathbf{C})$, and $M_n(\mathbf{C})$ with the addition forms a group (see Example 2.2.6) while $M_n(\mathbf{C})$ with the multiplication forms a semigroup (monoid in fact); and, the multiplication is left and right distributive for the addition. Clearly, in this example, we can replace $\mathbf{C}$ by $\mathbf{Q}$ or $\mathbf{R}$.

More generally, a subset $\mathbf{F}$ of the complex set $\mathbf{C}$ is said to be a *number field* if $\mathbf{F}$ is closed under the addition and multiplication and $(\mathbf{F}, +)$ is a subgroup of $(\mathbf{C}, +)$ and $(\mathbf{F}^*, \cdot)$ is a subgroup of $(\mathbf{C}^*, \cdot)$; where $(\mathbf{C}^*, \cdot)$ is the multiplicative group of non-zero complexes, see Exer.2.2.2(6), and $\mathbf{F}^* = \mathbf{F} - \{0\}$. Let $M_n(\mathbf{F})$ be the set of all matrices of dimension n with coefficients in $\mathbf{F}$. Then the system of $M_n(\mathbf{F})$ with the matrix addition and the matrix multiplication have the structure similar to $M_n(\mathbf{C})$.

3.1.1 Definition. Let R be a non-empty set and have two operations, one is called *addition* and denoted by "+", and the other is called *multiplication* and denoted by "$\cdot$"; such R is said to be a *ring* if the following three conditions hold:

R1) $(R, +)$ is an additive group, called *the additive group of the ring*, and its zero element 0_R (0 if there is no confusion) is also called the *zero element* of the ring;

R2) $(R, \cdot)$ is a semigroup, called the *semigroup of the ring*;

R3) the multiplication is both left and right distributive for the addition.

The cardinality of the underlying set of a ring R is called the *order* of the ring R and denoted by $|R|$. R is said to be a *finite ring* if $|R| < \infty$.

If the semigroup of a ring R is a monoid, then R is said to be a *unitary ring* and the unity 1_R (1 if there is no confusion) of the monoid is called the *unity* or *identity element* of the ring. Recall from Remark 2.2.2(2) that the unity of a unitary ring is unique.

If the semigroup of a ring R is commutative, then R is said to be a *commutative ring*. The mathematical area which discusses commutative rings is usually referred to as *Commutative Algebra*.

3.1.2 Remark. It is clear that there is exactly one ring of order 1: $R = \{0\}$ with $0 + 0 = 0$ and $0 \cdot 0 = 0$. By definition, it is a unitary ring and 0 is also its unity; and it is in fact the unique unitary ring whose zero element and unity coincide, see Exer.3. This ring is sometimes called the *zero ring*. To avoid this too trivial case and the unconvenient condition $0 = 1$, in the following we always take the agreement that a unitary ring R means $|R| > 1$ hence $0 \neq 1$.

So, the three systems mentioned just now are all rings.

3.1.3 Example. $(\mathbf{Z}, +, \cdot)$ is a commutative unitary ring which is called the *integer ring*.

$(\mathbf{Z}_m, +, \cdot)$ is a commutative unitary ring which is called the *ring of the residue classes modulo m*.

$(M_n(\mathbf{C}), +, \cdot)$ is a (non-commutative if $n > 1$) unitary ring which is called *the complex matrix ring*. Similarly, we have the *real matrix ring* $(M_n(\mathbf{R}), +, \cdot)$, the *rational matrix ring* $(M_n(\mathbf{Q}), +, \cdot)$, etc.

It is clear that the set $\mathbf{R}$ of all real numbers with the usual number addition and the usual multiplication forms a commutative unitary ring; and so do the $\mathbf{Q}$ and $\mathbf{C}$ etc., these are the special cases of dimension 1 of the matrix rings. Let $\mathbf{R}[x]$ be the set of all real polynomials; it is clear by the definition that, with the usual addition and multiplication of polynomials, $\mathbf{R}[x]$ is a commutative unitary ring which is called the *real polynomial ring*. Of course, here a real polynomial $f(x)$ is regarded as a real function. In fact, the set of all (continue) real functions is also a ring see Exer.1.

Similarly we have the *integral polynomial ring* $\mathbf{Z}[x]$, the *rational polynomial ring* $\mathbf{Q}[x]$, etc. Later we will show a formal way to construct the polynomial ring from a given ring in §3.8.

Another important example is as follows.

Example. Let $(A, +)$ be an additive group, let $\text{End}(A)$ be the set of all endomorphisms of A; define two operations on $\text{End}(A)$ as follows:

addition: $\qquad (f + g)(x) = f(x) + g(x) \qquad\qquad \forall x \in A,$

multiplication: $\qquad (f \cdot g)(x) = fg(x) \qquad\qquad \forall x \in A,$

for all $f, g \in \text{End}(A)$; then it is easy to see that $(\text{End}(A), +, \cdot)$ forms a unitary ring (except the trivial case that $A = \{0\}$), we call $\text{End}(A)$ the *endomorphism ring* of the additive group A. Note that here A can be any Abelian group even if its operation is written as multiplication, provided we define $(f+g)(x) = f(x)g(x)$. We will consider this example more carefully in (4.1.6).

(3.1.4) *Some elementary properties.* Let R be a ring. Since $(R, +)$ is an additive group, we can define "subtraction" as:

$$a - b = a + (-b) \qquad\qquad \text{for } a, b \in R,$$

and the distributive law holds for the subtraction:

$$a(b - c) = ab - ac \quad \text{and} \quad (b - c)a = ba - ca \qquad \text{for } a, b, c \in R;$$

and for $n \in \mathbf{Z}$ and $a \in R$ the *multiple* na is defined and, for $m, n \in \mathbf{Z}$ and $a, b \in R$, the following hold (see Remark 2.2.16 and Prop. 2.2.14):

$$n(a + b) = na + nb,$$
$$(m + n)a = ma + na,$$
$$(mn)a = m(na).$$

Since $(R, \cdot)$ is a semigroup, for positive integer n and $a \in R$ the *power* a^n is defined and, for positive integers m, n and $a \in R$, the following hold:

$$a^{m+n} = a^m a^n,$$
$$a^{mn} = (a^m)^n.$$

Note that, in the multiple formula $0a = 0$ (in fact this is by Definition 2.2.13), the 0 on the left is the integer 0 not the zero element of R while the 0 on the right is the zero element of R. For zero element, there is a formula of similar shape:

$$0a = 0 = a0 \qquad \text{for all } a \in R \text{ and the zero element } 0 \text{ of } R,$$

because $0a = (a - a)a = aa - aa = 0$. It is a known formula that $-(-a) = a$ (i.e. the additive version of formula $(a^{-1})^{-1} = a$). Further, from the above formula we have $ab + (-a)b = (a + (-a))b = 0b = 0$; thus we get:

$$(-a)b = -ab = a(-b) \qquad \text{for all } a, b \in R.$$

We remark that for integer n and $b \in R$ we also have $(-n)b = -nb$, but this is in fact a special case $m = -1$ of the formula $(mn)a = m(na)$. Moreover, we have

$$(-a)(-b) = a(-(-b)) = ab \qquad \text{for } a, b \in R.$$

At last, with the formulas on symbols and the distributive law, it is easy to verify that

$$n(ab) = (na)b = a(nb) \qquad \text{for } n \in \mathbf{Z} \text{ and } a, b \in R.$$

Of course, a^n for non-positive integer n is not defined for general rings; but it is defined for unitary rings and for the so-called invertible elements, we will discuss it in Def.3.2.4.

Many fundamentals for groups can be stated for rings.

3.1.5 Definition. A non-empty subset S of a ring R is said to be a *subring* of R, denoted $S \leq R$, if S is itself a ring under the operations of R; moreover, S is said to be a unitary subring if R is unitary and $1_R \in S$.

Note that, by Prop. 2.3.6, a subring S of a ring R must contain the zero element 0 of R; but S does not necessarily contain the unity 1 of R even if both R and S are unitary, see Exer.8.

From the corresponding facts for subgroups, we have some criterions for subrings at once; e.g. from Prop. 2.3.6 (and cf. Exer.2.3.8) we have

3.1.6 Proposition. *Let R be a ring and $S \subset R$. The following three statements are equivalent:*

i) $S \leq R$;

ii) $0 \in S$, *and* $a + b \in S$ *and* $-a \in S$ *and* $ab \in S$ *for any* $a, b \in S$;

iii) $S \neq \emptyset$ *and for any* $a, b \in S$ *we have* $a - b \in S$ *and* $ab \in S$. $\quad\square$

As an application, we see that $\mathbf{Z}$ is a unitary subring of $\mathbf{Q}$; $M_n(\mathbf{Q})$ is a unitary subring of $M_n(\mathbf{C})$; etc.

Similar to the normal subgroups for groups, the more important subrings are associated with the ring homormorphisms.

3.1.7 Definition. Let R and S be rings. A map $f : R \to S$ is called a *ring homomorphism* (or, a homomorphism for brevity) if

$$f(a + b) = f(a) + f(b) \quad \text{and} \quad f(ab) = f(a)f(b) \quad \forall a, b \in R;$$

and f is said to be *injective* (*surjective*) if it is injective (surjective); and f is said to be a *ring isomorphism* if it is bijective. If there is an isomorphism $f : R \to S$, then we say that R is isomorphic to S (through f), and denote $R \stackrel{f}{\cong} S$. A ring homomorphism $f : R \to S$ is said to be *unitary* if both R and S are unitary and $f(1_R) = 1_S$.

Of course, any properties of group homomorphisms are valid for ring homomorphisms restricted to the additive groups of rings; e.g. $f(0_R) = 0_S$, $f(-a) = -f(a)$ etc. But, it does not necessarily hold that $f(1_R) = 1_S$ even if both R and S are unitary, see Exer.9; this is the reason why we introduce the notation "unitary homomorphism".

Just like Corollary 2.4.4, it is easy to show that

3.1.8 Proposition. *Let $f : R \to S$ be a ring homomorphism, and* *define the* kernel *of f as*

$$\text{Ker}(f) = \{a \mid a \in R \text{ and } f(a) = 0\}.$$

Then the image $\text{Im}(f) \leq S$ *and the kernel* $\text{Ker}(f) \leq R$.   □

3.1.9 Example. Let $\mathbf{Z}$, $\mathbf{Z}_m$, $\mathbf{R}$ and $\mathbf{R}[x]$ be the rings in Example 3.1.3. Then it is easy to check that

$$\mathbf{Z} \longrightarrow \mathbf{Z}_m , \qquad n \longmapsto [n]$$

is a unitary surjective ring homomorphism, while

$$\mathbf{R} \longrightarrow \mathbf{R}[x] , \qquad r \longmapsto r$$

is a unitary injective ring homomorphism.

Similar to Theorem 2.4.11, any unitary ring can be represented as a unitary subring of the endomorphism ring of an additive group, see Exer.12.

Just like groups, the kernels of ring homomorphisms have their special features; we will discuss them in §3.3 later, and then discuss the ring homomorphisms further in §3.4.

Exercises 3.1

1. Let $C[0,1]$ be the set of all real functions defined on the interval $[0,1]$, and the addition and multiplication on $C[0,1]$ are the usual function addition and the usual function multiplication; then $C[0,1]$ is a unitary ring.

2. Let $(R, +)$ be an additive group; define a multiplication on R as follows: $ab = 0 \; \forall a, b \in R$; then R is a ring but not unitary ring.

3. If R is a unitary ring (remember that this means $|R| > 1$), then the unity 1 of R must be different from the zero element 0 of R.

4. Show that $2\mathbf{Z} = \{2n \mid n \in \mathbf{Z}\}$ equipped with the usual number addition and multiplication is a ring but not a unitary ring.

+5. An element a in a ring R is said to be an *idempotent* if $a^2 = a$. Prove: if every element of a ring R is an idempotent, then R is a commutative ring and $2a = 0$ for all $a \in R$.

6. Let R be a (unitary) ring.
    1) On $R \times R$ define: $(a, b) + (c, d) = (a + c, b + d)$, $(a, b) \cdot (c, d) = (ac, bd)$; then $R \times R$ is a (unitary) ring.
    2) On $S = R \times R$ define: $(a, b) + (c, d) = (a + c, b + d)$, $(a, b) \cdot (c, d) = (ac + bd, ad + bc)$; then S is a (unitary) ring.

7. Let R be a ring but is not unitary. Take $S = \mathbf{Z} \times R$ on which we define ($\forall m, n \in \mathbf{Z}$ and $a, b \in R$):

$$(m, a) + (n, b) = (m + n, \ a + b),$$

$$(m, a) \cdot (n, b) = (mn, \ mb + na + ab);$$

then $(S, +, \cdot)$ is a unitary ring and

$$R \longrightarrow S, \qquad a \longmapsto (0, a)$$

is an injective ring homomorphism.

$^{+}$8. Let $R = \left\{ \begin{pmatrix} a & 0 \\ 0 & 0 \end{pmatrix} \ \middle| \ a \in \mathbf{C} \right\}$. Show that R is a unitary ring, and it is a subring of $M_2(\mathbf{C})$ but not a unitary subring of $M_2(\mathbf{C})$.

9. Give an example to show that a ring homomorphism between two unitary rings is not necessarily unitary.

10. If $f : R \to S$ is a unitary homomorphism of unitary rings R and S, then $f(1_R) = 1_S$ and $f(a^{-1}) = f(a)^{-1}$ for any invertible element $a \in R$.

11. Let $f : R \to S$ be a surjective ring homomorphism.
    1) If R is unitary, then so is S and $f(1_R) = 1_S$.
    2) If R is commutative, then so is S.

$^{+}$12. Let R be a ring.
    1) For any given $a \in R$, the map $\lambda_a : R \to R$, $x \mapsto ax$ is a group homomorphism of the additive group $(R, +)$, i.e. $\lambda_a \in \mathrm{End}(R, +)$.
    2) The map $\lambda : R \to \mathrm{End}(R, +), a \mapsto \lambda_a$ is a ring homomorphism.
    3) Give an example to show that λ is not necessarily injective. (Hint: consider Exer.2.)
    4) λ is a unitary injective ring homomorphism if R is unitary.

13. (*Binomial Formula*). In a ring R, if $ab = ba$, then for any positive integer n we have (where $\binom{n}{i}$ is the usual binomial coefficients)

$$(a+b)^n = a^n + \binom{n}{1} a^{n-1} b + \binom{n}{2} a^{n-1} b^2 + \cdots + \binom{n}{n-1} ab^{n-1} + b^n.$$

3.2 Zero Divisors, Invertible Elements

3.2.1 Definition. Let R be a ring. For $S \subset R$ the set $\mathrm{Ann}_R^r(S) = \{x \mid x \in R \text{ and } sx = 0 \ \forall s \in S\}$ is called the *right annihilator* of S in R. It is clear that $0 \in \mathrm{Ann}_R^r(S)$. The *left annihilator* $\mathrm{Ann}_R^l(S)$ is defined similarly. If $S = \{a\}$, we write $\mathrm{Ann}_R^r(S) = \mathrm{Ann}_R^r(a)$ in short. An element a is called a *left zero divisor* of the ring R if $\mathrm{Ann}_R^r(a) \neq \{0\}$.

The *right zero divisors* are defined in a similar way. a is called a *zero divisor* if it is both a left zero divisor and a right zero divisor.

Obviously, 0 is a zero divisor of any ring of order > 1; a left (or right) zero divisor of a commutative ring is a zero divisor.

Example. The residue class [2] is a zero divisor of $\mathbf{Z}_6$, because $[3] \neq 0$ but $[2][3] = 0$.

Example. In $M_2(\mathbf{C})$, the matrix $\begin{pmatrix} 1 & 1 \\ 0 & 0 \end{pmatrix}$ is both a left and right zero divisor, hence a zero divisor. In fact, in matrix ring of finite dimension, a left (or a right) zero divisor must be a zero divisor, see Exer.1.

Example. Let V be a vector space over $\mathbf{R}$ of countable infinite dimension with a basis $\{b_1, b_2, \cdots\}$, let $\mathrm{End}_{\mathbf{R}}(V)$ be the set of all linear transformations of V. Then, with the usual addition and multiplication of linear transformations, it is easy to check that $\mathrm{End}_{\mathbf{R}}(V)$ is a unitary ring. Consider $\tau \in \mathrm{End}_{\mathbf{R}}(V)$ such that $\tau(b_i) = b_{i+1}$ for $i = 1, 2, \cdots$, and $\sigma \in \mathrm{End}_{\mathbf{R}}(V)$ such that $\sigma(b_1) = b_1$ and $\sigma(b_i) = 0$ for $i = 2, 3, \cdots$. Then $\sigma \neq 0$ and $\sigma\tau = 0$, i.e. τ is a right zero divisor; but, by Exer.9, τ is not a left zero divisor.

However, globally speaking, the existence of non-zero left zero divisors is equivalent to the existence of non-zero right zero divisors; and it is related to the *cancellation law*.

3.2.2 Proposition. *For a ring R the following are equivalent:*

  i) *R has no non-zero left zero divisor;*

  ii) *R has no non-zero right zero divisor;*

  iii) *the* left cancellation law *holds:*

$$(\forall a, b, c \in R)(ab = ac \text{ and } a \neq 0 \implies b = c);$$

  iv) *the* right cancellation law *holds:*

$$(\forall a, b, c \in R)(ba = ca \text{ and } a \neq 0 \implies b = c).$$

Proof. (i) $\Longleftrightarrow$ (ii). If $a \in R$ is a non-zero left zero divisor, then there is a $b \in R$ such that $b \neq 0$ and $ab = 0$; but $a \neq 0$, this also means that b is a non-zero right zero divisor. This proves (ii) $\Longrightarrow$ (i). The (i) $\Longrightarrow$ (ii) is proved in the same way.

(i) $\Longleftrightarrow$ (iii). Assume (i) holds and assume $ab = ac$ and $a \neq 0$, then $a(b - c) = 0$ hence $b - c = 0$ because a is not a left zero divisor by (i); i.e. $b = c$. Conversely, Assume (iii) holds and assume a is a left zero divisor, then there is a $b \neq 0$ such that $ab = 0 = a0$; but, by (iii), this means $b = 0$; a contradiction.

(ii) $\iff$ (iv). Similar to the above. $\qquad\qquad\qquad\qquad\qquad\square$

3.2.3 Definition. A ring R is called a *domain* (or *integral domain*) if R is a commutative unitary ring without non-zero zero divisor.

Example. The integer ring $\mathbf{Z}$, the real polynomial ring $\mathbf{R}[x]$ over $\mathbf{R}$ are domains.

Recall from (3.1.4) that the power a^n for non-positive integer n is not defined in general rings.

3.2.4 Definition. Assume that R is a unitary ring. It is reasonable to define $a^0 = 1$; and, an element a is said to be *invertible* and a' is called its *inverse element* if $aa' = 1 = a'a$ (such a is also called a *unit* of R); just like Remark 2.2.2(3), the inverse of an invertible element a is unique, hence denoted by a^{-1}. Thus, for an invertible element a and a negative integer n, it is reasonable to define $a^n = (a^{-1})^{-n}$.

This is jusk like the definition 2.2.13; and we can really understand it from the group situation because we have the following result.

3.2.5 Proposition and **Definition.** *Let R be a unitary ring, let R^* denote the set of all invertible elements of R. Then R^* with the multiplication forms a group, called the* multiplicative group *of the unitary ring.*

Proof. Exer.6. $\qquad\qquad\qquad\qquad\qquad\qquad\qquad\qquad\qquad\qquad\square$

Thus, all the definition and properties on powers in §2.2 are valid for the invertible elements of unitary rings, e.g. $(ab)^{-1} = b^{-1}a^{-1}$, etc.

It is clear that an invertible element is not a zero divisor (Exer.9). The converse is certainly false, e.g. in $\mathbf{Z}$ every non-zero element is not a zero divisor but only ± 1 are invertible.

The zero element 0 of a unitary ring R is not invertible because $0a = 0 \neq 1$, $\forall a \in R$. Consider an extreme case, we have

3.2.6 Definition. A unitary ring D is called a *division ring* (or *skew field*) if every non-zero element is invertible (equivalently, $R^* = R - \{0\}$ forms a multiplicative group, see Exer.12). A commutative division ring is called a *field*.

So, by Definitions 3.2.3 and 3.2.6, a field must be an integral domain. The converse is certainly false, e.g. **Z**. However, we will see in §3.7 that they are really related closely. For the moment we just state an easy fact in Exer.16.

Exercises 3.2

1. Prove: in $M_n(\mathbf{C})$, a left zero divisor must be a zero divisor.

2. An element a of a ring R is not a left zero divisor iff
$$(\forall b, c \in R)(ab = ac \implies b = c).$$

+3. An element a of a ring R is said to be *nilpotent* if $a^n = 0$ for a positive integer n. The idempotents are defined in Exer.3.1.5. Prove:

   1) A nilpotent element is a zero divisor.

   2) A non-zero idempotent is not nilpotent.

   3) In a unitary ring, an idempotent e other than 1 must be a zero divisor; on the other hand, $1 - a$ is invertible if a is a nilpotent element.

4. Show that the image of a ring homomorphism from a ring without non-zero zero divisors can have non-zero zero divisors, and the image of a ring homomorphism from a ring with non-zero zero divisors can have no non-zero zero divisors.
   (Hint: Consider **Z** and suitable $\mathbf{Z}_m$.)

5. A unitary subring of an integral domain is again an integral domain.

6. Prove Proposition 3.2.5.

+7. Let R be a ring. An element e of R is said to be a *left unity* if $ea = a$, $\forall a \in R$. The *right unity* is defined similarly.

   1) If R is unitary, then 1 is the unique left unity of R.

   2) If R have a unique left unity e, then R is unitary and e is the unity. (Hint: consider $(e + ae - a)b$ for $a, b \in R$.)

   3) Give an example to show that a ring can have more than one left unity. (Hint: consider $\left\{ \begin{pmatrix} a & b \\ 0 & 0 \end{pmatrix} \,\middle|\, a, b \in \mathbf{C} \right\}$.)

+8. Let R be a unitary ring. An element a' of R is said to be a *left inverse* of an element a of R if $a'a = 1$. The *right inverse* is defined similarly.

   1) If a is invertible, then a^{-1} is the unique left inverse of a.

   2) If a have a unique left inverse a', then a is invertible and $a^{-1} = a'$. (Hint: consider $(a' + aa' - 1)a$.)

   3) Give an example to show that an element can have more than one left inverse. (Hint: consider the third example after Def.3.2.1.)

+9. 1) An element having a left inverse is not a left zero divisor.

   2) An invertible element is not a zero divisor.

+10. In a unitary ring, if an element a have both a left inverse and a right inverse, then a is invertible.

   (Remark: cf. Remark 2.2.2(3). The concepts such as idempotent, left unity, left inverse etc. are in fact valid for monoids; this exercise is in fact for monoids. Also, one can find an easy similar exercise on left and right unities for semigroups, cf. Remark 2.2.2(2).)

11. In a ring with no non-zero zero divisor, prove:

   1) a left unity is the unity (cf. Exer.7(2));

   2) a left inverse of an element is the inverse of the element (cf. Exer.8(2)).

+12. Let R be a unitary ring. Prove:

   1) R has no non-zero zero divisor iff $R - \{0\}$ is a submonoid of the multiplicative monoid of R.

   2) R is a division ring iff, under the multiplication, $R - \{0\}$ forms a group.

13. Let R be a unitary ring and $a, b \in R$. If $1 - ab$ is invertible, then so is $1 - ba$.

   (Hint: let $(1 - ab)^{-1} = c$, prove $(1 - ba)^{-1} = 1 + bca$.)

14. An element a of a unitary ring R is invertible if one of the following holds:

   1) there is an element b such that $aba = a$ and $ba^2b = 1$.

   2) there is a unique element b such that $aba = a$. (Hint: $a(b+ab-1)a =?.$)

15. Let R be a non-zero ring. If for every $0 \neq a \in R$ there is a unique $b \in R$ such that $aba = a$, then the following hold:

   1) R has no non-zero zero divisor; (hint: $ax = 0 \implies a(b + x)a = a;$)

   2) if $aba = a$ and $a \neq 0$, then $bab = b$; (hint: $aba = a \implies ababa = a;$)

   3) R is unitary; (hint: $(\forall x \in R)(a(bax - x) = 0);$)

   4) R is a division ring.

16. A finite integral domain is a field.

   (Hint: for a finite integral domain R and $0 \neq a \in R$, the map $\lambda_a : R - \{0\} \to R - \{0\}, x \mapsto ax$ is injective, hence surjective since R is finite; so there is an $a' \in R - \{0\}$ such that $aa' = 1$.)

3.3 Ideals, Residue Rings

3.3.1 Definition. Let R be a ring. A non-empty subset I of R is called a *left ideal* of R if the following conditions (1) and (2) hold:

   1) I is an additive subgroup of $(R, +)$;

2) $a \in I$ and $r \in R \implies ra \in I$.

On the other hand, I is called a *right ideal* of R if the condition (1) and the following condition (2') hold:

2') $a \in I$ and $r \in R \implies ar \in I$.

We say that I is a *one-sided ideal* if I is either a left ideal or a right ideal. And I is called an *ideal* (or a *two-sided ideal*) of R if I is both a left ideal and a right ideal.

Thus, a one-sided ideal of a commutative ring must be an ideal.

It is clear that, in any ring R, $\{0\}$ and R are ideals, called the *zero ideal* and the *unity ideal* resp., and both are called the *trivial ideals* of R. An ideal other than R is said to be a *proper ideal*.

Example. 1). $m\mathbf{Z}$ is an ideal of the integer ring $\mathbf{Z}$.

2). For any ring R, let $C(R) = \{a \mid a \in R \text{ and } ax = xa \ \forall x \in R\}$ which is called the *center* of the ring R; then $C(R)$ is a subring of R, but not an ideal of R in general.

3.3.2 Definition. A ring R is called a *simple ring* if R has no non-trivial ideals.

3.3.3 Example. A division ring is a simple ring.

To see this, let I be a non-zero ideal, we can take $0 \neq a \in I$, then a is invertible, and for any $x \in R$ we have $x = (xa^{-1})a \in I$; so $I = R$.

In fact, the above argument shows that a division ring has no non-trivial one-sided ideals. The following simple ring has non-trivial one-sided ideals.

3.3.4 Example. $M_n(\mathbf{F})$ is a simple ring, where $\mathbf{F}$ is a number field.

To prove this, we note that a non-zero ideal $\mathcal{I}$ of $M_n(\mathbf{C})$ has a matrix $A \neq 0$; write it as: $A = \sum_{i,j} c_{ij} E_{ij}$ where E_{ij} is the matrix with 1 in (i,j)-entry and 0 elsewhere, then there is a $c_{ml} \neq 0$; so $E_{kk} = c_{ml}^{-1} E_{km} A E_{lk} \in \mathcal{I}$; and $1 = \sum_{k=1}^{k=n} E_{kk} \in \mathcal{I}$; thus $\mathcal{I} = M_n(\mathbf{C})$ by Exer.1.

From Prop.2.3.6 the following criterion follows immediately.

3.3.5 Proposition. *A non-empty set I of a ring R is an ideal if and only if $a - b \in I$ and $ra \in I$ and $ar \in I$ for any $a, b \in I$ and $r \in R$.* $\square$

Since a ring R has the additive group structure $(R, +)$, for $S, T \subset R$ by Notation 2.5.1 (but in additive version) we have

$$S + T = \{s + t \mid s \in S \text{ and } t \in T\}$$

and we appoint that

$$ST = \{\text{finite sum } \textstyle\sum_i s_i t_i \mid s_i \in S \text{ and } t_i \in T\}.$$

3.3.6 Proposition. *If A, B and C are ideals of a ring R, then so are A + B and AB and A ∩ B, and*

$$A(B + C) = AB + AC \qquad and \qquad (B + C)A = BA + CA.$$

Proof. Exer.2. □

3.3.7 Definition. Let R be a ring and $S \subset R$. Then the intersection of all the ideals of R which contains S is also an ideal by Prop. 3.3.6, hence it is the smallest ideal of R which contains S; we call it the *ideal generated by S* and denote it by $\langle S \rangle$. In addition, if $S = \{a_1, \cdots, a_n\}$, we denote $\langle S \rangle = \langle a_1, \cdots, a_n \rangle$ and call it a *finitely generated ideal*. In particular, the ideal $\langle a \rangle$ generated by one element a is called the *principal ideal* generated by a.

R is said to be a *principal ideal ring* if every ideal of R is principal. R is said to be a *principal ideal domain*, or *p.i.d.* in short, if R is a domain and every ideal is principal.

A typical example of p.i.d. is the integer ring, see Exer.3.

For principal ideals we have the following descriptions.

3.3.8 Proposition. *Let R be a ring and a ∈ R. Then*

$$\langle a \rangle = \{\textstyle\sum_i x_i a y_j + ra + as + ma \mid x_i, y_i, r, s \in R, \ m \in \mathbf{Z}\}$$

where $\sum_i x_i a y_i$ runs over the finite sums of such forms; in particular, if R is a commutative unitary ring, then $\langle a \rangle = \{ra \mid r \in R\} = Ra$.

Proof. It is clear that $\sum_i x_i a y_i + ra + as + ma \in \langle a \rangle$. Conversely, by Prop. 3.3.5, it is easy to verify that $\{\sum_i x_i a y_j + ra + as + ma \mid x_i, y_i, r, s \in R, \ m \in \mathbf{Z}\}$ is an ideal; and it contains a obviously as $a = 1a$ where $1 \in \mathbf{Z}$. Thus the desired equality is proved.

Further, if R is unitary, then ma for $m \in \mathbf{Z}$ can be expressed as a finite sum of $1a1$ or $-1a1$ according to m positive or negative resp. where $1 \in R$ is the unity, while $ra = ra1$ and $as = 1as$; thus

$$\langle a \rangle = \{\text{finite sum } \textstyle\sum_i x_i a y_j \mid x_i, y_i \in R\}.$$

In particular, when R is unitary and commutative, we have $\sum_i x_i a y_i = (\sum_i x_i y_i)a$, hence

$$\langle a \rangle = \{ra \mid r \in R\} = Ra. \qquad \qquad \square$$

3.3.9 Remark. If R is not commutative, $Ra = \{ra \mid r \in R\}$ is not necessarily an ideal; but it is easy to see that Ra is a left ideal of R. Coherently, Ra is said to be a *principal left ideal* of R. And, the notation *left ideals generated by sets*, *finitely generated left ideals* etc. can be defined similarly.

Example. Take $R = 2\mathbf{Z} = \{2n \mid n \in \mathbf{Z}\}$ (which is in fact a subring of $\mathbf{Z}$, but now we consider it as an independent ring), and $I = \{4x \mid x \in \mathbf{Z}\}$. Then I is a principal ideal, but note that $I \neq \langle 2 \rangle$ though $I = \{2x \mid x \in R\}$, because $2 \notin I$ and the $2x$ in the braces is the multiple, not a product; in fact, $I = \langle 4 \rangle$ by Prop.3.3.8. The set $I' = \{4r \mid r \in R\}$ is another principal ideal of R; however, since $4 \notin I'$, $I' \neq \langle 4 \rangle$; in fact, $I' = \langle 8 \rangle$.

3.3.10 Example. In the rational polynomial ring $\mathbf{Q}[x]$, the ideal $I = \langle 2, x \rangle$ generated by elements 2 and x is in fact the unity ideal by Exer.1 since 2 is invertible in $\mathbf{Q}[x]$, hence it is a principal ideal. Next, we consider the integral polynomial ring $\mathbf{Z}[x]$ and the ideal $I = \langle 2, x \rangle$ generated by 2 and x, then it is not a principal ideal. Otherwise, $\langle 2, x \rangle = \langle g(x) \rangle$ for some $g(x) \in \mathbf{Z}[x]$, and since $\mathbf{Z}[x]$ is unitary and commutative, by Prop.3.3.8 we have $2 = p(x)g(x)$ and $x = q(x)g(x)$ for some $p(x), q(x) \in \mathbf{Z}[x]$; from the former equality we have $g(x) = a \in \mathbf{Z}$, then from the latter we get $a = \pm 1$; hence $\langle 2, x \rangle = \langle \pm 1 \rangle = \langle 1 \rangle$. However, by Exer.7 and Prop.3.3.8, we have $\langle 2, x \rangle = \{2f(x) + xh(x) \mid f(x), h(x) \in \mathbf{Z}[x]\}$; in particular, $1 = 2f(x) + xh(x)$ for some $f(x), h(x) \in \mathbf{Z}[x]$; this is obviously impossible.

It is easy to show that the kernel of a ring homomorphism $R \to S$ is an ideal of R (Exer.8). In fact, any ideal of a ring is a kernel of a ring homomorphism.

Let R be a ring and I be an ideal of R. Since $(R, +)$ is an additive group with I being a subgroup, we have the quotient additive group $(R/I, +)$, and call the coset $a + I$ the *residue class* containing a and denote it by $[a]$ sometimes. With the further property of ideals, it is easy to show that

$$(a + I)(b + I) = ab + I \qquad \text{for } a + I, b + I \in R/I$$

is a well-defined multiplication on R/I; for, if $a' + I = a + I$ and $b' + I = b + I$, then $a' = a + u$ and $b' = b + v$ for some $u, v \in I$, and $a'b' = (a + u)(b + v) = ab + (av + ub + uv)$ and $av + ub + uv \in I$, i.e. $a'b' + I = ab + I$. Further, it is a routine to verify that R/I is a ring and the residue map $R \to R/I$, $x \mapsto x + I$ is a surjective ring homomorphism; it is obvious that the kernel of this homomorphism is just I.

3.3.11 Definition. Notations as above. The ring R/I is called the *residue ring* (or *quotient ring*) of the ring R modulo the ideal I, and the surjective homomorphism $R \to R/I$, $x \mapsto x + I$ is called the *natural (or canonical) homomorphism*.

3.3.12 Example. The ring $\mathbf{Z}_m$ is in fact the residue ring of the integer ring modulo the principal ideal $\langle m \rangle = m\mathbf{Z}$; Example 2.2.9 is in fact just the multiplicative group $\mathbf{Z}_m^*$ of this residue ring. It is clear that $[a] \in \mathbf{Z}_m^*$ iff a and m are coprime; hence $|\mathbf{Z}_m^*| = \varphi(m)$ (the Euler function in the number theory, it stands for the number of positive integers which are prime to m and $\leq m$). Thus, for any integer a prime to m we have $[a] \in \mathbf{Z}_m^*$ and $[a]^{\varphi(m)} = [1]$ by Cor.2.5.5; i.e. $a^{\varphi(m)} \equiv 1 \pmod{m}$. This fact is referred to as Euler's Theorem.

Exercises 3.3

+1. Let I be an ideal of a unitary ring R. The following are equivalent:
   i) I is the unity ideal;
   ii) I contains the unity 1 of R;
   iii) I contains an invertible element of R.
   (Remark: From this exercise, the unity ideal of a unitary ring is just the principal ideal generated by the unity of the ring.)

+2. Prove Proposition 3.3.6. Show that the intersection of any infinitely many ideals is again an ideal. Give an example to show that the union of two ideals is not necessarily an ideal. However, the union $\bigcup_{i=1}^{\infty} I_i$ of ideals I_i is again an ideal if $I_1 \subset I_2 \subset I_3 \subset \cdots$.

3. Prove: the integer ring $\mathbf{Z}$ is a principal ideal domain. And, in $\mathbf{Z}$, calculate: $\langle m \rangle + \langle n \rangle$, $\langle m \rangle \cap \langle n \rangle$, and $\langle m \rangle \cdot \langle n \rangle$.

+4. Let R be a commutative ring and $a, b \in R$. Prove: $\langle a \rangle \langle b \rangle \subset \langle ab \rangle$.

+5. A commutative simple unitary ring is a field.

+6. Let R be a unitary ring. The following are equivalent:

i) R is a division ring

ii) R has no non-trivial left ideals.

iii) R has no non-trivial principal left ideal.

(Hint: $Ra = R \implies a$ is left invertible; cf. Exer.2.2.9.)

7. In a ring, prove: $\langle S \cup T \rangle = \langle S \rangle + \langle T \rangle$; in particular, $\langle a_1, \cdots, a_n \rangle = \langle a_1 \rangle + \cdots + \langle a_n \rangle$.

+8. The kernel $\mathrm{Ker}(f)$ of a ring homomorphism $f : R \to S$ is an ideal of R.

+9. Let S be a subset of a ring R. Prove:

1) The left annihilator $\mathrm{Ann}_R^l(S)$ (see Definition 3.2.1) is a left ideal while the right annihilator $\mathrm{Ann}_R^r(S)$ is a right ideal.

2) If S is a left ideal, then $\mathrm{Ann}_R^l(S)$ is an ideal.

+10. Let R be a commutative ring and I be an ideal. The *radical* of I is defined to be $\mathrm{Rad}(I) = \{r \in R \mid (\exists$ a positive integer $n)(r^n \in I)\}$. Prove that $\mathrm{Rad}(I)$ is also an ideal.

+11. Let R be a commutative ring. Prove:

1) $\mathrm{Rad}(0)$ consists of all the nilpotent elements (cf. Exer.3.2.3).

2) If $N = \mathrm{Rad}(0)$, then R/N is a ring without non-zero nilpotent elements.

12. Let R be a ring and $\overline{R}$ be a residue ring of R.

1) If R is commutative (unitary resp.), then $\overline{R}$ is commutative (unitary resp.). (Remark. cf. Exer. 3.1.11.)

2) If any ideal of R is principal, then any ideal of $\overline{D}$ is principal.

13. Let $R = 2\mathbf{Z} = \{2n \mid n \in \mathbf{Z}\}$, and $I = \{4r \mid r \in R\}$.

1) Describe R/I;

2) Describe $R/\langle 4 \rangle$, is it a field?

14. A ring R is said to be a *von Neumann regular ring* if for every $a \in R$ there is an $a' \in R$ such that $aa'a = a$. Prove: in a von Neumann regular ring every principal left ideal (cf. Remark 3.3.9) is generated by an idempotent (see Exer.3.1.5).

3.4 Homomorphism Theorems

The homomorphism theorems for groups in §2.7 can be stated for rings; and the ring-theoretic versions are easy to verified since rings have additive group structures.

3.4.1 Theorem (Fundamental Theorem for Ring Homomorphisms). *Let $f : R \to S$ be a ring homomorphism with kernel K. Then K is an* *ideal of R and there exists a unique ring homomorphism $\overline{f} : R \rightarrow R/K$ such that $f = \overline{f} \circ \sigma$, where $\sigma : R \rightarrow R/K$ is the natural homomorphism; and such $\overline{f}$ is injective; moreover, $\overline{f}$ is an isomorphism provided f is surjective.*

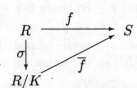

Proof. Note that the additive group of the residue ring R/K is just the quotient group of the additive group $(R, +)$ with respect to the subgroup $(K, +)$. As a group homomorphism, by Theorem 2.7.1, such $\overline{f}$ exists and is unique and it is in fact determined by $\overline{f}(a + K) = f(a)$ $\forall a + K \in R/K$; thus it remains only to prove that this $\overline{f}$ is indeed a ring homomorphism. For any $a + K, b + K \in R/K$, we do have

$$\overline{f}\big((a + K)(b + K)\big) = \overline{f}(ab + K) = f(ab) = f(a)f(b) = \overline{f}(aK)\overline{f}(bK). \square$$

3.4.2 Remark. Similar to Remark 2.7.2, we usually call the above $\overline{f}$ the homomorphism *induced by* f; and we have the following commutative diagram:

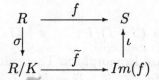

where ι is the inclusion homomorphism and $\widetilde{f}$ is an isomorphism and $\overline{f} = \iota\widetilde{f}$. In particular, we have

3.4.3 Corollary (First Isomorphism Theorem). *If $f : R \rightarrow S$ is a ring homomorphism, then $R/\mathrm{Ker}(f) \cong \mathrm{Im}(f)$; in particular, $R/\mathrm{Ker}(f) \cong S$ provided f is surjective.* $\square$

3.4.4 Theorem (Correspondence of Ideals). *Let $f : R \rightarrow \overline{R}$ be a ring surjective homomorphism with kernel $K = \mathrm{Ker}(f)$; let*

$$\mathcal{I}(\overline{R}) = \{\overline{I} \mid \overline{I} \ \text{is an ideal of} \ \overline{R}\},$$
$$\mathcal{I}_K(R) = \{I \mid I \ \text{is an ideal of} \ R \ \text{containing} \ K\}.$$

Then:

1) $\eta : \mathcal{I}_K(R) \rightarrow \mathcal{I}(\overline{R})$, $I \mapsto f(I)$ *is a bijective map;*

2) *for $I_1, I_2 \in \mathcal{I}_K(R)$, $I_1 \subset I_2$ iff $f(I_1) \subset f(I_2)$, and, in this case,* $|I_2 : I_1| = |f(I_2) : f(I_1)|$.

Proof. (1). $f(I)$ is a subgroup of $(\overline{R}, +)$ by Theorem 2.7.4; for $f(x) \in f(I)$ (i.e. $x \in I$) and $\overline{r} \in R$, there is $r \in R$ such that $f(r) = \overline{r}$, hence $\overline{r} \cdot f(x) = f(r) \cdot f(x) = f(rx) \in f(I)$ as $rx \in I$; similarly, $f(x) \cdot \overline{r} \in f(I)$. Thus $f(I) \in \mathcal{I}(\overline{R})$ and η is well-defined. Similarly, for $\overline{I} \in \mathcal{I}(\overline{R})$ we can show $f^{-1}(\overline{I}) \in \mathcal{I}_K(R)$; so the following map is also well-defined:

$$(3.4.5) \qquad \xi : \mathcal{I}(\overline{R}) \longrightarrow \mathcal{I}_K(R) , \qquad \overline{I} \longmapsto f^{-1}(\overline{I}).$$

Then, the same argument for Theorem 2.7.4 show that the two maps η and ξ are invertible to each other, hence both are bijective.

(2). Since ideals are subgroups of the additive groups of rings, this is in fact a consequence of Theorem 2.7.4(2). Or, one can argue this as the same as that for Theorem 2.7.4(2).   □

The following two results can be treated in the same way, we leave their proofs as exercises.

3.4.6 Corollary (Second Isomorphism Theorem). *Let R be a ring, I be an ideal and S be a subring. Then $S + I$ is a subring of R, and I is an ideal of $S + I$, and $I \cap S$ is an ideal of S, and we have a ring isomorphism:*

$$S/(I \cap S) \overset{\cong}{\longrightarrow} (S + I)/I , \qquad s + (I \cap S) \longmapsto s + I. \qquad □$$

3.4.7 Corollary (Third Isomorphism Theorem). *Let R be a ring and I, K be ideals of R and $I \subset K$. Then K/I is an ideal of R/I and we have a ring isomorphism:*

$$G/K \overset{\cong}{\longrightarrow} (G/I)\big/(K/I) , \qquad a + K \longmapsto (a + I) + (K/I). \qquad □$$

The homomorphism theorems give us effective tools; for the moment, we just exhibit an example to show how well they work.

Example. Prove: $\mathbf{C} \cong \mathbf{R}[x]/\langle x^2 + 1\rangle$.

Proof. It is easy to check that $\nu : \mathbf{R}[x] \to \mathbf{C}$, $f(x) \mapsto f(i)$ is a surjective ring homomorphism, where i stands for the imaginary unit; and $\text{Ker}(\nu) = \{f(x) \mid f(i) = 0\}$. But $f(i) = 0$ iff $x^2 + 1 \mid f(x)\}$, so $\text{Ker}(\nu) = \langle x^2 + 1\rangle$. Then the desired isomorphism follows from Theorem 3.4.1.   □

Exercises 3.4

1. State and prove a theorem for subrings similar to Theorem 3.4.4.
2. Prove Corollary 3.4.6.
3. Prove Corollary 3.4.7.
4. (Factorization Theorem for Ring Homomorphisms) Let $f : R \to X$ and $g : R \to S$ be ring homomorphisms and g be surjective. Then there is a homomorphism $h : S \to X$ such that $f = hg$ if and only if $\mathrm{Ker}(g) \subset \mathrm{Ker}(f)$; and, if this is the case, such h is unique and $\mathrm{Ker}(h) = g(\mathrm{Ker}(f))/g(\mathrm{Ker}(g))$.

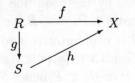

5. Prove: $\mathbf{R}[x]/\langle x \rangle \cong \mathbf{R}$.
6. Prove: the rational field $\mathbf{Q}$ has only identity and zero endomorphisms.
7. Prove: a non-zero homomorphism from a unitary ring to an integral domain must be unitary. (Hint: Exer.3.2.3.)
*8. Prove: the multiplicative group of a field is not isomorphic to the additive group of the field.

   (Hint: Let $f : (F^*, \cdot) \overset{\cong}{\to} (F, +)$; F must be infinite. If $-1 = 1$, then $2 \cdot 1 = 0$; otherwise, $d = f(-1) \neq 0$, and $2d = f((-1)^2) = f(1) = 0$. In short, there is $d \neq 0$ such that $2d = 0$. Hence $2a = 2d \cdot \frac{a}{d} = 0$ for any $a \in F$; consequently, for any $b \in F^*$ we have $b^2 = 1$, i.e. $(b+1)(b-1) = 0$. Then F has at most three elements $0, 1, -1$; a contradiction.)

3.5 Prime Ideals, Maximal Ideals

3.5.1 Definition. An ideal P of a ring R is said to be *prime* if $P \neq R$ and for any ideals A and B of R such that $AB \subset P$ we have either $A \subset P$ or $B \subset P$.

Example. It is easy to see that an ideal of $\mathbf{Z}$ (must be principal by Exer.3.3.3) is prime if and only if it is generated by a prime number.

This is in fact the original motivation of the prime ideals. In commutative rings, prime ideals take an important part; and they are characterized by a condition which is familiar in number theory.

3.5.2 Theorem. *Let P be a proper ideal of a ring R. Then P is prime if*

(P) $(\forall a, b \in R)\ \Big(\ (ab \in P)\ \Longrightarrow\ (\ \text{either}\ a \in P\ \text{or}\ b \in P\ )\ \Big).$

*Moreover, if R is commutative, then the condition **(P)** is also necessary for P being prime.*

Proof. First we assume the condition **(P)** holds. For any ideals A and B such that $AB \subset P$, if $A \not\subset P$, then there is an $a \in A - P$ such that $ab \in P$, $\forall b \in B$; but, by the condition **(P)** and the assumption that $a \notin P$, this means that $b \in P$, $\forall b \in B$; i.e. $B \subset P$. So P is prime.

Next, assume that R is commutative and P is prime. If $ab \in P$, then $\langle ab \rangle \subset P$; but, by Exer.3.3.4, $\langle a \rangle \langle b \rangle \subset \langle ab \rangle$; so $\langle a \rangle \langle b \rangle \subset P$. Thus, either $\langle a \rangle \subset P$ hence $a \in P$ or $\langle b \rangle \subset P$ hence $b \in P$. $\quad\square$

Example. Both $\langle x \rangle$ and $\langle 2, x \rangle$ are prime ideals of the integer polynomial ring $\mathbf{Z}[x]$. In fact, if $f(x)g(x) \in \langle 2, x \rangle$, then $f(x)g(x) = 2h(x) + xk(x)$ which is a polynomial with even constant term; thus one of the $f(x)$ and $g(x)$ has an even constant term, i.e. either $f(x) \in \langle 2, x \rangle$ or $g(x) \in \langle 2, x \rangle$. That is, by the above theorem, $\langle 2, x \rangle$ is a prime ideal.

3.5.3 Theorem. *An ideal P of a commutative unitary ring R is prime if and only if R/P is an integral domain.*

Proof. Assume tha P is prime. If $(a + P)(b + P) = P$, then $ab + P = P$, i.e. $ab \in P$; so either $a \in P$ or $b \in P$; consequently, either $a + P = P$ or $b + P = P$. Thus R/P has no non-zero zero divisors. It is clear that R/P is commutative and unitary, see Exer.3.3.12; hence it is an integral domain.

Conversely, assume that R/P is an integral domain. If $ab \in P$, then, in R/P, $(a + P)(b + P) = ab + P = P$; hence either $a + P = P$ or $b + P = P$; thus either $a \in P$ or $b \in P$. That is, P is prime. $\quad\square$

3.5.4 Corollary. *A commutative unitary ring is an integral domain if and only if $\{0\}$ is a prime ideal.* $\quad\square$

3.5.5 Corollary. *An ideal P of a commutative ring R is prime if and only if R/P has no non-zero zero divisors.* $\quad\square$

3.5.6 Definition. An ideal M of a ring R is said to be *maximal* if $M \neq R$ and for any ideal N of R such that $M \subset N$ we have either $N = R$ or $N = M$.

Example. An ideal $\langle p \rangle$ of $\mathbf{Z}$ is maximal if and only if p is prime.

3.5.7 Theorem. *Let R be a unitary ring. If an ideal $A \neq R$, then there is a maximal ideal M containing A. In particular, maximal ideals of R exist.*

Proof. We have to cite the Zorn's Lemma for the proof. Let $\mathcal{S} = \{I \mid A \subset I$ and I is a proper ideal of $R\}$. Then $\mathcal{S} \neq \emptyset$ since $A \in \mathcal{S}$. If $\mathcal{C} \subset \mathcal{S}$ is a chain, then $C = \bigcap_{I \in \mathcal{C}} I$ is an ideal containing A; and $C \neq R$, for, otherwise $1 \in C$ hence $1 \in I$ for some $I \in \mathcal{C}$, hence $I = \langle 1 \rangle = R$ which contradicts to that $I \in \mathcal{S}$. Thus $C \in \mathcal{S}$ is an upper bound of $\mathcal{C}$. By Zorn's Lemma there is a maximal member M in $\mathcal{S}$. Then M contains A and, by the definition of $\mathcal{S}$, M is a maximal ideal. $\square$

3.5.8 Remark. The above argument shows a way to get maximal ideals. If V is a non-empty subset of R consisting of invertible elements, and $\mathcal{S} = \{I \mid I$ is an ideal of R and $I \cap V = \emptyset\}$, then $\mathcal{S} \neq \emptyset$ and the Zorn's Lemma is applied; and the maximal members of $\mathcal{S}$ are maximal ideals of R. Exer.11. shows a similar way to get prime ideals of commutative rings.

3.5.9 Theorem. *An ideal M of a ring R is maximal if and only if R/M is a simple ring.*

Proof. Assume M is maximal. By Theorem 3.4.4, R/M has no non-trivial ideal; i.e. R/M is a simple ring.

Conversely, assume R/M is a simple ring. Then R/M has no non-trivial ideals; hence, by Theorem 3.4.4, R has no proper ideal which contains M properly; i.e. M is maximal. $\square$

3.5.10 Corollary. *An ideal M of a commutative unitary ring R is maximal if and only if R/M is a field.*

Proof. R/M is also a commutative unitary ring by Exer.3.3.12. The corollary follows from the above theorem and Exer.3.3.5 $\square$

3.5.11 Corollary. *A maximal ideal of a commutative unitary ring is a prime ideal.* $\square$

Exercises 3.5

1. An ideal P of a ring R is prime if and only if for $a, b \in R$ such that $aRb \subset P$ we have either $a \in P$ or $b \in P$.

$^{+}$2. If S is a subring of a ring R and P is a prime ideal of R, then $P \cap S$ is a prime ideal of S.

3. Let R be a commutative ring. If for any $a \in R$ there is a positive integer $n > 1$ such that $a^n = a$, then every prime ideal of R is a maximal ideal.

4. An ideal P of a ring R is prime if and only if
$$(\forall a, b \in R)\Big(\big(\langle a \rangle \langle b \rangle \subset P\big) \implies (a \in P \ \lor \ b \in P)\Big).$$

5. Find all the maximal ideals and the prime ideals of $\mathbf{Z}_m$.

6. Prove: a commutative unitary ring is a field if and only if $\{0\}$ is a maximal ideal. (Cf. Exer.3.3.5.)

7. Show that the ring $R = 2\mathbf{Z} = \{2n \mid n \in \mathbf{Z}\}$ has a maximal ideal M such that R/M is not a field.

8. A prime ideal P of a ring is said to be minimal if for any prime ideal $P' \subset P$ we have $P' = P$. Prove that a commutative unitary ring has minimal prime ideals. (Hint: use Zorn's Lemma.)

9. (Correspondence of Prime Ideals, cf. Theorem 3.4.4). Let $f : R \to \overline{R}$ be a surjective ring homomorphism with kernel $K = \mathrm{Ker}(f)$.
   1) If P is a prime ideal of R which contains K, then $f(P)$ is a prime ideal of $\overline{R}$;
   2) If $\overline{P}$ is a prime ideal of $\overline{R}$, then $f^{-1}(\overline{P})$ is a prime ideal of R which contains K.

10. (Prime Avoidance Lemma). Let R be a ring and I be an ideal. If $I \subset P_1 \cup \cdots \cup P_n$ for prime ideals $P_1, \cdots, P_n$, then $I \subset P_i$ for some i.
    (Hint: by induction on n and cf. Exer.2.3.5; if $I \not\subset (\cup P_i) - P_k$ for every k, then we have $x_k \in I$ and $x_k \in P_k$ but $x_k \notin P_i$ for any $i \neq k$, hence $x_1 + (x_2 \cdots x_n) \in I$ but $x_1 + (x_2 \cdots x_n) \notin P_i$ for any i.)

*11. Let R be a commutative ring and S be a subsemigroup of the multiplicative semigroup of R. If $\mathcal{D} = \{I \mid I$ is an ideal of R and $I \cap S = \emptyset\} \neq \emptyset$, then Zorn's Lemma can be applied and the maximal members of $\mathcal{D}$ are prime ideals of R. (Hint: let P be a maximal one; if $a_1 a_2 \in P$ but $a_i \notin P$, $i = 1, 2$, then both $\langle a_i \rangle + P$, $i = 1, 2$, contain P properly, hence meet S in non-empty sets; we have $s_i = x_i + p_i \in S$ with $x_i \in \langle a_i \rangle$ and $p_i \in P$; hence $s_1 s_2 - x_1 x_2 \in P$; but $x_1 x_2 \in P$ since $a_1 a_2 \in P$, cf. Prop.3.3.8; so $s_1 s_2 \in P \cap S$.)

12. Prove: in a commutative ring, there is at least one prime ideal which consists of zero divisors. (Hint: use Exer.11. to the subset of the elements which are not zero-divisors.)

*13. A left ideal L of a ring R is said to be *maximal* if $L \neq R$ and there is no proper left ideal which contains L properly. Prove that for any proper left ideal N of a unitary ring R there is a maximal left ideal L containing N. In particular, the maximal left ideals of R exist. (Hint: cf. the proof of Theorem 3.5.7.)

3.6 Direct Sums

We follow the ideas in §2.9 to consider the related topics for rings.

3.6.1 Definition. Let $R_1, R_2, \cdots, R_n$ be rings. From 2.9.3, we can define the additive group $R = R_1 \oplus \cdots \oplus R_n$ such that the addition is:

$$(a_1, \cdots, a_n) + (a'_1, \cdots, a'_n) = (a_1 + a'_1, \cdots, a_n + a'_n);$$

further, on R we define a multiplication:

$$(a_1, \cdots, a_n) \cdot (a'_1, \cdots, a'_n) = (a_1 a'_1, \cdots, a_n a'_n);$$

then it is easy to verify that R with the addition and the multiplication forms a ring, called the (*external*) *direct sum* of the rings $R_1, \cdots, R_n$, and denoted $R = R_1 \oplus \cdots \oplus R_n$.

3.6.2 Remark. The following facts can easily be verified:

1) The zero element of R is $0_R = (0_{R_1}, \cdots, 0_{R_n})$ with 0_{R_i} in the i-entry being the zero element of R_i.

2) If each R_i is unitary for $i = 1, \cdots, n$, then R is unitary with $1_R = (1_{R_1}, \cdots, 1_{R_n})$ where 1_{R_i} in the i-entry is the unity of R_i; and $e_i = (0, \cdots, 0, 1_{R_i}, 0, \cdots, 0)$, is an idempotent (cf. Exer.3.2.3) and $e_i e_j = 0$ for $i \neq j$ and $1_R = e_1 + \cdots e_n$.

Definition. An idempotent e is said to be *central* if e is a central element, i.e. $e \in C(R)$. Idempotents e_i and e_j are said to be *orthogonal* if $e_i e_j = 0 = e_j e_i$. A system $\{e_1, \cdots, e_n\}$ of idempotents is said to be *orthogonal* if they are orthogonal to each other, and it is said to be *complete* if $e_1 + \cdots + e_n = 1$. In these notations, the above $e_1, \cdots, e_n$ form a complete orthogonal system of central idempotents.

We continue the remark.

3) $R'_i = \{(0, \cdots, 0, x_i, 0, \cdots, 0) \mid x_i \in R_i\}$ is an ideal of R and, as rings, $R'_i \cong R_i$. Further, if every R_i is unitary, so is every R'_i and the above e_i is the unity of R'_i; in fact $R'_i = Re_i$.

4) In R the following condition holds:

$$R = R'_1 + \cdots + R'_n \quad \text{and} \quad R'_i \bigcap \left(\sum_{j \neq i} R'_j \right) = \{0\} \quad \text{for} \quad i = 1, \cdots, n;$$

in particular, the latter implies that $R'_i R'_j = \{0\}$, $\forall i \neq j$.

The condition in the above (4) is in fact a characterization of the direct sums of rings.

3.6.3 Theorem.   *Let R be a ring and $R_1, \cdots, R_n$ be ideals of R. The following two conditions are equivalent:*

i) $R = \sum_{i=1}^n R_i$ *and* $R_i \bigcap \left(\sum_{1 \leq j \neq i \leq n} R_j \right) = \{0\}$ *for* $i = 1, \cdots, n$;

ii) *For any $a \in R$ there are unique $a_i \in R_i$, $i = 1, \cdots, n$, such that* $a = a_1 + \cdots + a_n$.
Moreover, $R \cong R_1 \oplus \cdots \oplus R_n$ if this is the case.

Proof. The equivalence of the two conditions follows Theorem 2.9.9 at once. Assume they hold, then, by Theorem 2.9.9 again, the following

$$\sigma : R_1 \oplus \cdots \oplus R_n \longrightarrow R , \qquad (a_1, \cdots, a_n) \longmapsto a_1 + \cdots + a_n$$

is an additive group isomorphism; it remains to show that σ is a ring homomorphism. Note that

$$\sigma((a_1, \cdots, a_n)(a'_1, \cdots, a'_n)) = \sigma(a_1 a'_1, \cdots, a_n a'_n) = a_1 a'_1 + \cdots + a_n a'_n .$$

On the other hand, since R_i and R_j are ideals and $R_i \cap R_j = \{0\}$ for $i \neq j$, we have $a_i a'_j = 0$ because $a_i a'_j \in R_i \cap R_j$; so

$$(a_1 + \cdots + a_n)(a'_1 + \cdots + a'_n) = a_1 a'_1 + \cdots + a_n a'_n .$$

Thus

$$\sigma((a_1, \cdots, a_n)(a'_1, \cdots, a'_n)) = \sigma(a_1, \cdots, a_n)\sigma(a'_1, \cdots, a'_n). \qquad \square$$

3.6.4 Definition. A ring R is said to be an *internal direct sum* of its ideals $R_1, \cdots, R_n$, and denoted by $R = R_1 \oplus \cdots \oplus R_n = \bigoplus_{i=1}^n R_i$ too, if the conditions in Theorem 3.6.3 hold.   A ring is said to be *indecomposable* if it cannot be written as a direct sum of two non-zero ideals.

Theorem 3.6.3 shows that the external direct sums of rings and the internal direct sums of ideals have the same structures, hence it is reasonable to denote them by the same symbols.

Similarly to that for groups, if $f_i : R \to R_i$, $i = 1, \cdots, n$, are ring homomorphisms, then

$$f = (f_1, \cdots, f_n): \quad R \longrightarrow R_1 \oplus \cdots \oplus R_n, \quad a \longmapsto (f_1(a), \cdots, f_n(a))$$

is a ring homomorphism with kernel $\mathrm{Ker}(f) = \bigcap_{i=1}^n \mathrm{Ker}(f_i)$, cf. Prop. 2.9.5. As a consequence of Theorem 2.9.8 we immediately have the following fact.

3.6.5 Chinese Remainder Theorem (Ring-Theoretic Version). *Notations as above.* $f = (f_1, \cdots, f_n)$ *is surjective if and only if every* f_i *is surjective and*

$$\mathrm{Ker}(f_i) + \left(\bigcap_{j \neq i} \mathrm{Ker}(f_j) \right) = R \qquad \text{for all} \quad i = 1, \cdots, n. \qquad \square$$

3.6.6 Definition. Ideals A and B of a ring R are said to be *coprime* if $A + B = R$.

Example. In the integer ring $\mathbf{Z}$, ideals $\langle a \rangle$ and $\langle b \rangle$ are coprime iff the integers a and b are coprime in the number-theoretic sense.

Thus the above result can be restated as:

"$f = (f_1, \cdots, f_n)$ *is surjective if and only if every* f_i *is surjective and the ideals* $\mathrm{Ker}(f_i)$ *and* $\bigcap_{j \neq i} \mathrm{Ker}(f_j)$ *are coprime for* $i = 1, \cdots, n$."

3.6.7 Lemma. *In a unitary ring R, if every ideal I_i, $i = 1, \cdots, n$, is coprime to an ideal I, then both their product $I_1 \cdots I_n$ and their intersection $\bigcap_{i=1}^n I_i$ are coprime to I too.*

Proof. Since R is unitary, we have $R^2 = R$ hence $R = R^n$. By Prop.3.3.6 we have

$$R = R^n = (I_1 + I) \cdots (I_n + I) = (I_1 \cdots I_n) + \cdots ;$$

on the right hand side, every summand other than $I_1 \cdots I_n$ has I as a factor hence is contained in I. Thus $I + (I_1 \cdots I_n) = R$. Since $\bigcap_{i=1}^n I_i \supset I_1 \cdots I_n$ obviously, we also have $I + \bigcap_{i=1}^n I_i = R$. $\qquad \square$

The following is the usual version of the Chinese Remainder Theorem.

3.6.8 Chinese Remainder Theorem. *Let R be a unitary ring and $I_1, \cdots, I_n$ be ideals, let $I = \bigcap_{i=1}^n I_i$. Then the following*

$$\overline{f}: \quad R/I \longrightarrow R/I_1 \oplus \cdots \oplus R/I_n, \qquad a + I \longmapsto (a + I_1, \cdots, a + I_n)$$

is a ring isomorphism if and only if the ideals $I_1, \cdots, I_n$ are pairwise coprime.

Proof. The natural homomorphisms $f_i : R \to R/I_i$, $a \mapsto a + I_i$ induce a homomorphism

$$f : \ R \longrightarrow R/I_1 \oplus \cdots \oplus R/I_n, \qquad a \longmapsto (a + I_1, \cdots, a + I_n)$$

with kernel I since $\mathrm{Ker}(f_i) = I_i$. So the homomorphism $\overline{f}$ which is induced by f is injective. By 3.6.5, $\overline{f}$ is an isomorphism if and only if

$$I_i + \bigcap_{j \neq i} I_j = R, \qquad i = 1, \cdots, n.$$

By Lemma 3.6.7, these hold if $I_1, \cdots, I_n$ are pairwise coprime. Conversely, the condition $I_i + (\bigcap_{j \neq i} I_j) = R$ implies obviously that $I_i + I_j = R$ for $i \neq j$ since $I_j \supset \bigcap_{k \neq i} I_k$. □

As an application, we have a direct decomposition of $\mathbf{Z}_m$.

3.6.9 Corollary. *If $m_1, \cdots, m_n$ are pairwise coprime integers and $m = m_1 \cdots m_n$, then $\mathbf{Z}_m \cong \mathbf{Z}_{m_1} \oplus \cdots \oplus \mathbf{Z}_{m_n}$.*

Proof. $\langle m_1 \rangle, \cdots, \langle m_n \rangle$ are pairwise coprime. □

Thus, if $m_1, \cdots, m_n$ are pairwise coprime integers, then for any integers $r_1, \cdots, r_n$ there is a unique residue class $[a]_m \in \mathbf{Z}_m$ such that $a \equiv r_i \pmod{m_i}$, $i = 1, \cdots, n$. This is just the original number-theoretic version of the Chinese Remainder Theorem 2.9.1.

3.6.10 Corollary. *If an integer $m = p_1^{e_1} \cdots p_r^{e_r}$ with pairwise distinct prime numbers $p_1, \cdots, p_r$, then $\mathbf{Z}/m \cong \mathbf{Z}/\langle p_1^{e_1} \rangle \oplus \cdots \oplus \mathbf{Z}/\langle p_r^{e_r} \rangle$ and every $\mathbf{Z}/\langle p_i^{e_i} \rangle$ is indecomposable.*

Proof. Cor.3.6.9 and Exer.2.9.9. □

Exercises 3.6

1. If R_1 and R_2 are non-zero rings, prove: $R_1 \oplus R_2$ is not an integral domain.

2. Show that an integral domain is indecomposable; in particular, $\mathbf{Z}$, $\mathbf{R}[x]$ are indecomposable.

3. If R is a ring and R_1, R_2 are ideals and $R = R_1 \oplus R_2$, show that $R/R_1 \cong R_2$.

4. 1) If $R_1, \cdots, R_n$ are unitary rings and I is an ideal of $R_1 \oplus \cdots \oplus R_n$, then each $I_i = I \cap R_i$ is an ideal of R_i and $I = I_1 \oplus \cdots \oplus I_n$.

   2) Show that the above (1) is not necessarily true if R_i are not unitary.
   (Hint: Let $R_1 = R_2 = 2\mathbf{Z}$, see the example following 3.3.9; let $I =$

$\{(a, b) \in R |$ either $a \equiv b \equiv 0 (\mathrm{mod}\, 4)$ or $a \equiv b \equiv 2 (\mathrm{mod}\, 4)\}$; then $I \cap R_1 = I \cap R_2 = 4\mathbf{Z}$.)

5. If $R = R_1 \oplus \cdots \oplus R_n$ is a unitary ring, then so is every R_i. (Hint: consider the expression of 1_R.)

6. If e is an idempotent of a unitary ring R, then

    1) $1 - e$ is also an idempotent and $e, 1 - e$ form an orthogonal complete system of idempotents;

    2) $R = Re \oplus R(1 - e)$ is a direct sum of additive groups (in fact, of left ideals);

    3) if e is central, then so is $1 - e$ and the decomposition of (2) is a direct sum of rings.

7. Let R and $R_1, \cdots, R_n$ be unitary rings. The following statements are equivalent:

    i) $R \cong R_1 \oplus \cdots \oplus R_n$;

    ii) R has a complete orthogonal system $\{e_1, \cdots, e_n\}$ of central idempotents such that $Re_i \cong R_i$, $i = 1, \cdots, n$.

8. If $f : R \to R$ is an endomorphism of a ring R such that $f^2 = f$, then $R = \mathrm{Ker}(f) \oplus \mathrm{Im}(f)$ is a direct sum of additive groups; give an example to show that this is not an internal direct sum of ideals in general.

    (Hint: consider $\mathbf{R}[x]$ and the endomorphism $\gamma : \mathbf{R}[x] \to \mathbf{R}[x]$ such that $\gamma(f(x)) = $ the constant term of $f(x)$.)

3.7 Fraction Fields of Integral Domains

Following the idea of constructing the rational fields $\mathbf{Q}$ from the integer ring $\mathbf{Z}$, from any integral domain we can get a unique field.

3.7.1 Definition. Let D be an integral domain. A field F is called the *fraction field* of D if there is an injective ring homomorphism $\iota : D \to F$ and every $a \in F$ can be written as $a = \iota(b)\iota(c)^{-1}$ for some $b, c \in D$ and $c \neq 0$. Note that the ι must be unitary (Exer.3.4.8). It is convenient to identify the image $\iota(b)$ in F with b itself (just as identifying the fraction $n/1$ with the integer n); then, every $a \in F$ can be written as $a = bc^{-1}$ or $a = b/c$ for some $b, c \in D$ and $c \neq 0$.

The following result shows that not only the fraction fields of an integral domain D exist, and they are unique as well; hence we can say "the fraction field of D" instead of "a fraction field of D".

3.7.2 Theorem. *The fraction fields of an integral domain D exist and are unique up to isomorphism.*

Proof. Let $S = D \times (D - \{0\})$ be the Cartesian product of sets. Define a relation "$\sim$" on S:

$$(a,b) \sim (c,d) \quad \Longleftrightarrow \quad ad = bc,$$

which is an equivalence relation; for, the reflexity and the symmetry are obvious, and, if $(a,b) \sim (c,d)$ and $(c,d) \sim (e,f)$, then $ad = bc$ and $cf = de$, hence $adf = bcf = bde$; canceling d, we get $af = be$, i.e. $(a,b) \sim (e,f)$.

Take the quotient set $F = S/\sim$ whose elements $\overline{(a,b)}$ (the equivalence class containing (a,b)) are denoted by a/b. Note that $a/b = ac/bc$ for any $c \neq 0$. Define addition and multiplication on F as:

$$a/b + c/d = (ad + bc)/(bd), \qquad (a/b)(c/d) = (ac)/(bd).$$

If $a/b = a'/b'$ and $c/d = c'/d'$, then $ab' = a'b$ and $cd' = c'd$, hence

$$
\begin{aligned}
(a'd' + b'c')/(b'd') &= (a'bdd' + b'bdc')/(bdb'd') \\
&= (ab'dd' + b'bcd')/(bdb'd') \\
&= (ad + bc)/(bd);
\end{aligned}
$$

that is, the result of the addition is independent of the choice of representatives of the equivalence classes, so the addition is well-defined. Similarly, the multiplication is well-defined too.

Then, it is routine to verify that F is a field; we just point out that $0/b$ is the zero element $0 = 0_F$ of F (and $a/b = 0$ in F iff $a = 0$), and $1/1$ ($= b/b$ for any $b \neq 0$) is the unity $1 = 1_F$ of F, and $(a/b)^{-1} = b/a$ for $a/b \neq 0$ (which means $a \neq 0$).

The map $D \to F, b \mapsto b/1$, is clearly an injective ring homomorphism, since $(bc)/1 = (b \cdot c)/(1 \cdot 1) = (b/1)(c/1)$ and $b/1 = 0 \Longrightarrow b = 0$. As noted in Def.3.7.1, we identify $b/1 \in F$ with $b \in D$. For any $a = b/c \in F$, $a = (b/1)(1/c) = (b/1)(c/1)^{-1} = bc^{-1}$. Thus F is a fraction field of D.

For the uniqueness, we prove a more general conclusion:

(3.7.3). *If integral domains $D \overset{\varphi}{\cong} D'$ and F and F' are fraction fields of D and D' resp., then $\widetilde{\varphi} : F \to F'$, $a/b \mapsto \varphi(a)/\varphi(b)$ is an isomorphism.*

The proof for this is also routine. First we can calculate:

$$\tilde{\varphi}(a/b + c/d) = \tilde{\varphi}((ad + bc)/(bd)) = \varphi(ad + bc)/\varphi(bd)$$
$$= (\varphi(a)\varphi(d) + \varphi(b)\varphi(c))/(\varphi(b)\varphi(d))$$
$$= \varphi(a)/\varphi(b) + \varphi(c)/\varphi(d)$$
$$= \tilde{\varphi}(a/b) + \tilde{\varphi}(c/d);$$

and $\tilde{\varphi}(a/b \cdot c/d) = \tilde{\varphi}(a/b) \cdot \tilde{\varphi}(c/d)$ similarly; so $\tilde{\varphi}$ is a ring homomorphism. Then, by the definition of fraction fields and the surjectivity of φ we have $\tilde{\varphi}$ is surjective. If $\tilde{\varphi}(a/b) = \varphi(a)/\varphi(b) = 0$, then $\varphi(a) = 0$ hence $a = 0$ as φ is injective; so $\tilde{\varphi}$ is an isomorphism. $\square$

In fact, the above proof of (3.7.3) is based on a so-called *universal property* of the fraction fields.

3.7.4 Theorem. *Let D be an integral domain and $\iota : D \to F$ be a fraction field of D. If R is a commutative ring and $\varphi : D \to R$ is a ring homomorphism such that $f(a)$, $\forall 0 \neq a \in D$, is invertible, then exists a unique ring homomorphism $\tilde{\varphi} : F \to R$ such that $\varphi = \tilde{\varphi}\iota$, and such a $\tilde{\varphi}$ is injective.*

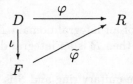

Proof. First we note that φ must be unitary since $\varphi(1_D)$ is an invertible idempotent hence must be the unity 1_R by Exer.3.2.3. Define (recall from 3.7.1 that we identify $\iota(a)$ with a for $a \in D$):

$$\tilde{\varphi} : F \longrightarrow R, \quad a/b \longmapsto \varphi(a)\varphi(b)^{-1}.$$

And $\tilde{\varphi}$ is a ring homomorphism, see (3.7.3). Further, if $\varphi(a)\varphi(b)^{-1} = 0$, then $\varphi(a) = 0$, so $a = 0$ (note that φ must be injective), hence $a/b = 0$. That is, $\tilde{\varphi}$ is injective. At last, if $\psi : F \to R$ is a ring homomorphism such that $\varphi = \psi\iota$, then for any $a/b \in F$ we have

$$\psi(a/b) = \psi(\iota(a)\iota(b)^{-1}) = \psi(\iota(a)) \cdot \psi(\iota(b))^{-1} = \varphi(a) \cdot \varphi(b)^{-1} = \tilde{\varphi}(a/b);$$

which shows $\psi = \tilde{\varphi}$. $\square$

As we suggested in the beginning of this section, one of the classical examples of fraction fields is the rational field from the integer ring. The following is another typical example.

3.7.5 Example. The fraction field of the real polynomial ring $\mathbf{R}[x]$ is the rational function field over $\mathbf{R}$:

$$\mathbf{R}(x) = \{f(x)/g(x) \mid f(x), g(x) \in \mathbf{R}[x] \text{ and } g(x) \neq 0\}.$$

Moreover, the fraction field of the real polynomial ring $\mathbf{R}[x_1, \cdots, x_n]$ of n variables is the rational function field of n variables:

$$\mathbf{R}(x_1, \cdots, x_n) =$$
$$\{f(x_1, \cdots, x_n)/g(x_1, \cdots, x_n) \mid f, g \in \mathbf{R}[x_1, \cdots, x_n] \text{ and } g \neq 0\}.$$

3.7.6 Remark. The idea of constructing fraction fields from integral domains can be extended; it leads to the so-called *localization theory* for commutative rings, even for non-commutative rings. The localization method (Exer.3) takes an important part in the commutative algebra.

Exercises 3.7

1. Find the fraction fields of the following integral domains:
    1) $\mathbf{Z}[i] = \{m + ni \mid m, n \in \mathbf{Z}\}$ where i is the imaginary unit.
    2) $\mathbf{Z}[x]$;

2. If F is a fraction field of an integral domain D and R is a subring of F such that $D \subset R \subset F$, then R is an integral domain and F is a fraction field of R.

+3. Let R be a commutative unitary ring and S is a submonoid of the multiplicative monoid of R. On the set $R \times S = \{(a, u) \mid a \in R \text{ and } u \in S\}$ define:

$$(a, u) \sim (b, v) \quad \Longleftrightarrow \quad (\exists w \in S)\Big((av - bu)w = 0\Big).$$

   1) "$\sim$" is an equivalence relation.
   2) Let $S^{-1}R = (R \times S)/\sim$ be the quotient set and denote the equivalence class of (a, u) by a/u; then the following operations

$$a/u + b/v = (av + bu)/uv \qquad \text{and} \qquad (a/u) \cdot (b/v) = (ab)/(uv)$$

   are well-defined on $S^{-1}R$.
   3) $(S^{-1}R, +, \cdot)$ is a commutative unitary ring.
   4) The map $\iota : R \to S^{-1}R, a \mapsto a/1$ is a ring homomorphism; and ι is injective if and only if S contains no zero divisors.
   5) If T is a commutative ring and $\varphi : R \to T$ is a ring homomorphism such that $\varphi(u)$, $\forall u \in S$, is invertible in T, then there exists a unique ring homomorphism $\widetilde{\varphi} : S^{-1}R \to T$ such that $\varphi = \widetilde{\varphi}\iota$.

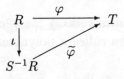

6) Take R to be an integal domain and $S = R - \{0\}$, then $S^{-1}R$ is the fraction field of R.

3.8 Polynomial Rings

For convenience, in this section we always adopt the following

3.8.0 Agreement. *Any ring is commutative and unitary, any subring and any ring homomorphism are unitary.*

Let R be a ring. As usual, we can regard $f(x) = a_0 + a_1 x + \cdots + a_n x^n$, with variable x and coefficients $a_i \in R$, as a function $f : R \to R, a \mapsto f(a) = a_0 + a_1 a + \cdots + a_n a^n$. Further, also as usual, we can define the addition and the multiplication of functions of R to itself. Then all such "polynomial functions" really form a ring. But, different from the real (or complex or rational etc.) ring, it may happen that $f(x) = a_0 + a_1 x + \cdots + a_n x^n$ and $g(x) = b_0 + b_1 x + \cdots + b_m x^m$ are distinct expressions (i.e. some $a_i \neq b_i$) but they are one and the same function. For example, take $R = \mathbf{Z}_2$ and $f(x) = 1 + x$ and $g(x) = 1 + x^2$, then $f(0) = 1 = g(0)$ and $f(1) = 0 = g(1)$.

On the other hand, If R is a subring of a ring R' and $u \in R'$, then it is easy to see that the subset $R[u] = \{f(u) = a_0 + a_1 u + \cdots + a_n u^n \mid n \geq 0, a_i \in R\}$ is a subring of R' which contains R and u, and it is in fact the smallest one of such subrings; see Exer.1. The following is the generalized description.

3.8.1 Proposition and Notation. *Let R be a subring of a ring R' and $u_1, \cdots, u_n \in R'$.*

1) *The intersection of all subrings of R' which contain R and $u_1, \cdots, u_n$ is the smallest subring containing R and $u_1, \cdots, u_n$; this subring is denoted by $R[u_1, \cdots, u_n]$ and called the subring of R' generated by $u_1, \cdots, u_n$ over R.*

2) $R[u_1, \cdots, u_n] = \{$*finite sums* $\sum_{i_1, \cdots, i_n} a_{i_1 \cdots i_n} u_1^{i_1} \cdots u_n^{i_n} \mid a_{i_1 \cdots i_n} \in R\};$ *such a finite sum is also called a* polynomial of $u_1, \cdots, u_n$ over R.

Proof. Exer.1 $\qquad\qquad\qquad\qquad\qquad\qquad\qquad\qquad\qquad\qquad$ □

Consider the case of $n = 1$. Even in very familiar cases, e.g. rational field and real field $\mathbf{Q} \leq \mathbf{R}$, it may happen that "polynomials" $f(u)$ and $g(u)$ over $\mathbf{Q}$ are in different expressions but $f(u) = g(u)$; e.g. if $u = \sqrt{2}$, then $u = u^2 + u - 2$. It is easy to see that this happens if and only if there is an expression $f(u) = a_0 + a_1 u + \cdots + a_n u^n = 0$ with at least one $a_i \neq 0$.

Let R be a ring. What is the expected "polynomial ring over R of one variable"? It should be a ring R' which contains R and has an element x such that $R' = R[x]$ whose element is *uniquely* (except for zero coefficients) expressed as $f(x) = a_0 + a_1 x + \cdots + a_n x^n$ for $n \geq 0$ and $a_i \in R$, or equivalently, if $a_0 + a_1 x + \cdots + a_n x^n = 0$ then $a_0 = \cdots = a_n = 0$. And it is expected that such rings are unique up to isomorphism if they exist.

In fact, the uniqueness is an easy consequence of its property, see Remark 3.8.6 below. For the existence, the wanted property suggests an "artificial" way to get such a ring.

Regarding x just as a symbol, called an "indeterminate", we consider all such expressions $a_0 + a_1 x + \cdots + a_n x^n$ with $n \geq 0$ and $a_i \in R$, and agree on that $a_0 + a_1 x + \cdots + a_n x^n = b_0 + b_1 x + \cdots + b_m x^m$ if and only if $a_i = b_i \; \forall i = 0, 1, \cdots$, where $a_i = 0$ ($b_i = 0$ resp.) if $i > n$ ($i > m$ resp.); let $R[x]$ be the set of all such expressions. Then $R[x]$ is a ring provided that we define (equalize the lengths by adjoining zero coefficients)

$$\Big(\sum_{i=0}^n a_i x^i \Big) + \Big(\sum_{i=0}^n b_i x^i \Big) = \sum_{i=0}^n (a_i + b_i) x^i$$

and (where $c_k = \sum_{i+j=k} a_i b_j$, for $k = 0, 1, \cdots, m+n$)

$$\Big(\sum_{i=0}^n a_i x^i \Big) \Big(\sum_{j=0}^m b_j x^j \Big) = \sum_{k=0}^{m+n} c_k x^k;$$

and $R \to R[x], a \mapsto a$, is an injective ring homomorphism, i.e. R can be identified with a subring of $R[x]$.

It is clear that any of the above expressions $f(x) = a_0 + a_1 x + \cdots +$

$a_n x^n \in R[x]$ is determined by the sequence $(a_0, a_1, \cdots)$ where almost $a_i \in R$ are zero. Thus, a more mathematical formulation of $R[x]$ is as follows.

3.8.2 Proposition and Definition. *Let R be a ring. Let*

$$R' = \Big\{ (a_0, a_1, \cdots) \Big| a_i \in R, \text{ and } a_i = 0 \text{ except for finite many indices } i \Big\},$$

and let $x = (0, 1, 0, 0, \cdots)$. Define two operations on R':

$$(a_0, a_1, \cdots) + (b_0, b_1, \cdots) = (a_0 + b_0, a_1 + b_1, \cdots);$$

$$(a_0, a_1, \cdots) \cdot (b_0, b_1, \cdots) = (c_0, c_1, \cdots), \text{ where } c_k = \sum_{i+j=k} a_i b_j.$$

Then R' is a ring and $\iota : R \to R', a \mapsto (a, 0, 0, \cdot)$, is an injective ring homomorphism; and, identifying R with its image in R', we have $R' = R[x]$; and $\sum_i a_i x^i = \sum_i b_i x^i$ iff $a_i = b_i$ $\forall i = 0, 1, \cdots$.

The element x is called an indeterminate *and the ring $R[x]$ is called the* polynomial ring *of x over R and elements of $R[x]$ are called* polynomials *over R (or R-polynomials) of x.*

Proof. It is routine to check that R' is a ring and ι is an injective ring homomorphism. An easy calculation shows that

$$x^i = (0, \cdots, 0, 1, 0, \cdots) \qquad \text{with 1 in the } i\text{-entry and 0 elsewhere;}$$

thus, obviously, $(a_0, a_1, \cdots) = \sum_i a_i x^i$. Then, the last conclusion is clear. □

Remark. In the above, $f(x) = \sum_i a_i x^i$ is in fact a finite sum of the terms $a_i x^i$, since almost $a_i = 0$. Thus, for $f(x) \neq 0$ we can write $f(x) = a_0 + a_1 x + \cdots + a_n x^n$ with $a_n \neq 0$, where n is called the *degree* of $f(x)$ and denoted by $\deg(f(x))$, and a_n is called the *leading coefficient* of $f(x)$. Note that it is an agreement that $\deg(0) = -\infty$.

3.8.3 Theorem (Universal Property of Polynomial Rings). *Let R and S be rings and $f : R \to S$ be a ring homomorphism and $u \in S$; let x be an indeterminate over R. Then there is a unique ring homomorphism $\tilde{f} : R[x] \to S$ such that $f = \tilde{f}|_R$ and $\tilde{f}(x) = u$.*

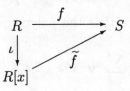

where ι is the inclusion map.

Proof. It is easy to verify that the following

$$\tilde{f}:\ R[x] \longrightarrow S,\ \ f(x) \longmapsto f(u)\,,$$

is a ring homomorphism and $\tilde{f}|R = f$. Such $\tilde{f}$ is clearly unique since any homomorphism from $R[x]$ is determined by its image of R and of x.   □

3.8.4 Corollary. *Let $R \le R'$ be rings and $u \in R'$, let x be an indeterminate over R. Then $R[u] \cong R[x]/I$ for an ideal I of $R[x]$.*

Proof. Let $\iota : R \to R'$ be the inclusion map. The above theorem shows a homomorphism $\tilde{\iota} : R[x] \to R'$ such that $\tilde{\iota}|_R = \iota$ and $\tilde{\iota}(x) = u$. Obviously, $\mathrm{Im}(\tilde{\iota}) = R[u]$. Let $I = \mathrm{Ker}(\tilde{\iota})$, then the corollary follows from the Fundamental Theorem on Homomorphisms.   □

It is clear that $I = \{f(x) \in R[x] \,|\, f(u) = 0\}$.

3.8.5 Definition. Notations as above. The polynomials $f(x)$ such that $f(u) = 0$ is called the *annihilating R-polynomials* of u. And u is called an *algebraic element* over R if it has non-zero annihilating R-polynomials; otherwise u is called a *transcendental element* over R.

3.8.6 Remark. An indeterminate x over R is of course transcendental over R; conversely, a transcendental element acts just as an indeterminate. In particular, for any two transcendental elements u and v we have a ring isomorphism $R[u] \to R[v], f(u) \mapsto f(v)$; see Exer.2.

Example. Let $\omega \in \mathbf{C}$ be a primitive third root of unity. Then $\mathbf{Z}[\omega] \cong \mathbf{Z}[x]/\langle x^2 + x + 1 \rangle$.

In fact, using $\tilde{\iota} : \mathbf{Z}[x] \to \mathbf{Z}[\omega]$ as in Cor.3.8.4, an integer polynomial $f(x) \in \mathrm{Ker}(\tilde{\iota})$ iff $f(\omega) = 0$, iff $x^2 + x + 1 | f(x)$.

Turn to the polynomial rings of several indeterminates over a ring R. A natural way is by induction. Let $R[x_1]$ be the polynomial ring of an indeterminate x_1, and x_2 be an indeterminate over $R[x_1]$; then the following is the polynomial ring over R of the two indeterminates x_1 and x_2:

$$R[x_1][x_2] = \Big\{ g_0(x_1) + g_1(x_1)x_2 + \cdots + g_n(x_1)x_2^n \ \Big|\ g_j(x_1) \in R[x_1] \Big\}.$$

Letting $g_j(x_1) = \sum_{i=0}^{m} a_{ij} x_1^i$, we see that the polynomial ring of two indeterminates over R is

$$R[x_1, x_2] = R[x_1][x_2] = \Big\{ \sum_{i,j} a_{ij} x_1^i x_2^j \ \Big|\ a_{ij} \in R \Big\};$$

and $\sum_{i,j} a_{ij}x_1^i x_2^j = 0$ iff every coefficient $a_{ij} = 0$.

3.8.7 Definition. By induction, the polynomial ring of n indeterminates over a ring R is defined as $R[x_1, \cdots, x_{n-1}][x_n]$ which is as follows:

$$R[x_1, \cdots, x_n] = \left\{ \sum_{i_1, \cdots, i_n} a_{i_1 \cdots i_n} x_1^{i_1} \cdots x_n^{i_n} \;\middle|\; a_{i_1 \cdots i_n} \in R \right\};$$

$$\sum_{i_1, \cdots, i_n} a_{i_1 \cdots i_n} x_1^{i_1} \cdots x_n^{i_n} = 0 \quad \Longleftrightarrow \quad \text{every coefficient } a_{i_1 \cdots i_n} = 0.$$

And elements $u_1, \cdots, u_n$ of a ring R' which contains R are said to be *algebraically dependent* over R if there exist finite many $a_{i_1 \cdots i_n} \in R$ which are not all zero such that $\sum_{i_1, \cdots, i_n} a_{i_1 \cdots i_n} u_1^{i_1} \cdots u_n^{i_n} = 0$. Otherwise, $u_1, \cdots, u_n$ are said to be *algebraically independent* over R. In particular, one element u is algebraically dependent over R iff u is an algebraic element over R.

The following two results are similar to 3.8.3 and 3.8.4.

3.8.8 Theorem (Universal Property of Polynomial Rings). *Let R and S be rings and $f : R \to S$ be a ring homomorphism and $u_1, \cdots, u_n \in S$. Then there is a unique ring homomorphism $\widetilde{f} : R[x_1, \cdots, x_n] \to S$ such that $f = \widetilde{f}|_R$ and $\widetilde{f}(x_i) = u_i$ for $i = 1, \cdots, n$.* $\qquad\square$

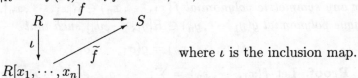

where ι is the inclusion map.

3.8.9 Corollary. *Let $R \leq R'$ be rings and $u_1, \cdots, u_n \in R'$. Then $R[u_1, \cdots, u_n] \cong R[x_1, \cdots, x_n]/I$ for an ideal I of $R[x_1, \cdots, x_n]$.* $\qquad\square$

In the polynomial ring $R[x_1, \cdots, x_n]$, by Def.3.8.7, the n indeterminates $x_1, \cdots, x_n$ are algebraically independent over R. Conversely, n elements $u_1, \cdots, u_n$ which are algebraically independent over R act just like the n indeterminates, see Exer.2.

A subtle point is that $u_1, \cdots, u_n$ are not necessarily algebraically independent over R even if every one of them is transcendental over R, see Exer.3. For this reason, in the case of polynomial rings of several indeterminates, the $x_1, \cdots, x_n$ are sometimes called *independent indeterminates*, instead of just "indeterminates".

Now we consider symmetric polynomials. For an intuitive inspiration we refer to Example 2.1.16.

3.8.10 Definition. Let G be a permutation group of degree n, i.e. $G \leq S_n$. For $\alpha \in G$ and $f(x_1, \cdots, x_n) = \sum_{i_1, \cdots, i_n} a_{i_1 \cdots i_n} x_1^{i_1} \cdots x_n^{i_n} \in R[x_1 \cdots, x_n]$, put

$$\alpha f(x_1, \cdots, x_n) = f(x_{\alpha(1)}, \cdots, x_{\alpha(n)}) = \sum_{i_1, \cdots, i_n} a_{i_1 \cdots i_n} x_{\alpha(1)}^{i_1} \cdots x_{\alpha(n)}^{i_n}.$$

If $\alpha f(x_1, \cdots, x_n) = f(x_1, \cdots, x_n)$, $\forall \alpha \in G$, then $f(x_1, \cdots, x_n)$ is said to be *G-invariant*. In particular, if $G = S_n$, then a G-invariant polynomial $f(x_1, \cdots, x_n)$ is called a *symmetric polynomial* in $x_1, \cdots, x_n$.

3.8.11 Example. Expansing the following polynomial in terms of the powers of x:

$$(x-x_1)(x-x_2)\cdots(x-x_n) = x^n - \sigma_1 x^{n-1} + \cdots + (-1)^{n-1}\sigma_{n-1}x + (-1)^n \sigma_n,$$

then

$$\sigma_1 = \sum_i x_i, \qquad \sigma_2 = \sum_{i<j} x_i x_j, \qquad \cdots, \qquad \sigma_n = \prod_i x_i$$

are clearly symmetric polynomials, which are called the *elementary symmetric polynomials* in $x_1, \cdots, x_n$.

3.8.12 Theorem (Fundamental Theorem on Symmetric Polynomials). *For any symmetric polynomial $f(x_1, \cdots, x_n) \in R[x_1, \cdots, x_n]$, there is a unique polynomial $g(y_1, \cdots, y_n) \in R[y_1, \cdots, y_n]$ such that*

$$f(x_1, \cdots, x_n) = g(\sigma_1, \cdots, \sigma_n).$$

Proof. Let $f(x_1, \cdots, x_n) = \sum_{i_1, \cdots, i_n} a_{i_1 \cdots i_n} x_1^{i_1} \cdots x_n^{i_n}$; and put the non-zero terms *alphabetically* as follows: $a_{i_1 \cdots i_n} x_1^{i_1} \cdots x_n^{i_n}$ before $b_{j_1 \cdots j_n} x_1^{j_1} \cdots x_n^{j_n}$ if $i_1 = j_1, \cdots, i_t = j_t$ and $i_{t+1} > j_{t+1}$.

Note that, in this alphabetical order, the set

$$\left\{ (i_1, \cdots, i_n) \ \middle| \ \text{all } i_k\text{'s are non-negative integers} \right\}$$

is a well-ordered set (just like the set of all non-negative integers), i.e. any non-empty subset of it has a least element; in particular, we can use the induction on it. In addition, it is easy to check the following fact:

(3.8.13). *The first term of $\sigma_1^{i_1} \sigma_2^{i_2} \cdots \sigma_n^{i_n}$ is $x_1^{i_1 + \cdots + i_n} x_2^{i_2 + \cdots + i_n} \cdots x_n^{i_n}$.*

For the *existence*, we first claim that the first term $a_0 x_1^{l_1} x_2^{l_2} \cdots x_n^{l_n}$, $a_0 \neq 0$, of the polynomial f must be such that $l_1 \geq l_2 \geq \cdots \geq l_n$.

If this is false, let $l_t < l_{t+1}$ be the first reverse pair; then, taking the transposition $\alpha = (t, t+1)$, by the symmetry of the polynomial we see that $f = \alpha f$ has a non-zero term $a_0 x_1^{l_1} \cdots x_t^{l_{t+1}} x_{t+1}^{l_t} \cdots x_n^{l_n}$ which is before $a_0 x_1^{l_1} \cdots x_t^{l_t} x_{t+1}^{l_{t+1}} \cdots x_n^{l_n}$; a contradiction.

Let $f_1 = f - a_0 \sigma_1^{j_1} \sigma_2^{j_2} \cdots \sigma_n^{j_n}$ where $j_k = l_k - l_{k+1}$ for $k < n$ and $j_n = l_n$. Then f_1 is again a symmetric polynomial, see Exer.7; but, since the first term of $a_0 \sigma_1^{j_1} \sigma_2^{j_2} \cdots \sigma_n^{j_n}$ is

$$a_0 x_1^{j_1 + \cdots + j_n} x_2^{j_2 + \cdots + j_n} \cdots x_n^{j_n} = a_0 x_1^{l_1} x_2^{l_2} \cdots x_n^{l_n},$$

the first term of f_1 is after the first term of f. Thus, if f_1 is not the zero polynomial, by induction we can get a $g_1(y_1, \cdots, y_n) \in R[y_1, \cdots, y_n]$ such that $f_1 = g_1(\sigma_1, \cdots, \sigma_n)$. Then, letting $g = g_1 + a_0 y_1^{j_1} \cdots y_n^{j_n}$, we have $f(x_1, \cdots, x_n) = g(\sigma_1, \cdots, \sigma_n)$.

For *uniqueness*, suppose

$$g(y_1, \cdots, y_n) \neq g'(y_1, \cdots, y_n) \in R[y_1, \cdots, y_n]$$

such that

$$g(\sigma_1, \cdots, \sigma_n) = g'(\sigma_1, \cdots, \sigma_n) \in R[x_1, \cdots, x_n].$$

Then, letting $h(y_1, \cdots, y_n) = g(y_1, \cdots, y_n) - g'(y_1, \cdots, y_n)$, we have a non-zero polynomial $h(y_1, \cdots, y_n)$ in $R[y_1, \cdots, y_n]$ such that

$$h(\sigma_1, \cdots, \sigma_n) = 0 \quad \text{in } R[x_1, \cdots, x_n].$$

However, for any two distinct non-zero terms $a y_1^{i_1} \cdots y_n^{i_n}$ and $b y_1^{j_1} \cdots y_n^{j_n}$ of $h(y_1, \cdots, y_n)$, from (3.8.13) we can see that, in $R[x_1, \cdots, x_n]$, the terms $a \sigma_1^{i_1} \cdots \sigma_n^{i_n}$ and $b \sigma_1^{j_1} \cdots \sigma_n^{j_n}$ have distinct first terms; thus it is impossible that $h(\sigma_1, \cdots, \sigma_n) = 0$ in $R[x_1, \cdots, x_n]$. This contradiction completes the proof. $\square$

Remark. The above proof for the existence gives a method to write a symmetric polynomial with elementary symmetric polynomials. For example, consider $f(x_1, x_2, x_3) = x_1^3 + x_2^3 + x_3^3$; its first term is x_1^3, hence set

$$f_1 = f - \sigma_1^{3-0} \sigma_2^{0-0} \sigma_3^0 = 3 \sum_{i<j} x_i^2 x_j + 6 x_1 x_2 x_3;$$

the first term of f_1 is $3 x_1^2 x_2$, so set

$$f_2 = f_1 - 3\sigma_1^{2-1} \sigma_2^1 = 6\sigma_3 - 9\sigma_3 = -3\sigma_3;$$

thus, $f = \sigma_1^3 + 3\sigma_1 \sigma_2 - 3\sigma_3$.

On the other hand, the uniqueness of Theorem 3.8.12 is equivalent to the algebraically independence of $\sigma_1, \cdots, \sigma_n$ over R; cf. Exer.8.

Return to the construction of the polynomial ring $R[x]$ in (3.8.2), where we consider the sequence $(a_0, a_1, \cdots)$ with only finitely many a_i non-zero. If this restriction is dropped, we then have

3.8.14 Proposition. *Let R be a ring, let*

$$R[[x]] = \Big\{ (a_0, a_1, \cdots) \ \Big| \ a_i \in R \Big\};$$

Define two operations on $R[[x]]$:

$$(a_0, a_1, \cdots) + (b_0, b_1, \cdots) = (a_0 + b_0, a_1 + b_1, \cdots),$$

$$(a_0, a_1, \cdots) \cdot (b_0, b_1, \cdots) = (c_0, c_1, \cdots) \qquad \text{where } c_k = \sum_{i+j=k} a_i b_j.$$

Then $R[[x]]$ is a ring containing (a copy of) R as a subring. $\qquad \square$

We call $R[[x]]$ the ring of the *formal power series* over R; let $x = (0, 1, 0, \cdots)$ then every element of $R[[x]]$ can be expressed as a formal power series $\sum_{i=0}^{\infty} a_i x^i$ and the operations defined above are the familiar operations on formal power series.

We show an interesting property of $R[[x]]$, one can compare it with Exer.6(2) and Exer.12.

3.8.15 Proposition. *An element $\sum_{i=0}^{\infty} a_i x^i \in R[[x]]$ is invertible in $R[[x]]$ if and only if a_0 is invertible in R.*

Proof. Note that the unity of $R[[x]]$ is $1 = (1, 0, 0, \cdots)$. If there is a $\sum_{i=0}^{\infty} b_i x^i \in R[[x]]$ such that $\sum_{i=0}^{\infty} a_i x^i \cdot \sum_{i=0}^{\infty} b_i x^i = 1$, then $a_0 b_0 + (a_0 b_1 + a_1 b_0)x + \cdots = 1$, hence $a_0 b_0 = 1$; i.e. a_0 is invertible in R.

Conversely, assume a_0 is invertible in R; we need to find b_i's such that $\sum_{i=0}^{\infty} a_i x^i \cdot \sum_{i=0}^{\infty} b_i x^i = 1$, or equivalently, to find b_i's satisfy the following equations in R:

$$a_0 b_0 = 1, \quad a_0 b_1 + a_1 b_0 = 0, \quad \cdots, \quad a_0 b_n + a_1 b_{n-1} + \cdots, + a_n b_0 = 0, \quad \cdots.$$

We can solve these equations recursively: $b_n = a_0^{-1}(-a_1 b_{n-1} - \cdots - a_n b_0)$ when $b_0, \cdots, b_{n-1}$ have been obtained. $\qquad \square$

Exercises 3.8

Note that the Agreement 3.8.0 is still adopted in the following exercises.

1. Prove Prop.3.8.1.

$^{+}$2. Let $R \le R'$ be rings, and x be an indeterminate over R. Then $u \in R'$ is transcendental over R if and only if $\sigma : R[x] \to R[u]$, $f(x) \mapsto f(u)$ is a ring isomorphism. Generalize this result to the polynomial rings of several indeterminates. (Hint: 3.8.4 and 3.8.9.)

3. It is well-known that the real number π is transcendental over the rational field $\mathbf{Q}$. Using this result, show that both π and 2π are transcendental $\mathbf{Q}$, but they are algebraically dependent over $\mathbf{Q}$.

4. Find the ideal I of $\mathbf{Q}[x]$ such that $\mathbf{Q}[\sqrt{2} + \sqrt{3}] \cong \mathbf{Q}[x]/I$, show that $\sqrt{2} + \sqrt{3}$ is an algebraic element over $\mathbf{Q}$.

*5. Let R be a ring and $f(x) = \sum_{i=0}^{n} a_i x^i \in R[x]$.

  1) $f(x)$ is invertible in $R[x]$ iff a_0 is invertible in R and all $a_1, \cdots, a_n$ are nilpotent in R. (Hint: the sum of an invertible element and a nilpotent element is invertible, cf. Exer.3.2.3(3).)

  2) If $f(x)$ is a nilpotent element in $R[x]$, then every a_i is a nilpotent element in R. (Hint: $1 + f(x)$ is invertible, use (1).)

  3) If $f(x)$ is a zero divisor in $R[x]$, then there is a $0 \ne b \in R$ such that $bf(x) = 0$. (Hint: choose $g = \sum_{i=0}^{m} b_i x^i$ of least degree m such that $fg = 0$; then $a_n b_m = 0$, hence $a_n g = 0$; by induction, $a_{n-r} g = 0$, $0 \le r \le n$.)

6. Let R be an integral domain.

  1) The polynomial ring $R[x_1, \cdots, x_n]$ is also an integral domain; and its fraction field is the field $R(x_1, \cdots, x_n)$ of the rational fractions in $x_1, \cdots, x_n$.

  $^{+}$2) An element of $R[x_1, \cdots, x_n]$ is invertible if and only if it is an element of R and invertible in R.

  3) For $f(x_1, \cdots, x_n) = \sum_{i_1, \cdots, i_n} a_{i_1 \cdots i_n} x_1^{i_1} \cdots x_n^{i_n}$, the sum $i_1 + \cdots + i_n$ is called the degree of the term $a_{i_1 \cdots i_n} x_1^{i_1} \cdots x_n^{i_n}$, and the maximum of the degrees of all non-zero terms of $f(x_1, \cdots, x_n)$ is called the *degree* of $f(x_1, \cdots, x_n)$ and denoted by $\deg(f)$. Prove:

  $$\deg(fg) = \deg(f) + \deg(g) \qquad \forall f, g \in R[x_1, \cdots, x_n].$$

  (*Remark:* $\deg(f)$ is not defined for $f = 0$, or one can agree on that $\deg(0) = -\infty$.)

7. Show that the sum (product) of two symmetric polynomials is again a symmetric polynomial.

$^{+}$8. Let R be a ring. Prove that, in the polynomial ring $R[x_1, \cdots, x_n]$, the elementary symmetric polynomials $\sigma_1, \cdots, \sigma_n$ are algebraically independent over R. (Hint: cf. the proof for the uniqueness of Theorem 3.8.12.)

9. Express $x_1^4 + x_2^4 + x_3^4 - 2x_1^2x_2^2 - 2x_2^2x_3^2 - 2x_3^2x_1^2$ in the elementary symmetric polynomials.

10. Let $\sigma_1, \cdots, \sigma_n$ be the elementary symmetric polynomials as in (3.8.11) and $s_k = \sum_{i=1}^n x_i^k$. Prove the Newton's formula:
    1) $s_k - s_{k-1}\sigma_1 + s_{k-2}\sigma_2 - \cdots + (-1)^k s_0\sigma_k = 0$, $(1 \le k \le n)$.
    2) $s_k - s_{k-1}\sigma_1 + s_{k-2}\sigma_2 - \cdots + (-1)^n s_{k-n}\sigma_n = 0$, $(k > n)$.

11. A ring R is an integral domain if and only if $R[[x]]$ is an integral domain.

12. Let R be an integral domain with fraction field F, let $F([x])$ denote the fraction field of $R[[x]]$. Prove: every element of $F([x])$ can be expressed as

$$x^n(a_0 + a_1x + a_2x^2 + \cdots), \qquad a_i \in F, \ a_0 \ne 0, \ n \in \mathbf{Z}.$$

3.9   Factorial Domains

In this section, the Agreement 3.8.0 is still adopted; in addition, we consider integral domains and assume that D is an integral domain. For convenience, we call an invertible element a *unit*.

We first show that many familiar arithmetic notations such as divisors, prime elements, least common divisors, etc. can be easily defined for any integral domains; next, we discuss when the familiar arithmetic property that *"any integer is uniquely written as a product of primes"* can hold for an integral domain.

3.9.1 Definition. Let $a, b \in D$.

1) We say that a *divides* b, denote $a|b$, if there is a $c \in D$ such that $b = ac$; and, in this case, we say that a is a *factor* or *divisor* of b and b is a *multiple* of a.

2) We say that a and b are *associates*, denote $a^{\bullet} \sim b$, if both $a|b$ and $b|a$. ($\sim$ is an equivalence relation, see Exer.1, and the equivalence class is called an *associate class*.)

3) Any $b \in D$ has two kinds of divisors: the units and the elements associate to b (see Exer.1), such divisors are said to be trivial; a nontrivial divisor of b is called a *proper divisor* of b.

Some easy properties about the above notations listed in Exer.1, which show why we assume that p is neither a zero element nor a unit in the following definition.

3.9.2 Definition. Let $p \in D$ be a non-zero and non-unit element. p is called a *prime element* if, for any $a, b \in D$, $p|ab \implies$ either $p|a$ or $p|b$. On the other hand, p is called an *irreducible element* if, for any $a, b \in D$, $p = ab \implies$ either a or b is a unit (i.e. p has no proper divisors).

Though an irreducible element is just a prime element in the integer ring, Exer.2 shows that the two concepts are really different in general case.

3.9.3 Theorem. 1) $p \in D$ *is a prime element if and only if* $\langle p \rangle$ *is a non-zero prime ideal.*

2) $c \in D$ *is an irreducible element if and only if* $\langle c \rangle$ *is maximal in the set of all the non-zero proper principal ideals.*

3) *A prime element is an irreducible element.*

Proof. (1) follows from Theorem 3.5.2.

(2). Assume c is irreducible; if $\langle c \rangle \subset \langle d \rangle \neq D$, then $d \mid c$ by Exer.1 and d is not a unit, hence, by the definition 3.9.2, d is associated to c; so $\langle c \rangle = \langle d \rangle$ by Exer.1. Conversely, assume $\langle c \rangle$ is maximal in the set of all non-zero proper principal ideals; if $d \mid c$, then $\langle c \rangle \subset \langle d \rangle$; thus either $\langle c \rangle = \langle d \rangle$ or $\langle d \rangle = D$, i.e. either d and c are associates or d is a unit; in other words, c has only trivial divisors.

3). Let $p \in D$ be prime and $a \mid p$, then $p = ab$, in particular, $p \mid ab$; so, by the primeness of p, either $p \mid a$ or $p \mid b$; in the formal case $a \sim p$; in the latter case $p \sim b$ hence a is a unit by Exer.1(9); thus a is a trivial divisor of p. $\qquad\Box$

The following concepts also come from the number theory.

3.9.4 Definition. c is called a *common divisor* of a and b if c is a divisor of both a and b. d is said to be a *greatest common divisor*, abbreviated by g.c.d., of a and b, if d is a common divisor of a and b and, for any common divisor c of a and b, we have $c \mid d$. It is easy to see that the g.c.d.'s of a and b (if they exist) form an associate class, which is denoted by (a, b). Moreover, if (a, b) exists and equal to the class of the units, then we denote $(a, b) = 1$ for convenience and say that a and b are *coprime*, (or a is prime to b, or b is prime to a).

The common divisors and the g.c.d.'s of $a_1, \cdots, a_n$ are defined similarly, the g.c.d.'s, also denoted by $(a_1, \cdots, a_n)$, form an associate class. Moreover, $a_1, \cdots, a_n$ are said to be *coprime*, denoted by $(a_1, \cdots, a_n) = 1$, if the g.c.d.'s of $a_1, \cdots, a_n$ are the class of units.

Exer.3 shows that the g.c.d.'s do not necessarily exist in general cases. On the other hand, Exer.5 lists some easy properties about g.c.d.'s when they exist.

The concept dual to g.c.d. is exhibited in Exer.6.

3.9.5 Definition. An integral domain D is said to be *factorial* if any non-zero non-unit $a \in D$ has a *unique factorization*, i.e. the following two conditions hold for a:

1) $a = p_1 \cdots p_n$ for finitely many irreducible elements $p_1, \cdots, p_n$;

2) if $a = p_1 \cdots p_n = q_1 \cdots q_m$ where all p_i and all q_j are irreducible, then $m = n$ and, up to a reindexing, $q_i \sim p_i$, $i = 1, \cdots, n$.

In this case, $a = p_1 \cdots p_n$ is called an *irreducible factorization* of a; further, collecting the irreducible factors which are associates to each other, we can rewrite it as $a = up_1^{k_1} \cdots p_s^{k_s}$, called a *reduced irreducible factorization*, where u is a unit and p_i's are pairwisely distinct irreducible elements and k_i's are positive integers.

3.9.6 Lemma. *If D is a factorial domain and $a = up_1^{k_1} \cdots p_s^{k_s}$ is a reduced irreducible factorization, then c is a divisor of a if and only if $c = vp_1^{l_1} \cdots p_s^{l_s}$ with v being a unit and $0 \le l_i \le k_i$ for $i = 1, \cdots, s$.*

Proof. If $a = cd$; let $c = wq_1^{l_1} \cdots q_r^{l_r}$ and $d = w'(q_1')^{l_1'} \cdots (q_{r'}')^{l_{r'}'}$ be irreducible factorizations of c and d resp., then

$$a = up_1^{k_1} \cdots p_s^{k_s} = ww'q_1^{l_1} \cdots q_r^{l_r}(q_1')^{l_1'} \cdots (q_{r'}')^{l_{r'}'}.$$

By the uniqueness of factorizations, reindexing if necessary, we have that $r \le s$, and $q_i = \varepsilon_i p_i$ and $l_i \le k_i$ for $1 \le i \le r$; letting $v = w\varepsilon_1 \cdots \varepsilon_r$ which is a unit and $l_i = 0$ for $r < i \le s$, we get $c = vp_1^{l_1} \cdots p_s^{l_s}$.

Conversely, if $c = vp_1^{l_1} \cdots p_s^{l_s}$, then $a = c \cdot (v^{-1}up_1^{k_1-l_1} \cdots p_s^{k_s-l_s})$. $\square$

With this lemma, we can get some properties similar to the integers; see Exer.7.

3.9.7 Definition. We refer to three conditions on D as follows:

divisor chain condition, means that there is no infinite sequences $a_1, a_2, \cdots$ such that every a_{i+1} is a *proper* divisor of a_i;

primeness condition, means that any irreducible element is a prime element;

g.c.d. condition, means that any two elements have g.c.d.'s.

3.9.8 Theorem. *For an integral domain D, the following three are equivalent:*

   i) *D is factorial;*

   ii) *the divisor chain condition and the primeness condition hold;*

   iii) *the divisor chain condition and the g.c.d. condition hold.* $\square$

Proof. (i) $\Longrightarrow$ (iii). Suppose $a_1, a_2, \cdots$ is a sequence such that every a_{i+1} is a proper divisor of a_i; then $a_i = a_{i+1}a'_{i+1}$ and a'_{i+1} is also a proper divisor of a_i; combining a factorization of a_{i+1} and a'_{i+1}, we get a factorization of a_i which length is at least 2; recursing from a_i back to a_1, we see that a_1 has a factorization of length at least i; because i is arbitrary, this contradicts to that a_1 has a unique factorization. Therefore, the divisor chain condition holds. For any $a, b \in D$, if one of them is zero or a unit, then (a, b) exists by Exer.5; otherwise (a, b) exists by Exer.7.

(iii) $\Longrightarrow$ (ii). Assume p is an irreducible element and $p \nmid a$ and $p \nmid b$, then $(p, a) = 1$ and $(p, b) = 1$ by Exer.5(2); so, by Exer.5(6), $(p, ab) = 1$, that is, $p \nmid ab$; thus p is prime.

(ii) $\Longrightarrow$ (i). Let $a \in D$ be non-zero non-unit. If a is reducible, then $a = a_1 b_1$ with proper divisor a_1 and b_1 of a; similarly, $a_1 = a_2 b_2$ with proper divisor a_2 and b_2 of a_1 provided a_1 is reducible; in this way, because of the divisor chain condition, we stop at an irreducible divisor p_1 of a. So $a = p_1 a'$ with irreducible p_1 hence a' is a proper divisor of a. If a' is reducible, we have an irreducible p_2 such that $a' = p_2 a''$ hence a'' is a proper divisor of a' and $a = p_1 p_2 a''$. Continuing this process, we must stop after finite steps because of the divisor chain condition, and get $a = p_1 p_2 \cdot p_n$ with all p_i's being primes.

Next, assume $a = p_1 \cdots p_n = q_1 \cdots q_m$ where all p_i and all q_j are irreducible; we prove the uniqueness by induction on n. If $n = 1$, then $a = p_1$ is irreducible, and it has to be the case that $m = 1 = n$. Assume $n > 1$. By the primeness condition, p_1 is prime. By Definition 3.9.2 and cf. Exer.4, there is a q_i, say q_1, such that $p_1 \mid q_1$; but q_1 has no non-trivial divisors, hence $p_1 \sim q_1$ and there is a unit u_1 such that $q_1 = u_1 p_1$. So $a = p_1 p_2 \cdots p_n = u_1 p_1 q_2 \cdots q_m$. Applying induction to $a_1 = p_2 \cdots p_n = (u_1 q_2) \cdots q_m$, we get that $m - 1 = n - 1$ hence $m = n$

and, up to a reindexing, $u_1 q_2 \sim p_2$ (i.e. $q_2 \sim p_2$), $q_2 \sim p_2, \cdots, q_n \sim p_n$. □

We exhibit two sorts of factorial domains.

3.9.9 Theorem. *A principal ideal domain is factorial.*

Proof. Assume D is a principal ideal domain. Let $a_1, a_2, \cdots$ be a sequence such that $a_{i+1} \mid a_i$ for all $i \geq 0$. Then, by Exer.1(1) we have

$$\langle a_1 \rangle \subset \langle a_2 \rangle \subset \cdots$$

hence $\bigcup_{i=1}^{\infty} \langle a_i \rangle$ is an ideal of D by Exer.3.3.2; so $\bigcup_{i=1}^{\infty} \langle a_i \rangle = \langle b \rangle$. Then $b \in \langle a_k \rangle$ for some k, i.e. $\langle b \rangle \subset \langle a_k \rangle$ hence

$$\langle a_k \rangle = \langle a_{k+1} \rangle = \cdots = \langle b \rangle ;$$

in other words, by Exer.1(3), all $a_k, a_{k+1}, \cdots$ are associates to each other. Thus the divisor chain condition holds in D. Next, for any $a, b \in D$, the ideal $\langle a, b \rangle = \langle d \rangle$ for $d \in D$; consequently, $d \mid a$ and $d \mid b$ by Exer.1(1); moreover, if $d' \mid a$ and $d' \mid b$, then $a \in \langle d' \rangle$ and $b \in \langle d' \rangle$, hence $\langle d \rangle = \langle a, b \rangle \subset \langle d' \rangle$; thus $d' \mid d$ by Exer.1(1) again; that is, d is a g.c.d. of a and b. The g.c.d. condition holds in D. Therefore, D is a factorial domain by Theorem 3.9.8. □

The converse of 3.9.9 is certainly false; a counterexample is the Example 3.3.10: $\mathbf{Z}[x]$ is not a principal ideal domain since $\langle 2, x \rangle$ is not a principal ideal, but it is a factorial domain by Theorem 3.10.6 below.

3.9.10 Corollary. *If D is a principal ideal domain and d is a g.c.d. of $a, b \in D$ and m is a l.c.m.(see Exer.6), then $\langle a \rangle + \langle b \rangle = \langle d \rangle$ and $\langle a \rangle \cap \langle b \rangle = \langle m \rangle$; in particular, there are $s, t \in D$ such that $as + bt = d$.* □

3.9.11 Corollary (of Theorem 3.9.3). *If D is a principal ideal domain, then any prime ideal is a maximal ideal* (note that the converse is always true, see Cor.3.5.10).

Proof. Let $P = \langle p \rangle$ be a prime ideal of D, then p is a prime element by 3.9.3(1), hence p is irreducible by 3.9.3(3), consequently, $P = \langle p \rangle$ is maximal ideal by 3.9.3(2) since all ideals of D are principal. □

The integer ring $\mathbf{Z}$ is a principal ideal domain, see Exer.3.3.3; hence $\mathbf{Z}$ is a factorial domain. The more fundamental fact is that the additive group of $\mathbf{Z}$ is a cyclic group and any subgroup is also cyclic, see Lemma 2.8.1. The argument is based on the Euclidean Division.

3.9.12 Definition. An integral domain is called a *Euclidean domain* if there is a non-negative integral function φ defined on $D - \{0\}$ and, for any $a, b \in D$ with $b \neq 0$, there are $q, r \in D$ such that $a = bq + r$ and either $r = 0$ or $\varphi(r) < \varphi(b)$.

3.9.13 Theorem. *A Euclidean domain is a principal ideal domain, hence a factorial domain.*

Proof. The argument is the same as the proof of Lemma 2.8.1(1). Let D be a Euclidean domain and I be a non-zero ideal of D (the zero ideal $\langle 0 \rangle$ is obviously principal). Then $\{\varphi(a) \mid 0 \neq a \in I\}$ is a non-empty set of non-negative integers, hence there is a $b \in I$ such that $\varphi(b)$ is the minimal one in this set. For any $a \in I$ we have $q, r \in D$ such that $a = bq + r$ and either $r = 0$ or $\varphi(r) < \varphi(b)$; suppose $r \neq 0$, then $r = a - bq \in I$ but $\varphi(r) < \varphi(b)$ which is impossible; thus $r = 0$. So $a = bq \in \langle b \rangle$; i.e. $I = \langle b \rangle$ is a principal ideal. $\square$

3.9.14 Corollary. *The integer ring $\mathbf{Z}$ is a Euclidean domain, hence it is a principal domain, hence it is a factorial domain.*

Proof. Take $\varphi(n) = |n|$ (the absolute value of n) for $0 \neq n \in \mathbf{Z}$. $\square$

3.9.15 Corollary. *The polynomial ring over a field is a Euclidean domain, hence it is a principal domain, and thus it is a factorial domain.*

Proof. Let F be a field and $F[x]$ be the polynomial ring in indeterminate x; for $0 \neq f(x) \in F[x]$, let $\varphi(f(x)) = \deg(f(x))$ be the degree of $f(x)$. Then, by Exer.14, $F[x]$ is a Euclidean domain. $\square$

Exercises 3.9

Note that the agreements at the beginning of this section are still adopted in the following exercises.

$^{+}$1. 1). $a \mid b$ in D if and only if the principal ideal $\langle b \rangle \subset \langle a \rangle$.

2) 0 is not a divisor of any non-zero element.

3) $a \sim b$ if and only if there is a unit u such that $a = ub$, if and only if $\langle a \rangle = \langle b \rangle$.

4) The relation of divisability "$\mid$" is reflexive and transitive, but not symmetric.

5) The relation of associateness "$\sim$" is an equivalence relation; the equivalence class including 1 is the subset of all the units.

6) If $a \mid b$ and $a' \sim a$, then $a' \mid b$ (i.e. the divisors of b are partitioned into associates classes).

7) Any non-zero non-unit $b \in D$ has two kinds of divisors: the units and the elements associate to b itself (such divisors are said to be *trivial*).

8) $u \in D$ is a unit if and only if $a|1$, if and only if u has only trivial divisors.

9) If $b = ac$ and $b \sim a$, then c is a unit, hence $a = bc^{-1}$.

2. Take $D = \mathbf{Z}[\sqrt{-5}] = \left\{ a + b\sqrt{-5} \mid a, b \in \mathbf{Z} \right\}$, which is an integral domain since it is a subring of $\mathbf{C}$ (see Exer.3.2.5).

1) $(a + b\sqrt{-5}) \mid (c + d\sqrt{-5})$ in $D \implies (a^2 + 5b^2) \mid (c^2 + 5d^2)$ in $\mathbf{Z}$.

(Hint: $(c + d\sqrt{-5}) = (a + b\sqrt{-5})(a' + b'\sqrt{-5}) \implies c^2 + 5d^2 = (a^2 + 5b^2)(a'^2 + 5b'^2)$.)

2) Only ± 1 are the units of D.

3) $3 \in D$ is an irreducible element.

4) $3 \mid (2 + \sqrt{-5})(2 - \sqrt{-5})$ in D, but $3 \nmid (2 + \sqrt{-5})$ and $3 \nmid (2 - \sqrt{-5})$ in D; thus 3 is not a prime element in D.

3. Take $D = \mathbf{Z}[\sqrt{-5}]$ as above. Let $x = 9$ and $y = 3(2 + \sqrt{-5})$.

1) The only common divisors of x and y are ± 1, ± 3, and $\pm(2 + \sqrt{-5})$.

(Hint: if $a + b\sqrt{-5}$ is a common divisor of x and y, then $a^2 + 5b^2 \mid 81$.)

2) x and y has no g.c.d.

$^+$4. If p is a prime element and $p \mid a_1 \cdots a_n$, then $p \mid a_i$ for some $1 \leq i \leq n$.

$^+$5. Assume that the g.c.d.'s (a, b) exist for any $a, b \in D$. Then the g.c.d.'s $(a_1, \cdots, a_n)$ exist; and the following hold:

1) $(0, a) \sim a$ for any $a \in D$;

2) $(a, p) = 1$ if p is irreducible and $\nmid a$;

3) $(u, a) = 1$ if u is a unit;

4) $(a, (b, c)) = ((a, b), c) = (a, b, c)$;

5) $c(a, b) \sim (ca, cb)$;

6) $(a, b) = 1 \lor (a, c) = 1 \implies (a, bc) = 1$;

7) A common divisor d of $a_1, \cdots, a_n$ is a g.c.d. $\iff (a_1/d, \cdots, a_n/d) = 1$.

$^+$6. m is called a least common multiple (l.c.m. for short) of a and b if $a \mid m$ and $b \mid m$ and $m \mid m'$ provided $a \mid m'$ and $b \mid m'$. Prove that all the l.c.m.'s of a and b form an associate class provided they exist.

$^+$7. Assume D is a factorial domain and $a = up_1^{k_1} \cdots p_s^{k_s}$ and $b = vp_1^{h_1} \cdots p_s^{h_s}$ are reduced irreducible factorizations.

1) $p_1^{l_1} \cdots p_s^{l_s}$ with $l_i = \min(k_i, h_i)$ (the minimum of k_i and h_i) is a g.c.d. of a and b;

2) $p_1^{g_1} \cdots p_s^{g_s}$ with $g_i = \max(k_i, h_i)$ (the maximum of k_i and h_i) is a l.c.m. of a and b;

3) $ab \sim [a, b](a, b)$.

(Hint: use Lemma 3.9.6.)

8. Give an example to show that a subring of a factorial ring is not necessarily a factorial ring.

9. Prove: $\mathbf{Z}[i] = \{a + bi \,|\, a, b \in \mathbf{Z}\}$, where i is the imaginary unit, is a Euclidean domain.

(Hint: take $\varphi(a + bi) = a^2 + b^2$; for $a + bi \neq 0$ and $c + di$, if $c + di = (a + bi)(u + vi) + r$, then

$$r = (c + di) - (a + bi)(u + vi) = (a + bi)((\tfrac{ac+bd}{a^2+b^2} - u) + (\tfrac{ad-bc}{a^2+b^2} - v)i);$$

one can choose u and v such that $|\tfrac{ac+bd}{a^2+b^2} - u| \leq \tfrac{1}{2}$ and $|\tfrac{ad-bc}{a^2+b^2} - u| \leq \tfrac{1}{2}$.)

10. Notations as above.

1) What are the units in $\mathbf{Z}[i]$?

2) Show that $1 - 2i$ is a prime in $\mathbf{Z}[i]$.

3) Determine all the primes of $\mathbf{Z}[i]$.

11. Prove $\mathbf{Z}[\sqrt{2}] = \{a + b\sqrt{2} \,|\, a, b \in \mathbf{Z}\}$ is a Euclidean domain.

(Hint: take $\varphi(a + b\sqrt{2}) = |a^2 - 2b^2|$.)

12. Let D be a factorial domain and $0 \neq a \in D$. Then there are finitely many principal ideals which contain $\langle a \rangle$.

13. Let D be a principal ideal domain. Then every proper ideal I can be written as $I = P_1 \cdots P_n$ with prime ideals $P_1, \cdots, P_n$ and such expression is unique up to order.

+14. Let R be an arbitrary commutative unitary ring and $f(x), g(x) \in R[x]$. If $g(x) \neq 0$ and its leading coefficient is a unit of R, then there are $q(x), r(x) \in R[x]$ such that $f(x) = g(x)q(x) + r(x)$ and either $r(x) = 0$ or $\deg(r(x)) < \deg(g(x))$. (Hint: if $f(x) = a_n x^n + a_{n-1}x^{n-1} + \cdots + a_0$ and $g(x) = b_m x^m + m_{m-1}x^{m-1} + \cdots + b_0$ and $m \leq n$, then $f(x) = -b_m^{-1}a_n x^{n-m} \cdot g(x) + f_1(x)$ and $\deg(f_1(x)) < n$; then use induction on $\deg(f(x))$.)

3.10   Polynomial Rings over Factorial Rings

In this section, we *always* assume

3.10.0 Assumption. *D is a factorial ring and F is the fraction field of D, hence $D \subset F$; and $D[x]$ and $F[x]$ are the polynomial rings in indeterminate x, hence $D[x] \subset F[x]$.*

We will prove that $D[x]$ is also a factorial ring, see Theorem 3.10.6.

3.10.1 Definition. $f(x) = a_0 + a_1 x + \cdots + a_n x^n \in D[x]$ is called a *primitive polynomial* if $f(x) \neq 0$ and $a_0, \cdots, a_n$ are coprime, i.e. $(a_1, \cdots, a_n) = 1$.

3.10.2 Proposition.   1) *A D-polynomial which is associate to a primitive D-polynomial is also primitive.*

2) *Any non-zero D-polynomial can be expressed as a product of a primitive polynomial and an element of D, and such expression is unique up to associate relation.*

Proof. (1). The g.c.d.'s of the coefficients of a D-polynomial form an associate class in D; and, by Exer.3.8.6(2), the units of $D[x]$ are just the units of D.

(2). Let $f(x) = a_0 + a_1 x + \cdots + a_n x^n \in D[x]$ and d be a g.c.d. of $a_0, a_1, \cdots, a_n$; then $f(x) = dg(x)$ and $g(x)$ is a primitive D-polynomial, see Exer.3.9.5(7). If $f(x) = dg(x) = d'g'(x)$ with $g'(x)$ being primitive D-polynomial too, then both d and d' are g.c.d. of $a_0, a_1, \cdots, a_n$, hence they are associates, i.e. there is a unit $u \in D$ such that $d' = du$; hence $dg(x) = dug'(x)$, consequently, $g(x) = ug'(x)$, i.e. $g(x)$ and $g'(x)$ are associates in $D[x]$. $\Box$

3.10.3 Lemma (Gauss Lemma). *The product of two D-polynomials is primitive if and only if both the D-polynomials are primitive.*

Proof. The necessity is clear. Let $f_1(x) = a_0 + a_1 x + \cdots + a_n x^n$ and $f_2(x) = b_0 + b_1 x + \cdots + b_m x^m$ be primitive D-polynomials. Suppose that $f_1(x) \cdot f_2(x) = c_0 + c_1 x + \cdots + c_{m+n} x^{m+n}$ is not primitive, then there is a prime p of D which divides every c_i. But, p does not divide every coefficient of $f_i(x)$, $i = 1, 2$; i.e. there are $0 \leq k \leq n$ and $0 \leq h \leq m$ such that $p \mid a_i$ for $0 \leq i < k$ but $p \nmid a_k$, and $p \mid b_i$ for $0 \leq i < h$ but $p \nmid a_h$. Observing $c_{k+h} = \sum_{i+j=k+h} a_i b_j$, we see that, except for $a_k b_h$, p divides every term of this equality; but $p \nmid a_k b_h$ since p is a prime; a contradiction. $\Box$

3.10.4 Lemma.   1) *Two primitive D-polynomials are associates in $D[x]$ if and only if they are associates in $F[x]$.*

2) *For any $0 \neq f(x) \in F[x]$ there are $0 \neq c \in F$ and a primitive $g(x) \in D[x]$ such that $f(x) = c \cdot g(x)$, and such $g(x)$ is unique up to associateness.*

Proof. (1).   Assume two primitive $f(x), g(x)$ of $D[x]$ are associates in $F[x]$, then there are non-zero $a, b \in D$ such that $f(x) = (a/b) \cdot g(x)$, then $bf(x) = ag(x)$; thus, by Prop.3.10.2(2), there is a unit $u \in D$ such that $a = ub$; hence $f(x) = ug(x)$. The converse is obviously true.

(2).   Let $f(x) = \frac{a_0}{b_0} + \frac{a_1}{b_1}x + \cdots + \frac{a_n}{b_n}x^n$; then $f(x) = \frac{1}{c}f_1(x)$ with $c = b_0 b_1 \cdots b_n$ and $f_1(x) \in D[x]$; by Prop.3.10.1(2), $f_1(x) = dg(x)$ with $d \in D$ and primitive $g(x) \in D[x]$; so $f(x) = (d/c) \cdot g(x)$. Further, if $f(x) = (d/c) \cdot g(x) = (a/b) \cdot h(x)$ where $a, b \in D$ and $h(x) \in D[x]$ is primitive too, then $g(x)$ and $h(x)$ are associates in $F[x]$, hence, by (1), they are also associates in $D[x]$. $\qquad\square$

3.10.5 Corollary.   Let $0 \neq f(x) \in D[x]$.

1)   If $f(x) = f_1(x) \cdots f_n(x)$ with all $f_i(x) \in F[x]$, then there are primitive D-polynomials $g_i(x)$ associate to $f_i(x)$ in $F[x]$ for $1 \leq i \leq n$ and $a \in D$ such that $f(x) = ag_1(x) \cdots g_n(x)$.

2) $f(x)$ is irreducible in $D[x]$ if and only if either $\deg(f(x)) = 0$ and it is an irreducible element in D or $\deg(f(x)) > 0$ and it is primitive in $D[x]$ and irreducible in $F[x]$.

Proof. (1).   By Prop.3.10.2(2), $f(x) = bg(x)$ with primitive $g(x) \in D[x]$ and $b \in D$. By Lemma 3.10.4(2), $f_i(x) = c_i g_i(x)$ with $0 \neq c_i \in F$ and $g_i(x) \in D[x]$. Then $bg(x) = f(x) = (c_1 \cdots c_n)(g_1(x) \cdots g_n(x))$; but, by Lemma 3.10.3, $g_1(x) \cdots g_n(x)$ is primitive; hence, by Lemma 3.10.4(1), there is a unit $u \in D$ such that $g(x) = ug_1(x) \cdots g_n(x)$; therefore, $f(x) = bg(x) = bug_1(x) \cdots g_n(x)$.

(2).   It is clear that an irreducible element of D is irreducible in $D[x]$. Assume that $\deg(f(x)) > 0$ and $f(x)$ is primitive in $D[x]$ and irreducible in $F[x]$; if $f(x) = f_1(x)f_2(x)$ with $f_1(x), f_2(x) \in D[x]$, then, by the irreducibility in $F[x]$, one of $f_i(x)$, say $f_2(x)$, is of degree 0; i.e. $a = f_2(x) \in D$ and $f(x) = af_1(x)$; by the primitivity of $f(x)$ in $D[x]$, a must be a unit of D, hence a is a unit of $D[x]$.

Conversely, assume $f(x) \in D[x]$ is irreducible in $D[x]$. If $f(x) = a \in D$, then it is irreducible in D clearly. If $\deg(f(x)) > 0$, then, by Cor.3.10.5(1), $f(x)$ is irreducible in $F[x]$; further, taking a g.c.d of the coefficients of $f(x)$, we have $f(x) = df_1(x)$ with $f_1(x) \in D[x]$; then d must be a unit of D, otherwise d is a proper divisor of $f(x)$ in $D[x]$. $\qquad\square$

3.10.6 Theorem. $D[x]$ *is a factorial ring.*

Proof. Let $f(x) \in D[x]$ be non-zero and non-unit. If $\deg(f(x)) = 0$, then $f(x) \in D$ can be written as a product of irreducible elements of D which are also irreducible elements of $D[x]$ by Cor.3.10.5(2). Assume $\deg(f(x)) > 0$. Since $F[x]$ is a factorial ring (see Cor.3.9.15), $f(x) = f_1(x) \cdots f_n(x)$ with irreducible $f_i(x) \in F[x]$. By Cor.3.10.5(1), there are primitive $g_i(x) \in D[x]$ associates to $f_i(x)$ for $1 \leq i \leq n$ and $a \in D$ such that $f(x) = a g_1(x) \cdots g_n(x)$. Every $g_i(x)$ is irreducible in $F[x]$ since $f_i(x)$ is irreducible; thus every $g_i(x)$ is irreducible in $D[x]$ by Cor.3.10.5(2). Expressing $a \in D$ as a product of irreducible elements of D, we write $f(x)$ as a product of irreducible elements of $D[x]$.

Next assume that

$$f(x) = (p_1 \cdots p_k)(g_1(x) \cdots g_n(x)) = (q_1 \cdots q_l)(h_1(x) \cdots h_m(x)),$$

where all p_i and q_j are irreducible elements of D and every $g_i(x)$ and every $h_j(x)$ are irreducible in $D[x]$; then every $g_i(x)$ and every $h_j(x)$ are primitive, hence both $g_1(x) \cdots g_n(x)$ and $h_1(x) \cdots h_m(x)$ are primitive by Lemma 3.10.3. By Prop.3.10.2(2), there is a unit $u \in D$ such that $g_1(x) \cdots g_n(x) = u h_1(x) \cdots h_m(x)$, hence $u p_1 \cdots p_k = q_1 \cdots q_l$. Since D is factorial, we have that $k = l$ and, reindexing suitably, we have p_i and q_i are associates in D hence in $D[x]$ for $i = 1, \cdots, k$. On the other hand, every $g_i(x)$ and every $h_j(x)$ are irreducible in $F[x]$ by Cor.3.10.5(2), and $F[x]$ is factorial, thus $m = n$ and, up to reindexing, $g_i(x)$ and $h_i(x)$ are associates in $F[x]$ hence in $D[x]$ by Lemma 3.10.4(1). □

3.10.7 Corollary. *The polynomial ring* $D[x_1, \cdots, x_n]$ *in independent indeterminates* $x_1, \cdots, x_n$ *is factorial.* □

To conclude this section, we mention several facts about how to test the irreducibility. In general, it is difficult to determine whether a polynomial is reducible. The following is a general fact.

3.10.8 Proposition. *Let* R *be an* arbitrary *commutative unitary ring. Let* $f(x) \in R[x]$ *and* $a \in R$. *Then* $x - a \mid f(x)$ *if and only if* $f(a) = 0$ *(recall that* $a \in R$ *such that* $f(a) = 0$ *is called a* root *of* $f(x)$*).*

Proof. By Exer.3.9.14, there are $q(x) \in R[x]$ and $r \in R$ such that $f(x) = (x - a)q(x) + r$. Thus $f(a) = r$; and $x - a \mid f(x)$ if and only if $f(a) = r = 0$. □

3.10.9 Corollary. *A D-polynomial $f(x)$ of degree $n > 0$ has at most n roots.* □

Exer.4 shows that the above statement is false for general rings.

3.10.10 Theorem (Eisenstein's Criterion). *Let $f(x) = a_0 + a_1 x + \cdots + a_n x^n$ be a D-polynomial. If there is a prime $p \in D$ such that $p \mid a_i$ for $0 \le i < n$ and $p \nmid a_n$ and $p^2 \nmid a_0$, then $f(x)$ is irreducible in $F[x]$.*

Proof. Suppose $f(x)$ is reducible in $F[x]$, then, by Cor.3.10.5(1), there are D-polynomials $g(x) = b_0 + \cdots + b_r x^r$ and $h(x) = c_0 + \cdots + c_s x^s$ such that $r + s = n$ and $0 < s < n$ and $f(x) = g(x)h(x)$. Since $a_0 = b_0 c_0$, by the hypothesis we can assume that $p \nmid b_0$ and $p \mid c_0$; on the other hand, we have both $p \nmid b_r$ and $p \nmid c_s$ since $p \nmid a_n$. Assume $p \mid c_i$ for $0 \le i < k$ and $p \nmid c_k$, then $0 < k \le s$. Observing $a_k = b_0 c_k + b_1 c_{k-1} + \cdots + b_k c_0$, we see that $p \nmid a_k$ since p divides all the summands on the right-hand side except the first one. However, $p \mid a_k$ by the hypothesis since $k \le s < n$. The contradiction completes the proof. □

Example. Let p be a prime integer and n a positive integer. Then, by Theorem 3.10.10, $x^n - p$ is irreducible in the rational polynomial ring $\mathbf{Q}[x]$.

Example. The real polynomial $f(x,y) = y^3 + x^2 y^2 + x^3 y + x$ can be regarded as a polynomial over $\mathbf{R}[x]$. Considering the prime element x of $\mathbf{R}[x]$, by Theorem 3.10.10 we see that $f(x,y)$ is irreducible in $\mathbf{R}[x,y]$.

Exercises 3.10

Note that the agreements at the beginning of this section are still valid in the following exercises.

1. A D-polynomial of degree 0 is primitive if and only if it is a unit of D.

2. Show that a prime element of D is also a prime element of $D[x]$.

3. If $f(x) \in D[x]$ is *monic* (i.e. its leading coefficient is 1) and $f(x) = g(x)h(x)$ for $g(x), h(x) \in F[x]$ and $g(x)$ is monic, then $g(x) \in D[x]$. (Hint: a monic F-polynomial which is associate in $F[x]$ to a primitive D-polynomial must be a D-polynomial.)

4. Show that $(x - [1])(x - [2]) \in \mathbf{Z}_6[x]$ has four distinct roots in $\mathbf{Z}_6$.

5. Determine whether the following polynomials are irreducible in the rational polynomial ring $\mathbf{Q}[x]$:

    1) $x^p + px + 1$, where p is a prime integer;

    2) $x^5 + x^3 + 3x^2 - x + 1$.

6. Find all the irreducible polynomials of degree ≤ 3 in $\mathbf{Z}_3[x]$.

7. Let $f(x) = a_0 + a_1 x + \cdots + a_n x^n \in \mathbf{Z}[x]$ of degree n and p be a prime integer. If there is a $0 < k < n$ such that $p \mid a_i$ for $i = 0, \cdots, k-1$ and $p \nmid a_k$ and $p \nmid a_n$ and $p^2 \nmid a_0$, then $f(x)$ has an irreducible divisor of degree at least k.

$^{+}$8. Let p be a prime integer. Prove: $\varphi(x) = (x^p - 1)/(x - 1) = x^{p-1} + x^{p-2} + \cdots + 1$ is irreducible in $\mathbf{Q}[x]$. (Hint: take $x = y + 1$, $\varphi(x) = \varphi(y+1) = y^{p-1} + \dbinom{p}{1} y^{p-2} + \cdots + \dbinom{p}{p-1}$; where $\dbinom{p}{i}$ is the binomial coefficient.)

9. Let p be a prime integer and $\mathbf{Z}[x] \to \mathbf{Z}_p[x], f(x) \mapsto \overline{f}(x)$ be the canonical homomorphism induced by the canonical homomorphism $\mathbf{Z} \to \mathbf{Z}_p, a \mapsto \overline{a} = [a]$, i.e. $\overline{f}(x) = \sum \overline{a}_i x^i$ if $f(x) = \sum a_i x^i$.

    1) If $f(x) \in \mathbf{Z}[x]$ is monic and $\overline{f}(x)$ is irreducible, then $f(x)$ is irreducible.

    2) Give an example to show that the above (1) is false if $f(x)$ is not monic.

    3) Extend (1) to $D[x]$.

Chapter 4

Modules

Note: *in this chapter, all rings are unitary, all homomorphisms of rings are unitary.*

4.1 Modules

Recall that a vector space V over a field F is an additive group V equipped with a map (called the *scalar multiplication*)

$$F \times V \longrightarrow V, \quad (a, v) \longmapsto av \text{ (the image of } (a, v) \text{ is denoted by } av),$$

and the following four conditions hold:

1) $a(v_1 + v_2) = av_1 + av_2, \ \forall \ a \in F \text{ and } v_1, v_2 \in V$;
2) $(a_1 + a_2)v = a_1v + a_2v, \ \forall \ a_1, a_2 \in F \text{ and } v \in V$;
3) $(a_1 a_2)v = a_1(a_2 v), \ \forall \ a_1, a_2 \in F \text{ and } v \in V$;
4) $1_F v = v, \ \forall \ v \in V$, where 1_F is the unity of F.

The image av of $(a, v) \in F \times V$ is called the *scalar product* of a and v.

On the other hand, an additive group in fact has a similar structure. Let $\mathbf{Z}$ denote the integer ring as usual, and V be an additive group. Then we have a map

$$\mathbf{Z} \times V \longrightarrow V, \quad (n, v) \longmapsto nv$$

where nv is defined in 2.2.13 (but in additive notation as remarked in 2.2.16), and the following hold (see 2.2.14 but in additive notation):

1) $n(v_1 + v_2) = nv_1 + nv_2, \ \forall \ n \in \mathbf{Z} \text{ and } v_1, v_2 \in V$;

123

2) $(n_1 + n_2)v = n_1 v + n_2 v$, $\forall n_1, n_2 \in \mathbf{Z}$ and $v \in V$;

3) $(n_1 n_2)v = n_1(n_2 v)$, $\forall n_1, n_2 \in \mathbf{Z}$ and $v \in V$;

4) $1v = v$, $\forall v \in V$, where 1 is the integer 1, i.e. the unity of $\mathbf{Z}$.

Generalizing such algebraic structures, we introduce modules.

4.1.1 Definition. Let R be a ring. We say that M is a *left module over* R (or a *left R-module*) if M is an additive group equipped with a map (called the *module operation* of the module M, or *scalar multiplication*)

$$R \times M \longrightarrow M, \quad (r, x) \longmapsto rx \text{ (the image of } (r, x) \text{ is denoted by } rx),$$

and the following four conditions hold:

1) $r(x_1 + x_2) = rx_1 + rx_2$, $\forall r \in R$ and $x_1, x_2 \in M$;

2) $(r_1 + r_2)x = r_1 x + r_2 x$, $\forall r_1, r_2 \in R$ and $x \in M$;

3) $(r_1 r_2)x = r_1(r_2 x)$, $\forall r_1, r_2 \in R$ and $x \in M$;

4) $1x = x$, $\forall x \in M$, where 1 is the unity of R.

The image rx of $(r, x) \in R \times M$ is called the *scalar product* of r and x.

Remark. The module operation of M is in fact a biadditive function from $R \times M$ to M; for, if we denote the image of (r, x) by $\langle r, x \rangle$ as the usual notation in the linear algebra, then condition (2) means that the function is additive in the first variable r while condition (1) means that it is additive in the second variable x.

The *right R-modules* are defined similarly, provided the scalar multiplication is put on the right:

$$M \times R \longrightarrow M, \quad (x, r) \longmapsto xr$$

and the four conditions are modified correspondingly; e.g. 3) is rewritten as

3'). $x(r_1 r_2) = (x r_1)r_2$, $\forall r_1, r_2 \in R$ and $x \in M$.

Agreement. *In this book, "module" usually means "left module" unless otherwise stated.*

4.1.2 Example. If R is a field, then R-modules are just vector spaces over R.

4.1.3 Example. Any additive group can be regarded as a $\mathbf{Z}$-module.

4.1.4 Example. Let R be a ring and S be a unitary subring of R. Then R be an additive group, and multiplying by every element $s \in S$

on the left of the ring R gives a map: $S \times R \to R, (s, x) \mapsto sx$; it is easy to check that, in this way, R is a left S-module, denoted by $_S R$. In particular, R is a left R-module, called the *left regular module* of R and denoted by $_R R$,

On the other hand, multiplying by every element $s \in S$ on the right of the ring R gives the right S-module R_S. In particular, we have the *right regular* R-module: $R \times R \to R, (x, r) \mapsto xr$, denoted by R_R.

Moreover, the above two module structures "commute" with each other: $(sx)r = s(xr)$, $\forall s, x, r \in R$.

4.1.5 Definition. Let R and S be rings. A *left R- right S-bimodule* M, denoted by $_R M_S$ also, means that M is both a left R-module and a right S-module and $(rx)s = r(xs)$ for all $x \in M$ and $r \in R$ and $s \in S$.

A left R- left S-module M, denoted by $_{R-S} M$ also, means that M is both a left R-module and a left S-module and $s(rx) = r(sx)$ for all $x \in M$ and $r \in R$ and $s \in S$.

The right R- right S-modules are defined similarly.

Such algebraic structures which admit two module operations compatible to each other are called *bimodules*. For a bimodule, e.g. $_R M_S$, since the two scalar multiplications are compatible, we can write rxs instead of $(rx)s = r(xs)$.

Example. For any ring R, it follows from Example 4.1.4 that R is a left R- right R-bimodule, denoted by $_R R_R$.

Now we give another point of view to consider the modules.

4.1.6 Definition. Let U, V be additive groups and denote

$$\text{Hom}(U, V) = \{\text{homomorphisms } f : U \to V\}.$$

For $f_1, f_2 \in \text{Hom}(U, V)$, the sum $f_1 + f_2$ is defined by:

$$(f_1 + f_2)(u) = f_1(u) + f_2(u), \qquad \forall \, u \in U.$$

Then it is easy to check that $f_1 + f_2 \in \text{Hom}(U, V)$ and, with this addition, $\text{Hom}(U, V)$ is an additive group (see Exer.2), called the *homomorphism group* from U to V.

Further, denote $\text{End}(V) = \text{Hom}(V, V)$ for an additive group V. And, besides the addition, we have a multiplication on $\text{End}(V)$ as follows: for $f_1, f_2 \in \text{End}(V)$,

$$(f_1 f_2)(v) = f_1(f_2(v)), \qquad \forall \, v \in V.$$

Then End(V) becomes a ring (Exer.3), we denote it by End$^l(V)$ and call it the *left endomorphism ring* of the additive group V. The attributive "left" means that the symbol f representing the homomorphism is put on the left: $f(v)$, for $v \in V$.

If we put f on the right: $(v)f$, for $v \in V$, then it does not matter for the addition of homomorphisms; however, the multiplication of homomorphisms appears like this: $(v)(f_1 f_2) = ((v)f_1)f_2$ for $v \in V$. In this way, End(V) also becomes a ring, we denote it by End$^r(V)$ and call it the *right homomorphism ring* of the additive group V.

4.1.7 Proposition. *Let R be a ring and V is an additive group.*

1) *If V is a left R-module, then for $r \in R$ the map $\rho_r : V \to V, v \mapsto rv$ is a homomorphism of the additive group V, and $\rho : R \to$ End$^l(V), r \mapsto \rho_r$, is a ring homomorphism.*

2) *If $\rho : R \to$ End$^l(V), r \mapsto \rho_r$, is a ring homomorphism, then, with the map $R \times V, (r, v) \mapsto \rho_r(v)$, V is a left R-module.*

Proof. (1). From condition (1) of Definition 4.1.1, ρ_r is a group homomorphism; conditions (2), (3) and (4) imply that $\rho : R \to$ End$^l(V)$ is a ring homomorphism.

(2). Since ρ_r is a group homomorphism, $\rho_r(v_1+v_2) = \rho_r(r_1)+\rho(v_2)$, this is condition (1) of Definition 4.1.1; further, conditions (2), (3) and (4) follow from the assumption that ρ is a ring homomorphism. □

For right modules, we have the following similar characterization.

4.1.7′ Proposition. *Let R be a ring and V be an additive group.*

1) *If V is a right R-module, then for $r \in R$ the map $\tau_r : V \to V, v \mapsto vr$ is a homomorphism of the additive group V, and $\tau : R \to$ End$^r(V), r \mapsto \tau_r$, is a ring homomorphism.*

2) *If $\tau : R \to$ End$^r(V), r \mapsto \tau_r$, is a ring homomorphism, then, with the map $V \times R, (v, r) \mapsto (v)\tau_r$, V is a right R-module.* □

4.1.8 Remark. Note that, for an additive group V, both End$^l(V)$ and End$^r(V)$ have the same underlying additive group End(V); but their multiplications are different. The difference is the order of the composition of the mappings f_1 and f_2 on $v \in V$: in End$^l(V)$, $f_1 f_2$ maps v first by f_2 then by f_1; while in End$^r(V)$, $f_1 f_2$ maps v first by f_1 then by f_2. From this observation we have the following fact.

4.1.9 Proposition. *If R is a commutative ring, then any left R-module is also a right R-module, and* vice versa; *hence, any R-module is an R-bimodule.* □

Some more related observations stated in Exer.5—7.

Exercises 4.1

1. Write down the definitions of right R-modules.

+2. Check that the sum of two homomorphisms of additive groups U and V is again a homomorphism and $\text{Hom}(U, V)$ is an additive group.

    (Remark: the zero element of $\text{Hom}(U, V)$ is the *zero homomorphism* $0 : U \to V$ which sends every $u \in U$ to the zero element 0 of V.)

+3. Check that the $\text{End}^l(V)$ of an additive group V is a ring.

    (Remark: the unity element of $\text{End}^l(V)$ is the identity endomorphism id_V; we denote it by 1 if there is no confusion.)

4. Prove Prop.4.1.9.

5. Let R be a ring with addition "+" and multiplication "·". On R we define another multiplication "∘" as follows: $r \circ s = s \cdot r$ for $r, s \in R$. Then, the set R with the addition "+" and the new multiplication "∘" is also a ring.

    (**Definition:** the new ring constructed as above is denoted by R^o and called the *opposite ring* of the original ring R.)

6. An *anti-homomorphism* $f : R \to S$ from a ring R to a ring S is an additive homomorphism such that $f(rr') = f(r')f(r)$ for all $r, r' \in R$. Show that, if $f : R \to S, r \mapsto f(r)$ is a homomorphism from a ring R to a ring S and R^o denotes the opposite ring as in Exer.5 above, then both $f : R \to S^o, r \mapsto f(r)$ and $f : R^o \to S, r \mapsto f(r)$ are anti-homomorphisms of rings; while $f : R^o \to S^o, r \mapsto f(r)$ is a homomorphism of rings.

7. Let R be a ring and M be an Abelian group, let $\text{End}^l(M)$ be the left endomorphism ring of M. If $\alpha : R \to \text{End}^l(M)$ is an anti-homomorphism of rings, then, with the map $M \times R$, $(x, r) \mapsto \alpha(r)(x)$, M is a right R-module.

4.2 Homomorphisms

Keeping the fundamentals about the vector spaces or the additive groups in mind, we can easily get some basic facts about modules.

4.2.1 Definition. Let M be a module over a ring R. N is called a *submodule* of M, denoted by $N \leq M$, if N is an additive subgroup of M and is closed under the scalar multiplication, i.e. $ry \in N$ for all $y \in N$ and $r \in R$.

Submodules of right modules and bimodules are defined similarly.

Thus, if R is a field, then a submodule U of an R-module V is just a subspace of the vector space V over R.

If an additive group V is regarded as a **Z**-module, then a submodule of V is just a subgroup of the additive group V.

4.2.2 Example. Let R be a ring. A submodule L of the left regular R-module $_RR$ is a subgroup of the additive group R such that $ry \in L$ for any $y \in L$ and $r \in R$; in other words, a submodule L of $_RR$ is just a left ideal of the ring R. Similarly, a submodule N of the right regular module R_R is just a right ideal of the ring R; and a submodule I of the bimodule $_RR_R$ is just an ideal of the ring R.

It is easy to get a criterion for submodules, see Exer.1. Then it is easy to see that the intersection of submodules of a module is still a submodule. On the other hand, we can define the sum of submodules.

4.2.3 Definition. Let M be an R-module and N_i, $i \in I$, be submodules of M. The sum of N_i, $i \in I$, denoted by $\sum_{i \in I} N_i$, is the set of all finite sums $\sum x_i$ with $x_i \in N_i$ and i running on a finite subset of I; obviously, the sum $\sum_{i \in I} N_i$ is a submodule, see Exer.2.

4.2.4 Definition. Let M be a module over a ring R and $T \subset M$ be a subset. The intersection of the submodules which contain T is a submodule, called the *submodule generated by* T, and denoted by $\langle T \rangle$. If $T = \{x_1, \cdots, x_n\}$, then denote $\langle T \rangle = \langle x_1, \cdots, x_n \rangle$. In particular, the submodule $\langle x \rangle$ generated by one element $x \in M$ is called a *cyclic submodule*.

If $M = \langle T \rangle$, then T is called a *generator set* of the R-module M. If $M = \langle x_1, \cdots, x_n \rangle$ for finite number of elements, then M is said to be *finitely generated*. In particular, if $M = \langle x \rangle$, then M is called a *cyclic module*.

It is easy to show that $\langle x \rangle = \{rx \mid r \in R\}$ which we denote by Rx, and $\langle T \rangle = \sum_{x \in T} Rx$; see Exer.3. Hence, if M is a cyclic R-module generated by x, then $M = Rx$.

4.2.5 Definition. Let R be a ring and M and M' be R-modules. We say that a map $f : M \to M'$ is an *R-module homomorphism* (or *R-homomorphism*, or *R*-linear map) if the following two conditions hold:
$$f(x_1 + x_2) = f(x_1) + f(x_2) \qquad \forall \; x_1, x_2 \in M$$
$$f(rx) = rf(x) \qquad \forall \; r \in R, \;\; x \in M \,.$$
The *kernel* of an R-homomorphism f, denoted by $\ker(f)$, is the inverse image in M of the zero submodule $\{0\}$ of M'.

An R-homomorphism $f : M \to M'$ is called an *isomorphisms* if it is both injective and surjective. A homomorphism (isomorphism resp.) from a module to itself is called *endomorphisms* (*automorphisms* resp.).

The first condition says that f is a homomorphism of additive groups. The right module homomorphisms and the bimodule homomorphisms are defined similarly, provided the second condition is replaced by the following resp. For right R-modules:
$$f(xr) = f(x)r\,, \qquad \forall \; r \in R, \;\; x \in M\,;$$
and for left R- right S-modules:
$$f(rxs) = rf(x)s\,, \qquad \forall \; r \in R, \;\; s \in S, \;\; x \in M\,.$$

For example, if R is a field, then an R-homomorphism is just a usual R-linear maps of vector spaces. If $R = \mathbf{Z}$ and regard additive groups as $\mathbf{Z}$-modules, then an additive group homomorphism can be regarded as a $\mathbf{Z}$-homomorphism.

It is clear that the kernel $\mathrm{Ker}(f)$ of an R-homomorphism $f : M \to M'$ is a submodule of M while the image $\mathrm{Im}(f) = f(M)$ is a submodule of M'; see Exer.4.

4.2.6 Definition. Let M be a module over a ring R and $N \le M$ be a submodule. Then we have the quotient group M/N of the additive group M; it is clear that $R \times M/N \to M/N$, $(r, x + N) \mapsto rx + N$, is a well defined mapping and M/N becomes an R-module and $\sigma : M \to M/N, x \mapsto x + N$ is an R-homomorphism with kernel N (Exer.5). The R-module M/N is called the *quotient module* of M with respect to the submodule N, and σ is called the *natural* (or *canonical*) homomorphism.

All the homomorphism theorems for additive groups (see §2.7, but the notations should be replaced by additive versions) can be easily extended to modules; we state them but leave their proofs as exercises.

4.2.7 Theorem (Fundamental Theorem for Module Homomorphisms). Let $f : M \to N$ be a homomorphism with kernel K of R-modules M and N. Then there is a unique R-homomorphism $\overline{f} : M \to M/K$ such that $f = \overline{f} \circ \sigma$, where $\sigma : M \to M/K$ is the natural homomorphism; such $\overline{f}$ is injective; moreover, $\overline{f}$ is an isomorphism if f is surjective. $\square$

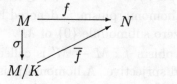

4.2.8 Remark. We refer the above $\overline{f}$ as the homomorphism *induced by f*. And, if $f : M \to N$ is not surjective, we consider the inclusion homomorphism $\iota : Im(f) \to N$; then the induced homomorphism $\overline{f} : M/K \to N$ can be written as a composition $M/K \to Im(f) \to N$ as follows, where $\tilde{f} : M/K \to Im(f)$ is always an isomorphism:

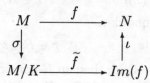

In particular, we have

4.2.9 Corollary (First Isomorphism Theorem). *If $f : M \to N$ is a homomorphism of modules over a ring R, then $M/\mathrm{Ker}(f) \cong Im(f)$; in particular, $M/\mathrm{Ker}(f) \cong N$ provided f is surjective.* $\square$

4.2.10 Theorem (Correspondence of Submodules).
Let R be a ring and $f : M \to \overline{M}$ be a surjective R-module homomorphism and $K = \mathrm{Ker}(f)$ be its kernel; let $\mathcal{L}(\overline{M}) = \{\overline{N} \mid \overline{N} \leq \overline{M}\}$ while $\mathcal{L}_K(M) = \{N \mid N \leq M \text{ and } N \supset K\}$. Then

    1) *$\eta : \mathcal{L}_K(M) \to \mathcal{L}(\overline{M})$, $N \mapsto f(N)$ is a bijective map;*

    2) *for $N_1, N_2 \in \mathcal{L}_K(M)$, $N_1 \subset N_2$ iff $f(N_1) \subset f(N_2)$.* $\square$

4.2.11 Corollary (Second Isomorphism Theorem). *Let M be a module over a ring R and both N and K be submodules of M. Then we have an R-isomorphism:*

$$N/(N \cap K) \xrightarrow{\cong} (N+K)/K , \qquad y + (N \cap K) \longmapsto y + K . \qquad \square$$

4.2.12 Corollary (Third Isomorphism Theorem). *Let M be a module over a ring R and both $N \subset K$ submodules of M. Then we have an R-isomorphism:*

$$M/K \xrightarrow{\cong} (M/N)\big/(K/N) , \qquad x + K \longmapsto (x+N) + (K/N) . \quad \square$$

4.2.13 Remark. For R-modules M, N, let

$$\mathrm{Hom}_R(M, N) = \{ \ R\text{-homomorphisms from } M \text{ to } N \ \};$$

and let $\mathrm{End}_R(M) = \mathrm{Hom}_R(M, M)$. Then, just like (4.1.6), it is easy to see that $\mathrm{Hom}_R(M, N)$ is an additive group, and there are two ring structures on $\mathrm{End}_R(M)$, i.e. the *left endomorphism ring* $\mathrm{End}_R^l(M)$ and the *right endomorphism ring* $\mathrm{End}_R^r(M)$; see Exer.11–12.

Exercises 4.2

$^+$1. Let M be a module over a ring R and $N \subset M$. The following two are equivalent:
   i) $N \leq M$;
   ii) $N \neq \emptyset$ and for all $x, y \in N$ and $r \in R$ we have $x - y \in N$ and $rx \in N$.

$^+$2. Prove: the sum $\sum_{i \in I} N_i$ of submodules N_i, $i \in I$ (see Definition 4.2.3), is a submodule of M.

$^+$3. Let M be a module over a ring R. Prove:
   1) The cyclic submodule generated by $x \in M$ is $Rx = \{rx \mid r \in R\}$.
   2) The submodule $\langle T \rangle = \sum_{x \in T} Rx$.

$^+$4. Let R be a ring and $f : M \to M'$ be an R-homomorphism of R-modules M and M'. Prove that $\ker(f) \leq M$ and $\mathrm{Im}(f) \leq M'$.

5. Notations are as in Definition 4.2.6. Prove: $r(x + N) = rx + N$ is a well-defined scalar multiplication on M/N, and M/N is an R-module, and $\tau : M \to M/N, x \mapsto x + N$ is a surjective R-homomorphism.

6. Prove the results 4.2.7 — 4.2.12.

7. (Factorization Theorem for Module Homomorphisms) Let R be a ring and $f : M \to X$ and $g : M \to N$ be R-module homomorphisms and g be surjective. Then there is a homomorphism $h : N \to X$ such that $f = hg$ if and only if $\mathrm{Ker}(g) \subset \mathrm{Ker}(f)$; and, if this is the case, such an h is unique and $\mathrm{Ker}(h) = g(\mathrm{Ker}(f))$.

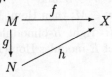

8. Let M be a module over a ring R. For any left ideal L of R, let LM denote the subset of M consisting of all the finite sums $\sum a_i x_i$ with $a_i \in L$ and $x_i \in M$.

   1) Show that LM is a submodule of M.

   2) M/IM becomes a module over the quotient ring R/I if I is an ideal of R.

9. Let M be a module over a ring R. Prove:

   1) The union $N_1 \cup N_2$ of two submodules N_1 and N_2 is still a submodule if and only if one of them is contained in the other.

   2) Let N_i, $i \in I$, be submodules of M. If for any two $i, j \in I$ there is a $k \in I$ such that both $N_i, N_j \subset N_k$, then $\bigcup_{i \in I} N_i$ is a submodule of M.

$+$10. (*Glueing of homomorphisms*). Let M, N be modules over a ring R, and $M_1, M_2 \le M$ such that $M = M_1 + M_2$. If $f_i : M_i :\to N$, $i = 1, 2$, are R-homomorphisms such that the restriction maps $f_1|_{M_1 \cap M_2} = f_2|_{M_1 \cap M_2}$, then $f : M \to N$ with $f(x_1 + x_2) = f_1(x_1) + f_2(x_2)$ for $x_i \in M_i$ is a well-defined R-homomorphism.

$+$11. Let M, N be modules over a ring R.

   1) Define an addition on $\mathrm{Hom}(M, N)$ as: for $f_1, f_2 \in \mathrm{Hom}(M, N)$,

   $$(f_1 + f_2)(x) = f_1(x) + f_2(x) \qquad \forall x \in M.$$

   Then $\mathrm{Hom}(M, N)$ is an additive group.

   2) If $\varphi : M \to M'$ is an R-homomorphism, then

   $$\varphi^* : \ \mathrm{Hom}(M', N) \longrightarrow \mathrm{Hom}(M, N), \quad f' \longmapsto f'\varphi$$

   is an additive group homomorphism.

   3) If $\varphi : N \to N'$ is an R-homomorphism, then

   $$\varphi_* : \ \mathrm{Hom}(M, N) \longrightarrow \mathrm{Hom}(M, N'), \quad f \longmapsto \varphi f$$

   is an additive group homomorphism.

$+$12. Notations as above Exer.11. Consider $\mathrm{End}_R(M)$.

   1) Define the left multiplication on $\mathrm{End}_R(M)$ as: $(f_1 f_2)(x) = f_1(f_2(x))$ for $f_1, f_2 \in \mathrm{End}_R(M)$ and $x \in M$; then we get a ring $\mathrm{End}_R^l(M)$.

   2) Define the right multiplication on $\mathrm{End}_R(M)$ as: $(x)(f_1 f_2) = ((x)f_1)f_2$ for $f_1, f_2 \in \mathrm{End}_R(M)$ and $x \in M$; then we get a ring $\mathrm{End}_R^r(M)$.

$+$13. 1) Notations as in above Exer.11; further, assume M is a left R- right S-bimodule. Then for $s \in S$ and $f \in \mathrm{Hom}(M, N)$, define $s \cdot f$ as $(s \cdot f)(m) = f(ms)$, $\forall m \in M$; then $\mathrm{Hom}(M, N)$ is an S-module.

   2) Consider R as a left R- right R-bimodule. For any R-module N we have an isomorphism of R-modules $\mathrm{Hom}(R, N) \to N, f \mapsto f(1_R)$.

4.3 Direct Products, Direct Sums

Just as vector spaces and as additive groups (see §2.9), we can define the direct sums and direct products of modules.

Let R be a ring and M_i, $i \in I$, be R-modules. Let $M = \prod_{i \in I} M_i = \{(x_i)_{i \in I} \mid x_i \in M_i\}$ be the Cartesian product (cf. §1.1). On M we define

$$(x_i)_{i \in I} + (y_i)_{i \in I} = (x_i + y_i)_{i \in I} \qquad \text{for } (x_i)_{i \in I}, (y_i)_{i \in I} \in M;$$

$$r(x_i)_{i \in I} = (r x_i)_{i \in I} \qquad \text{for } r \in R \text{ and } (x_i)_{i \in I} \in M;$$

in other words, operations are made componentwise. Then it is easy to check that M is an R-module, and the map

$$\rho_j : \quad \prod_{i \in I} M_i \longrightarrow M_j, \qquad (x_i)_{i \in I} \longmapsto x_j$$

is an R-homomorphism for every $j \in I$.

4.3.1 Definition. The R-module $M = \prod_{i \in I} M_i$ constructed as above is called the *direct product* of M_i, $i \in I$; and the module homomorphism ρ_j is called the *canonical projection* to the jth-component.

4.3.2 Proposition. *Notations as above. For any R-module N and any R-homomorphisms $f_j : N \to M_j$ where $j \in I$, there is a unique R-homomorphism $f : N \to \prod_{i \in I} M_i$ such that $f_j = \rho_j f$ for all $j \in I$.*

Proof. This is in fact the module-version of Prop.2.9.5. Define

$$f : \quad N \longrightarrow \prod_{i \in I} M_i, \qquad y \longmapsto (f_i(y))_{i \in I};$$

then f is an R-homomorphism satisfying the desired condition. If $f' : N \to \prod_{i \in I} M_i$ also satisfies the condition $f_j = \rho_j f'$ for all $j \in I$, for any $y \in N$ let $f'(y) = (x_i)_{i \in I}$, then $x_j = \rho_j f'(y) = f_j(y)$ for all $j \in I$; thus $f'(y) = (f_i(y))_{i \in I} = f(y)$; i.e. $f' = f$. □

In the direct product $\prod_{i \in I} M_i$ of the R-modules M_i, let $\bigoplus_{i \in I} M_i$ be the subset consisting of all the elements $(x_i)_{i \in I}$ with only finitely many components x_i non-zero (i.e. almost $x_i = 0$); then it is easy to see from Exer.4.2.1 that $\bigoplus_{i \in I} M_i$ is a submodule of $\prod_{i \in I} M_i$; hence $\bigoplus_{i \in I} M_i$ is itself an R-module. Further, for every $j \in I$ we have an R-homomorphism $\iota_j : M_j \longrightarrow \bigoplus_{i \in I} M_i$, mapping $x_j \in M_j$ to the element whose jth-component is x_j and all the other components are zero.

4.3.3 Definition. The R-module $\bigoplus_{i \in I} M_i$ obtained above is called the *direct sum* of M_i, $i \in I$; and the R-homomorphism ι_j defined as above is called the *canonical injection* to the jth-component.

For any $(x_i)_{i \in I} \in \bigoplus_{i \in I} M_i$, almost all components $x_i = 0$ except for finitely many indices; so $\iota_i(x_i) = 0$ for almost $i \in I$. For convenience we can write

(4.3.4) $(x_i)_{i \in I} = \sum_{i \in I} \iota_i(x_i);$

provided we keep in mind the following

4.3.5 Agreement: The sum $\sum_{i \in I} x_i$ for any $(x_i)_{i \in I} \in \bigoplus_{i \in I} M_i$ is in fact a finite sum over the finite subset of the indices $j \in I$ such that $x_j \neq 0$.

4.3.6 Proposition. *Notations as above. For any R-module N and any R-homomorphisms $g_j : M_j \to N$ where $j \in I$, there is a unique R-homomorphism $g : \bigoplus_{i \in I} M_i \to N$ such that $g_j = g \iota_j$ for all $j \in I$.*

Proof. With the agreement above, we can define a map

$$g : \bigoplus_{i \in I} M_i \longrightarrow N, \qquad (x_i)_{i \in I} \longmapsto \sum_{i \in I} \iota_i(x_i).$$

Obviously, g is an R-homomorphism satisfying the desired condition (in fact, this is the module-version of the group homomorphism (2.9.10)). If $g' : \bigoplus_{i \in I} M_i \to N$ also satisfies that $g_j = g' \iota_j$ for all $j \in I$, then by (4.3.4) we have

$$g'((x_i)_{i \in I}) = g'(\sum_{i \in I} \iota_i(x_i)) = \sum_{i \in I} g' \iota_i(x_i) = \sum_{i \in I} g_i(x_i) = g((x_i)_{i \in I}).$$

That is, $g' = g$. $\square$

Let M_i be as above again; let $M_i' = \mathrm{Im}(\iota_i)$. Then $M_i' \leq \bigoplus_{i \in I} M_i$ and $M_i' \cong M_i$, and (4.3.4) shows that $\bigoplus_{i \in I} M_i = \sum_{i \in I} M_i'$; moreover, it is easy to see that the two conditions of the following proposition are satisfied for M', $i \in I$, in $\bigoplus_{i \in I} M_i$.

4.3.7 Proposition. *Let M be a module over a ring R and $N_i \leq M$, $i \in I$. The following three conditions are equivalent:*

i) $M = \sum_{i \in I} N_i$ *and* $N_j \bigcap (\sum_{i \neq j} N_i) = 0$ *for all $j \in I$.*

ii) *Any $x \in M$ can be written as $x = x_{i_1} + \cdots + x_{i_k}$ for a finite subset $\{i_1, \cdots, i_k\} \subset I$ and $x_{i_t} \in N_{i_t}$; and such expression is unique except for zeros; that is, if $x = x_{i_1} + \cdots + x_{i_k} = x'_{i'_1} + \cdots + x'_{i'_{k'}}$ and all $x_{i_t} \in M_{i_t}$ and all $x'_{i'_t} \in M_{i'_t}$ are non-zero, then $\{i_1, \cdots, i_k\} = \{i'_1, \cdots, i'_{k'}\}$ (hence $k' = k$) and $x_{i_t} = x'_{i_t}$ for $t = 1, \cdots, k$.*

iii) $\iota: \bigoplus_{i \in I} N_i \longrightarrow M$, $(x_i) \longmapsto \sum_{i \in I} x_i$ *is an R-isomorphism.*

Proof. (i) $\Longrightarrow$ (ii). The existence of the expression $x = x_{i_1} + \cdots + x_{i_k}$ follows from the condition $M = \sum_{i \in I} N_i$. Suppose $x = x_{i_1} + \cdots + x_{i_k} = x'_{i'_1} + \cdots + x'_{i'_{k'}}$ without zero summands. If $i'_1 \notin \{i_1, \cdots, i_k\}$, then

$$x'_{i'_1} = x_{i_1} + \cdots + x_{i_k} - x'_{i'_2} - \cdots - x'_{i'_{k'}} \in N_{i'_1} \bigcap (\sum_{i \neq i'_1} N_i) = 0;$$

which contradicts to that $x'_{i'_1} \neq 0$. Thus $i'_1 \in \{i_1, \cdots, i_k\}$. Similarly, every $i'_t \in \{i_1, \cdots, i_k\}$, and every $i_t \in \{i'_1, \cdots, i'_{k'}\}$. Thus $\{i_1, \cdots, i_k\} = \{i'_1, \cdots, i'_{k'}\}$. If $x_{i_1} \neq x'_{i_1}$, then

$$0 \neq x'_{i_1} - x_{i_1} = x_{i_2} + \cdots + x_{i_k} - x'_{i_2} - \cdots - x'_{i_k} \in N_{i_1} \bigcap (\sum_{i \neq i_1} N_i) = 0;$$

which is impossible. Therefore $x_{i_t} = x'_{i_t}$ for $t = 1, \cdots, k$.

(ii) $\Longrightarrow$ (iii). The existence of the expression $x = x_{i_1} + \cdots + x_{i_k}$ show that ι is surjective; and the uniqueness of the expressions show that ι is injective.

(iii) $\Longrightarrow$ (i). The surjectivity of ι implies that $M = \sum_{i \in I} N_i$. If $0 \neq x_j \in N_j \bigcap (\sum_{i \neq j} N_i) \neq 0$, then there are indices $i_1, \cdots, i_n$ which are all different from j such that $x_j = x_{i_1} + \cdots + x_{i_n}$ with $x_{i_t} \in N_{i_t}$; thus the element of $\bigoplus_{i \in I} N_i$, which jth-component is $-x_j$ and i_tth-component is x_{i_t} for $1 \leq t \leq n$ and the other components are zero, is non-zero but mapped by ι to zero; this contradicts to the injectivity of ι. $\square$

These suggest the following

4.3.8 Definition. Notations are as in Prop.4.3.7. We say that M is the *(internal) direct sum* of its submodules N_i, $i \in I$, denoted by $\bigoplus_{i \in I} N_i$ again, if the conditions of Prop.4.3.7 hold.

4.3.9 Remark. If we consider a finite number of modules, then the direct product is the same as the direct sum, and the internal direct sum is similar to the internal product in a group defined in (2.9.11). The result for modules similar to Theorem 2.9.8 can be obtained easily.

On the other hand, Theorem 2.9.9 and Definition 2.9.11 can be, by the similar argument, extended to an infinite number of normal subgroups. Further, the following can also be stated for groups.

4.3.10 Proposition. *If M_i and N_i, $i \in I$, are modules over a ring R and $f_i : M_i \to N_i$, $i \in I$, are R-homomorphisms. Then*

1) $\prod_{i\in I} f_i : \prod_{i\in I} M_i \longrightarrow \prod_{i\in I} N_i$, $(x_i)_{i\in I} \longmapsto (f_i(x_i))_{i\in I}$, is an R-homomorphism, and

$$\text{Ker}(\prod_{i\in I} f_i) = \prod_{i\in I} \text{Ker}(f_i), \qquad \text{Im}(\prod_{i\in I} f_i) = \prod_{i\in I} \text{Im}(f_i).$$

In particular, $\prod_{i\in I} f_i$ is injective (surjective, isomorphism) if and only if all f_i, $i \in I$, are injective (surjective, isomorphism).

2) $\bigoplus_{i\in I} f_i : \bigoplus_{i\in I} M_i \longrightarrow \bigoplus_{i\in I} N_i$, $(x_i)_{i\in I} \longmapsto (f_i(x_i))_{i\in I}$, is an R-homomorphism, and

$$\text{Ker}(\bigoplus_{i\in I} f_i) = \bigoplus_{i\in I} \text{Ker}(f_i), \qquad \text{Im}(\bigoplus_{i\in I} f_i) = \bigoplus_{i\in I} \text{Im}(f_i).$$

In particular, $\bigoplus_{i\in I} f_i$ is injective (surjective, isomorphism) if and only if all f_i, $i \in I$, are injective (surjective, isomorphism).

Proof. Exer.5.   □

Exercises 4.3

1. Let $M_1, \cdots, M_n$ be modules over a ring R, notations be as in 4.3.1 and 4.3.3. Prove that:

$$\sum_{i=1}^{n} \iota_i\rho_i = \text{id}_{M_1\oplus\cdots\oplus M_n} \quad \text{and} \quad \rho_j\iota_i = \begin{cases} \text{id}_{M_i} & \text{if } i=j \\ 0 & \text{if } i\neq j \end{cases}.$$

2. Let M and $M_1, \cdots, M_n$ be modules over a ring R. Prove: $M \cong M_1 \oplus \cdots \oplus M_n$ if and only if there are R-homomorphisms $\varphi_i : M \to M_i$ and $\tau_i : M_i \to M$ for $1 \leq i \leq n$ such that

$$\sum_{i=1}^{n} \tau_i\varphi_i = \text{id}_M \quad \text{and} \quad \varphi_j\tau_i = \begin{cases} \text{id}_{M_i} & \text{if } i=j \\ 0 & \text{if } i\neq j \end{cases}.$$

3. If the left regular module $_RR$ of a ring R is decomposed into a direct sum of left ideals $_RR = L_1 \oplus \cdots \oplus L_n$, then there are $e_1, \cdots, e_n \in R$ such that

$$1_R = \sum_{i=1}^{n} e_i \quad \text{and} \quad e_je_i = \begin{cases} e_i & \text{if } i=j \\ 0 & \text{if } i\neq j \end{cases},$$

   and $L_i = Re_i$ for $1 \leq i \leq n$. Conversely, if there are $e_1, \cdots, e_n \in R$ satisfy the above two conditions, then $_RR$ is the direct sum $_RR = Re_1 \oplus \cdots \oplus Re_n$.

4. State and prove the result for right regular module R_R similar to Exer.3.

5. Prove Prop.4.3.10.

6. Let M_i, $i \in I$ be modules over a ring R and $N_i \leq M_i$, $i \in I$. Prove:

   1) $\prod_{i\in I} M_i \big/ \prod_{i\in I} N_i \cong \prod_{i\in I}(M_i/N_i)$.

   2) $\bigoplus_{i\in I} M_i \big/ \bigoplus_{i\in I} N_i \cong \bigoplus_{i\in I}(M_i/N_i)$.

7. Assume that an R-module $M = M_1 \oplus M_2$.

   1) If $M_i = \bigoplus_{j \in J_i} M_{ij}$, $i = 1, 2$, are direct sums of their submodules resp., then $M = \bigoplus_{i=1}^{2} \bigoplus_{j \in J_i} M_{ij}$.

   2) If $M_i = \prod_{j \in J_i} M_{ij}$, $i = 1, 2$, are direct products of their submodules resp., then $M = \prod_{i=1}^{2} \prod_{j \in J_i} M_{ij}$.

8. Let $M_1, \cdots, M_m$ and $M_1', \cdots, M_n'$ be R-modules. Then there is an additive group isomorphism:

$$\mathrm{Hom}_R(\bigoplus_i M_i, \ \bigoplus_j M_j') \cong \bigoplus_{i,j} \mathrm{Hom}_R(M_i, \ M_j')$$

(Hint: let $\iota_k : M_k \to \bigoplus_i M_i$ and $\iota_k' : M_k' \to \bigoplus_j M_k'$ be the canonical injections, and $\rho_k : \bigoplus_i M_i \to M_k$ and $\rho_k' : \bigoplus_j M_j' \to M_k'$ be the canonical projections; the following is an isomorphism (called the *canonical isomorphism*):

$$\mathrm{Hom}_R(\bigoplus_i M_i, \ \oplus_j M_j') \longrightarrow \bigoplus_{i,j} \mathrm{Hom}_R(M_i, \ M_j'), \quad f \longmapsto (\rho_j' f \iota_i)_{i,j} \, .)$$

4.4   Exact Sequences of Homomorphisms

4.4.1 Definition. Let R be a ring, and $M_i \xrightarrow{d_i} M_{i-1}$ be homomorphisms of R-modules. The sequence of R-homomorphisms (finite or infinite)

$$\cdots \longrightarrow M_{i+1} \xrightarrow{d_{i+1}} M_i \xrightarrow{d_i} M_{i-1} \xrightarrow{d_{i-1}} \cdots$$

is said to be *exact at the point* M_i if $\mathrm{Im}(d_{i+1}) = \mathrm{Ker}(d_i)$; Moreover, we say that the sequence is *exact* if it is exact at every M_i. The following exact sequence is said to be a *short exact sequence*:

$$0 \longrightarrow M' \longrightarrow M \longrightarrow M'' \longrightarrow 0 \, .$$

4.4.2 Proposition.

   1) $0 \longrightarrow M' \xrightarrow{f} M$ *is exact if and only if f is injective;*

   2) $M \xrightarrow{g} M'' \longrightarrow 0$ *is exact if and only if g is surjective;*

   3) $0 \longrightarrow M' \xrightarrow{f} M \xrightarrow{g} M'' \longrightarrow 0$ *is exact if and only if f is injective and g is surjective and $\mathrm{Im}(f) = \mathrm{Ker}(g)$.*

Proof. Exer.1.   □

Example. If $M' \leq M$, then $0 \to M' \to M \to M/M' \to 0$ is exact. If M_1 and M_2 are two modules, then the following sequence is exact:

$$0 \to M_1 \to M_1 \oplus M_2 \to M_2 \to 0 \, .$$

4.4.3 Example. For an R-module homomorphism $f : M \to N$, recall that $\mathrm{Ker}(f) = f^{-1}(0)$; further, the *cokernel* of f, denoted by $\mathrm{Coker}(f)$, is defined to be the quotient module $N/\mathrm{Im}(f)$. Then it is easy to check that the following sequence is exact:

$$0 \longrightarrow \mathrm{Ker}(f) \longrightarrow M \stackrel{f}{\longrightarrow} N \longrightarrow \mathrm{Coker}(f) \longrightarrow 0.$$

4.4.4 Lemma (Five Lemma). *Assume that the following is a commutative diagram of R-module homomorphisms with two exact rows.*

$$
\begin{array}{ccccccccc}
M_1 & \stackrel{d_1}{\longrightarrow} & M_2 & \stackrel{d_2}{\longrightarrow} & M_3 & \stackrel{d_3}{\longrightarrow} & M_4 & \stackrel{d_4}{\longrightarrow} & M_5 \\
\downarrow{\scriptstyle f_1} & & \downarrow{\scriptstyle f_2} & & \downarrow{\scriptstyle f_3} & & \downarrow{\scriptstyle f_4} & & \downarrow{\scriptstyle f_5} \\
M_1' & \underset{d_1'}{\longrightarrow} & M_2' & \underset{d_2'}{\longrightarrow} & M_3' & \underset{d_3'}{\longrightarrow} & M_4' & \underset{d_4'}{\longrightarrow} & M_5'
\end{array}
$$

1) f_3 *is injective if f_1 is surjective and f_2 and f_4 are injective.*

2) f_3 *is surjective if f_2 and f_4 are surjective and f_5 is injective.*

3) f_3 *is an isomorphism if f_1, f_2, f_4 and f_5 are isomorphisms.*

Proof. (1). Let $x_3 \in M_3$ such that $f_3(x_3) = 0$. Then $f_4 d_3(x_3) = d_3' f_3(x_3) = 0$. Since f_4 is injective, we have $d_3(x_3) = 0$; hence $x_3 \in \mathrm{Ker}(d_3) = \mathrm{Im}(d_2)$. So we have an $x_2 \in M_2$ such that $x_3 = d_2(x_2)$. Then $d_2' f_2(x_2) = f_3 d_2(x_2) = f_3(x_3) = 0$; thus $f_2(x_2) \in \mathrm{Ker}(d_2') = \mathrm{Im}(d_1')$; and we have an $x_1' \in M_1'$ such that $d_1'(x_1') = f_2(x_2)$. Since f_1 is surjective, there is an $x_1 \in M_1$ such that $f_1(x_1) = x_1'$; hence $f_2 d_1(x_1) = d_1' f_1(x_1) = d_1'(x_1') = f_2(x_2)$. But, f_2 is injective, so $d_1(x_1) = x_2$. Thus, we get $x_3 = d_2(x_2) = d_2 d_1(x_1) = 0$.

The argument like the above is referred to as *"diagram chasing"*.

(2). Proved by a diagram chasing similar to the above.

(3). An immediate consequence of (1) and (2).   □

4.4.5 Lemma. *Assume that the following two rows are exact sequences of R-homomorphisms:*

$$
\begin{array}{ccccccc}
M' & \stackrel{f}{\longrightarrow} & M & \stackrel{g}{\longrightarrow} & M'' & \longrightarrow & 0 \\
\downarrow{\scriptstyle \alpha} & & \downarrow{\scriptstyle \beta} & & \downarrow{\scriptstyle \gamma} & & \\
0 & \longrightarrow & N' & \underset{h}{\longrightarrow} & N & \underset{k}{\longrightarrow} & N''
\end{array}
$$

1) *If there are R-homomorphisms α and β such that the left square is commutative, then there is a unique R-homomorphism γ such that the diagram is commutative (it is not necessary that h is injective here).*

2) *If there are R-homomorphisms* β *and* γ *such that the right square is commutative, then there is a unique R-homomorphism* α *such that the diagram is commutative (here* g *is not necessarily surjective).*

3) (Snake Lemma) *If R-homomorphisms* α, β, γ *make the diagram commutative, then there is an R-homomorphism* ∂ : $\mathrm{Ker}(\gamma) \to$ $\mathrm{Coker}(\alpha)$ *(referred to as the* connecting homomorphism*) such that*

$$\mathrm{Ker}(\alpha) \xrightarrow{f} \mathrm{Ker}(\beta) \xrightarrow{g} \mathrm{Ker}(\gamma) \xrightarrow{\partial} \mathrm{Coker}(\alpha) \xrightarrow{\overline{h}} \mathrm{Coker}(\beta) \xrightarrow{\overline{k}} \mathrm{Coker}(\gamma)$$

is an exact sequence of R-homomorphisms, where f *and* g *are restriction maps and* $\overline{h}$ *and* $\overline{k}$ *are induced maps.*

Proof. (1). γ is constructed by a diagram chasing as follows and the uniqueness can be seen from the construction. Let $x'' \in M''$, select an $x \in M$ such that $g(x) = x''$; if another $x_1 \in M$ with $g(x_1) = x''$, then there is an $x' \in M'$ such that $x_1 = x + f(x')$, hence

$$k\beta(x_1) = k\beta(x + f(x')) = k\beta(x) + k\beta f(x') = k\beta(x) + kh\alpha(x') = k\beta(x);$$

thus it is well-defined that $\gamma(x'') = k\beta(x)$ with $x \in g^{-1}(x'')$. Obviously, γ is an R-homomorphism and the diagram is commutative.

(2) is proved similarly.

(3). By (2), the first two restriction maps make sense, i.e.

$$f(\mathrm{Ker}(\alpha)) \subset \mathrm{Ker}(\beta), \qquad g(\mathrm{Ker}(\beta)) \subset \mathrm{Ker}(\gamma).$$

Similarly, by (1), we can obtain the $\overline{h}$ and $\overline{k}$. The map ∂ is constructed as follows: for $x'' \in \mathrm{Ker}(\gamma)$ there is an $x \in M$ such that $f(x) = x''$, so $k\beta(x) = 0$; hence there is a $y' \in N'$ such that $h(y') = \beta(x)$; then define $\partial(x'') = y' + \mathrm{Im}(\alpha) \in \mathrm{Coker}(\alpha)$. Then one can show that such ∂ is a well-defined R-homomorphism and the sequence is exact. The details are left as Exer.2. □

4.4.6 Definition. A surjective R-homomorphism $g : M \to M''$ is said to be *split* if there is an R-homomorphism $s : M'' \to M$ such that $gs = \mathrm{id}_{M''}$; in this case, s is called a *splitting* for g.

An injective R-homomorphism $f : M' \to M$ is said to be *split* if there is an R-homomorphism $t : M \to M'$ such that $tf = \mathrm{id}_{M'}$; in this case, t is called a *splitting* for f.

The condition for split homomorphisms of modules can be weakened; see Exer.3.

4.4.7 Proposition.　Let $0 \longrightarrow M' \xrightarrow{f} M \xrightarrow{g} M'' \longrightarrow 0$ be an exact sequence of R-homomorphisms. Then the following are equivalent:

i) g is split;

ii) f is split;

iii) There is a direct decomposition $M = \operatorname{Im}(f) \oplus N$, hence $N \cong M''$ (of course $\operatorname{Im}(f) \cong M'$).

iv) There is a commutative diagram of R-homomorphisms:

$$
\begin{array}{ccccccccc}
0 & \longrightarrow & M' & \xrightarrow{\iota'} & M' \oplus M'' & \xrightarrow{\rho''} & M'' & \longrightarrow & 0 \\
 & & \alpha \downarrow & & \beta \downarrow & & \downarrow \gamma & & \\
0 & \longrightarrow & M' & \xrightarrow[f]{} & M & \xrightarrow[g]{} & M'' & \longrightarrow & 0
\end{array}
$$

where ι' and ρ'' are canonical injection and canonical projection resp. and α, β and γ are isomorphisms.

Proof. (i) $\implies$ (iii). Let $s : M'' \to M$ be a splitting for g. For $x \in M$, we have $x = sg(x) + (x - sg(x))$, and $g(x - sg(x)) = g(x) - gsg(x) = g(x) - g(x) = 0$; that is, $x - sg(x) \in \operatorname{Ker}(g) = \operatorname{Im}(f)$; hence $M = \operatorname{Im}(f) + \operatorname{Im}(s)$. If $x \in \operatorname{Im}(f) \cap \operatorname{Im}(s)$, then $x \in \operatorname{Ker}(g)$ and $x = s(y)$ for a $y \in M''$, hence $x = s(y) = sgs(y) = sg(x) = 0$. Thus $M = \operatorname{Im}(f) \oplus \operatorname{Im}(s)$.

(iii) $\implies$ (iv). Let $\varphi : M'' \to N$ be an isomorphism; and let

$$\tau : M' \oplus M'' \longrightarrow M, \quad (x', x'') \longmapsto f(x') + \varphi(x'').$$

Then τ is an isomorphism by Prop.4.3.7; and the following left square is clearly commutative.

$$
\begin{array}{ccccccccc}
0 & \longrightarrow & M' & \xrightarrow{\iota'} & M' \oplus M'' & \xrightarrow{\rho''} & M'' & \longrightarrow & 0 \\
 & & \operatorname{id}_{M'} \downarrow & & \tau \downarrow & & \downarrow \gamma & & \\
0 & \longrightarrow & M' & \xrightarrow[f]{} & M & \xrightarrow[g]{} & M'' & \longrightarrow & 0
\end{array}
$$

Hence, by Lemma 4.4.5, there is an R-homomorphism γ which makes the diagram commutative; and by Five Lemma, γ is also an isomorphism.

(iv) $\implies$ (i). Set $h = \beta \iota'' \gamma^{-1}$ where $\iota'' : M'' \to M' \oplus M''$ is the canonical injection; then, note that $g\beta = \gamma \rho''$ and $\rho'' \iota'' = \operatorname{id}_{M''}$, we have

$$gh = g\beta \iota'' \gamma^{-1} = \gamma \rho'' \iota'' \gamma^{-1} = \gamma \gamma^{-1} = \operatorname{id}_{M''}.$$

(ii) $\implies$ (iii) and (iv) $\implies$ (ii) can be proved similarly.　　□

Remark. From the above proof we can see that, if s is a splitting for g and t is a splitting for f, then $0 \longleftarrow M' \overset{t}{\longleftarrow} M \overset{s}{\longleftarrow} M'' \longleftarrow 0$ is also an exact sequence.

4.4.8 Definition. The short exact sequences satisfying the conditions of Prop. 4.4.7 are said to be *split*.

Exercises 4.4

1. Prove Prop.4.4.2.
2. Complete the proof of the Snake Lemma 4.4.5(3).
3. 1) A surjective R-homomorphism $g : M \to M''$ is split if and only if there is an R-homomorphism $s : M'' \to M$ such that gs is an isomorphism of M''. (Hint: if gs is an isomorphism; set $s' = s(gs)^{-1}$, then $gs' = \mathrm{id}_{M''}$.)
   2) An injective R-homomorphism $f : M' \to M$ is split if and only if there is an R-homomorphism $t : M \to M'$ such that tf is an isomorphism of M'.
4. In the following diagrams of R-homomorphisms, if α and β are isomorphisms, then the left square is commutative if and only if the right square is commutative.

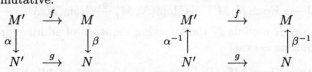

5. Assume the following is a commutative diagram of R-homomorphisms

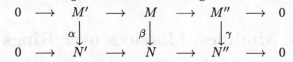

and α, β, γ all are isomorphisms. Then the top row is (split) exact if and only if the bottom row is (split) exact.
6. Give an example to show that a short exact sequence $0 \to M' \to M \to M'' \to 0$ such that $M \cong M' \oplus M''$ is not necessarily split.

   (Hint: let $T = \oplus_{i=1}^{\infty} \mathbf{Z}_2$, then $0 \to \mathbf{Z} \overset{f}{\to} \mathbf{Z} \oplus T \to T \to 0$ is an exact but not split sequence of $\mathbf{Z}$-modules, where $f(1) = (2, 0)$.)
7. Let $0 \to M' \overset{f}{\to} M \overset{g}{\to} M'' \to 0$ be an exact sequence of R-homomorphisms. The following are equivalent:
   i) The sequence is split.
   ii) For any R-homomorphism $h : M' \to N$ there is an R-homomorphism $\widetilde{h} : M \to N$ such that $h = \widetilde{h}f$.

iii) For any R-homomorphism $h : N \to M''$ there is an R-homomorphism $\overline{h} : N \to M$ such that $h = g\overline{h}$.

$+$8. Notations as in Exer.4.2.13.

1) If $0 \to N' \xrightarrow{f} N \xrightarrow{g} N''$ is an exact sequence of R-homomorphisms and M is an R-module, then the following sequence of additive group homomorphisms is exact:

$$0 \longrightarrow \mathrm{Hom}(M, N') \xrightarrow{f_*} \mathrm{Hom}(M, N) \xrightarrow{g_*} \mathrm{Hom}(M, N'').$$

2) If $M' \xrightarrow{f} M \xrightarrow{g} M'' \to 0$ is an exact sequence of R-homomorphisms and N is an R-module, then the following sequence of additive group homomorphisms is exact:

$$0 \longrightarrow \mathrm{Hom}(M'', N) \xrightarrow{g^*} \mathrm{Hom}(M, N) \xrightarrow{f^*} \mathrm{Hom}(M'', N).$$

9. Let $0 \to M' \xrightarrow{f} M \xrightarrow{g} M'' \to 0$ be an exact sequence of R-homomorphisms. The following are equivalent:

i) The exact sequence is split.

ii) For any R-module N the following sequence of additive group homomorphisms is exact:

$$0 \longrightarrow \mathrm{Hom}(N, M') \xrightarrow{f_*} \mathrm{Hom}(N, M) \xrightarrow{g_*} \mathrm{Hom}(N, M'') \longrightarrow 0.$$

iii) For any R-module N the following sequence of additive group homomorphisms is exact :

$$0 \longrightarrow \mathrm{Hom}(M'', N) \xrightarrow{g^*} \mathrm{Hom}(M, N) \xrightarrow{f^*} \mathrm{Hom}(M', N) \longrightarrow 0.$$

4.5   Free Modules, Matrices over Rings

Though we introduce modules by observing vector spaces, modules are different from vector spaces. The kind of modules which are like vector spaces is the so-called free modules. Let us begin with the linear relations.

4.5.1 Definition. Let R be a ring and M be an R-module. The expression $r_1x_1 + \cdots + r_nx_n$ for $r_i \in R$ and $x_i \in M$ is said to be an R-*linear combination* (or, R-*combination* for short) of $x_1, \cdots, x_n$. We say that $x_1, \cdots, x_n$ are *linearly independent* over R (or R-*linearly independent*, or R-*independent* for short) if

$$(\forall r_1, \cdots, r_n \in R)\Big(\big(\ r_1x_1 + \cdots + r_nx_n = 0\ \big) \implies \big(\ r_1 = \cdots = r_n = 0\ \big)\Big);$$

otherwise, we say that $x_1, \cdots, x_n$ are *linearly dependent* over R (or, *R-dependent* for short). An infinite subset X of M is said to be R-linearly independent if every finite subset of X is R-linearly independent. A subset X of M is said to be an *R-basis* of M if X is R-linearly independent and $M = \langle X \rangle$, i.e. X generates M. An R-module F is called a *free R-module* if it has an R-basis.

4.5.2 Remark. 1) It is clear that the above condition for elements $x_1, \cdots, x_n$ being R-independent is equivalent to the following:

$$(r_1 x_1 + \cdots + r_n x_n = r'_1 x_1 + \cdots + r'_n x_n) \implies (r_i = r'_i \text{ for } i = 1, \cdots, n).$$

2) If F is a free R-module with an R-basis B, then, by the above (1), every $f \in F$ is uniquely written as a sum $\sum_{b \in B} f(b)b$, where $f(b) \in R$ is zero for almost $b \in B$ hence the sum is in fact finite; in other words, every $f \in F$ is in fact a function $f : B \to R$ whose values $f(b)$ are almost zero; cf. (4.3.5). Conversely, if B is a set; let $F = \{f : B \to R \mid f(b) = 0 \text{ for almost } b \in B\}$; define an addition on F as $(f_1 + f_2)(b) = f_1(b) + f_2(b)$, $\forall b \in B$; and a scalar multiplication as $(rf)(b) = r \cdot f(b)$, $\forall b \in B$. Then it is clear that F is an R-module. For every $b \in B$ we have a $b' \in F$ such that $b'(c) = 1$ if $c = b$, and 0 otherwise. Let $B' = \{b' \mid b \in B\} \subset F$; then $B \to B', b \mapsto b'$ is a bijection; and it is easy to check that every $f \in F$ is uniquely written as $f = \sum_{b \in B} f(b)b'$; that is, F is a free R-module with R-basis B'. The following Example 3.5.3 provides another way to construct a free module with a given basis.

Many basic facts on the linear relation for vector spaces are false for general modules; Exer.1 exhibits some examples, where we see that **Q** is not a free **Z**-module and a **Z**-module with an element of finite order is not a free **Z**-module; but, the regular **Z**-module is a free **Z**-module. In fact, these are two general facts; see Exer.2 and the following example.

4.5.3 Example. For any ring R the regular module $_R R$ is a free module which has a basis $\{1\}$. This is clear because it is trivial that $r1 = 0 \implies r = 0$ and $R = R \cdot 1$. Moreover, for any cardinality n, we can construct a free module which has a basis of cardinality n as follows. By $R^{(n)}$ we denote the direct sum of n copies of the regular module $_R R$. Let $b_i = (0, \cdots, 0, 1, 0, \cdots, 0)$ with 1 in the ith-component and 0 elsewhere. Then it is easy to check that $b_1, \cdots, b_n$ form an R-basis of $R^{(n)}$, hence $R^{(n)}$ is a free R-module. Obviously, this is still true even if n is infinite.

A well-known fact for vector spaces is that any map defined on a basis determines a unique linear map.

4.5.4 Theorem. *Let F be a module over a ring R and $\emptyset \neq B \subset F$. The following are equivalent:*

i) *F is a free module with an R-basis B.*

ii) *For any R-module M and any map $\xi : B \to M$, there is a unique R-homomorphism $\widetilde{\xi} : F \to M$ such that $\widetilde{\xi}|_B = \xi$.*

Proof. (i) $\Longrightarrow$ (ii). Since every $f \in F$ is uniquely written as a finite sum $\sum_{b \in B} f(b)b$, we have a well-defined map

$$(4.5.5) \qquad \widetilde{\xi} : \quad F \longrightarrow M, \quad \sum_{b \in B} f(b)b \longmapsto \sum_{b \in B} f(b)\xi(b).$$

It is easy to check that $\widetilde{\xi}$ is an R-homomorphism and $\widetilde{\xi}|_B = \xi$. If $\varphi : F \to M$ is an R-homomorphism such that $\varphi|_B = \xi$, then

$$\varphi\Big(\sum_{b \in B} f(b)b \Big) = \sum_{b \in B} f(b)\varphi(b) = \sum_{b \in B} f(b)\xi(b) = \widetilde{\xi}\Big(\sum_{b \in B} f(b)b \Big).$$

(ii) $\Longrightarrow$ (i). Let B' be a set such that $B \xrightarrow{\xi} B'$ is a bijection, let F' be a free R-module with R-basis B'; cf. Remark 4.5.2. Then we have an R-homomorphism $\widetilde{\xi} : F \to F'$ such that $\widetilde{\xi}|_B = \xi$. So the R-independence of B in F follows from the R-independence of B' in F'. Suppose $\langle B \rangle \lneqq F$, then $F/\langle B \rangle \neq 0$, hence both the natural homomorphism and the zero homomorphism from F to $F/\langle B \rangle$ send B to $\{0\}$ but they are distinct homomorphisms, this contradicts condition (ii). $\square$

From the theorem and the proof, we have the following three corollaries; the details of their proofs are left as Exer.3.

4.5.6 Corollary. *Notations as in Theorem 4.5.4. Then the unique R-homomorphism in 4.5.4(ii) must be $\widetilde{\xi} : F \to M$ in (4.5.5), and the following hold:*

1) *$\widetilde{\xi}$ is injective if and only if ξ is injective and the set $\xi(B)$ is R-independent.*

2) *$\widetilde{\xi}$ is surjective if and only if $\langle \xi(B) \rangle = M$.*

3) *$\widetilde{\xi}$ is an isomorphism if and only if ξ is injective and the set $\xi(B)$ is a basis of M (hence M is free).* $\square$

4.5.7 Corollary. *Any R-module generated by a set of cardinality n (n may be infinite) is a homomorphism image of any free R-module with a basis of cadinality $\geq n$.*

Sketch. *Define a surjective map from the basis of the free module onto the generator set of the module.* □

4.5.8 Corollary. *Two free modules are isomorphic if and only if they have basis of the same cardinality. In particular, any free R-module F with a basis of cardinality n (n may be infinite) is isomorphic to $R^{(n)}$.*

Sketch. It is interesting to show an expilicit isomorphism from F onto $R^{(n)}$ for finite cardinality n (this is not a crucial assumption, just for convenience): let $\{b_1, \cdots, b_n\}$ be a basis of F; any element of F is uniquely expressed as $f = \sum_{i=1}^{n} f_i b_i$ with $f_i \in R$; then the following

$$F \longrightarrow R^{(n)}, \quad f \longmapsto (f_1, \cdots, f_n)$$

is an R-isomorphism.

4.5.9 Remark. For an n-dimensional vector space V and a basis B, it is well-known that any linear transformation of V corresponds to an $n \times n$ matrix, and *vice versa*; the addition and the multiplication resp. of transformations correspond to the addition and the multiplication resp. of matrices. It is the same for free modules with finite basis. But note that a ring R is not necessarily commutative. So, for *left* free R-modules we write R-homomorphisms on the *right* for convenience, and obtain results (4.5.10) — (4.5.15); while for *right* free R-modules, we write R-homomorphisms on the *left*, and state the dual results but leave the details as Exer.11.

Let F and E be free left R-modules with basis $B = \{b_1, \cdots, b_n\}$ and basis $A = \{a_1, \cdots, a_m\}$ resp. Let $\operatorname{Hom}_R(F, E) = \{R\text{-homomorphisms}$ from F to $E\}$, which is an additive group, see Remark 4.2.13. For $\alpha \in \operatorname{Hom}_R(F, E)$, since $(b_i)\alpha \in E$, there are $\alpha_{ij} \in R$ such that

$$(4.5.10) \qquad (b_i)\alpha = \sum_{j=1}^{m} \alpha_{ij} a_j \qquad i = 1, \cdots, n;$$

so a unique $n \times m$ R-matrix $(\alpha_{ij})_{n \times m}$ is determined. By Theorem 4.5.4, α is determined by the equalities (4.5.10), that is, determined by the matrix $(\alpha_{ij})_{n \times m}$ as follows: for any $\sum_{i=1}^{n} r_i b_i \in F$,

(4.5.11)

$$\Big(\sum_{i=1}^{n} r_i b_i \Big) \alpha = \sum_{i=1}^{n} r_i \cdot (b_i) \alpha = \sum_{i=1}^{n} r_i \sum_{j=1}^{m} \alpha_{ij} a_j = \sum_{j=1}^{m} \Big(\sum_{i=1}^{n} r_i \alpha_{ij} \Big) a_j .$$

Conversely, an $n \times m$ R-matrix $(\alpha_{ij})_{n \times m}$ determines, with Formula (4.5.10) and (4.5.11), a unique R-homomorphism $\alpha \in \mathrm{Hom}_R(F, E)$. Further, if $\beta \in \mathrm{Hom}_R(F, E)$ corresponds to R-matrix $(\beta_{ij})_{n \times m}$, then a calculation shows that $\alpha + \beta$ corresponds to the usual *matrix sum*:

(4.5.12)

$$(\alpha_{ij})_{n \times m} + (\beta_{ij})_{n \times m} = (\alpha_{ij} + \beta_{ij})_{n \times m}, \quad \text{with } (i, j)\text{-entries } \alpha_{ij} + \beta_{ij}.$$

By $M_{n \times m}(R)$ we denote the set of all $n \times m$ R-matrices; with the above addition, it is easy to check that $M_{n \times m}(R)$ is an additive group, see Exer.6. Thus we get:

4.5.13 Proposition.   *For left free R-modules F and E with basis $B = \{b_1, \cdots, b_n\}$ and basis $A = \{a_1, \cdots, a_m\}$ resp., the map $\alpha \mapsto (\alpha_{ij})_{n \times m}$ determined by (4.5.10) is an additive group isomorphism from $\mathrm{Hom}_R(F, E)$ onto $M_{n \times m}(R)$.*   □

Moreover, if P is another free left R-module with a basis $C = \{c_1 \cdots, c_k\}$ and $\beta : E \to P$ is an R-homomorphism corresponding to the R-matrix $(\beta_{ij})_{m \times k}$, then, with (4.5.10) and (4.5.11) it is easy to calculate that the composition homomorphism $\alpha\beta : F \to P$ (remember that here homomorphisms are written on the right) corresponds to the usual *matrix product*:

(4.5.14) $(\alpha_{ij})_{n \times m} \cdot (\beta_{ij})_{m \times k} = (\gamma_{ij})_{n \times k}, \quad \text{where } \gamma_{ij} = \sum_{t=1}^{m} \alpha_{it} \beta_{tj}.$

Thus, considering the right endomorphism ring $\mathrm{End}_R^r(F)$, the corresponding $M_n(R) = M_{n \times n}(R)$ is not only an additive group, and, under the matrix multiplication (4.5.14) and the *identity matrix* I with 1 in (i, i)-entries and 0 elsewhere, it is a ring as well (Exer.6), called the *R-matrix ring*.

4.5.15 Theorem.   *Let R be a ring and F be a left free R-module with an R-basis $B = \{b_1, \cdots, b_n\}$. Then the map $\alpha \mapsto (\alpha_{ij})$ determined by (4.5.10) is a ring isomorphism from $\mathrm{End}_R^r(F)$ onto $M_n(R)$.*   □

4.5.16 Remark. Extending the matrix multiplication (4.5.14) to scalar multiplications, the formula (4.5.10) and (4.5.11) can be rewritten as matrix version:

$$(4.5.10\mathrm{m}) \quad \begin{pmatrix} b_1 \\ b_2 \\ \vdots \\ b_n \end{pmatrix} \alpha = \begin{pmatrix} (b_1)\alpha \\ (b_2)\alpha \\ \vdots \\ (b_n)\alpha \end{pmatrix} = \begin{pmatrix} \alpha_{11} & \alpha_{12} & \cdots & \alpha_{1m} \\ \alpha_{21} & \alpha_{22} & \cdots & \alpha_{2m} \\ \cdots & \cdots & \cdots & \cdots \\ \alpha_{n1} & \alpha_{n2} & \cdots & \alpha_{nm} \end{pmatrix} \begin{pmatrix} a_1 \\ a_2 \\ \vdots \\ a_m \end{pmatrix};$$

(4.5.11m)

$$(r_1\, r_2\, \cdots\, r_n) \begin{pmatrix} b_1 \\ b_2 \\ \vdots \\ b_n \end{pmatrix} \alpha = (r_1\, r_2\, \cdots\, r_n) \begin{pmatrix} \alpha_{11} & \alpha_{12} & \cdots & \alpha_{1m} \\ \alpha_{21} & \alpha_{22} & \cdots & \alpha_{2m} \\ \cdots & \cdots & \cdots & \cdots \\ \alpha_{n1} & \alpha_{n2} & \cdots & \alpha_{nm} \end{pmatrix} \begin{pmatrix} a_1 \\ a_2 \\ \vdots \\ a_m \end{pmatrix}.$$

On the other hand, if F, E and P are *right* free R-modules, all the formulas are nearly the same except for some expressions. The formulas (4.5.10) and (4.5.11) become

$$(4.5.10^*) \qquad \alpha(b_j) = \sum_{i=1}^{m} a_i \alpha_{ij} \qquad j = 1, \cdots, n;$$

$$(4.5.11^*) \qquad \alpha\Big(\sum_{j=1}^{n} b_j r_j\Big) = \sum_{i=1}^{m} a_i \Big(\sum_{j=1}^{n} \alpha_{ij} r_j\Big).$$

Or, in matrix form, they appear as:

$$(4.5.10^*\mathrm{m}) \quad \alpha(b_1\, b_2\, \cdots\, b_n) = (a_1\, a_2\, \cdots\, a_m) \begin{pmatrix} \alpha_{11} & \alpha_{12} & \cdots & \alpha_{1n} \\ \alpha_{21} & \alpha_{22} & \cdots & \alpha_{2n} \\ \cdots & \cdots & \cdots & \cdots \\ \alpha_{m1} & \alpha_{m2} & \cdots & \alpha_{mn} \end{pmatrix};$$

(4.5.11*m)

$$\alpha(b_1\, b_2\, \cdots\, b_n) \begin{pmatrix} r_1 \\ r_2 \\ \vdots \\ r_n \end{pmatrix} = (a_1\, a_2\, \cdots\, a_m) \begin{pmatrix} \alpha_{11} & \alpha_{12} & \cdots & \alpha_{1n} \\ \alpha_{21} & \alpha_{22} & \cdots & \alpha_{2n} \\ \cdots & \cdots & \cdots & \cdots \\ \alpha_{m1} & \alpha_{m2} & \cdots & \alpha_{mn} \end{pmatrix} \begin{pmatrix} r_1 \\ r_2 \\ \vdots \\ r_n \end{pmatrix}.$$

Formulas (4.5.12) and (4.5.14) and Prop.4.5.13 are the same for right free modules provided we remember that the homomorphisms are written on the left; and Theorem 4.5.15 is stated for the left endomorphism ring:

4.5.15* Theorem.   *Let R be a ring and F be a right free R-module with an R-basis $B = \{b_1, \cdots, b_n\}$. Then the map $\alpha \mapsto (\alpha_{ij})$ determined by ($4.5.10^*$) is a ring isomorphism from $\text{End}^l_R(F)$ onto $M_n(R)$.*   □

Unfortunately, however, some basic concepts cannot be introduced for general free modules; for example, "dimensions", and "determinants". It is not very difficult to show that

4.5.16 Proposition.   *If a free module has a basis of infinite cardinality, then its any two bases have the same cardinality.*

Proof. See Exer.15.   □

However, there are rings R such that a free R-module can have two bases of different cardinalities; see Exer.14.

4.5.17 Definition. A ring R is said to be of *invariant basis number* (be an *IBN ring*, for short) if any two bases of every free R-module have the same cardinality; moreover, if it is the case, the same cardinality of a free R-module is called the *rank* (or *R-rank* precisely) of the free module. And by rank(F) (or $\text{rank}_R(F)$ precisely) we denote the rank of a free R-module F.

In the next two sections we discuss free modules over two types of IBN rings: division rings, and commutative rings; for them, more fundamental facts of linear algebra are extended.

Exercises 4.5

+1. Consider **Z**-modules, i.e. additive groups.
   1) An element x of a **Z**-module M is **Z**-linearly independent if and only if the order of x is infinite. Show a **Z**-module M and an $0 \neq x \in M$ such that $\{x\}$ is a **Z**-linearly dependent set.
   2) If a **Z**-module M has a basis, then every element of M has infinite order. Thus a **Z**-module with some elements of finite order has no basis.
   3) Prove that the regular **Z**-module (i.e. the additive group of integers) has a basis $\{1\}$. But $\{2\}$ is not a basis of **Z** though it is a **Z**-linearly independent set.
   4) Find all the bases of the regular **Z**-module.
   5) Consider the **Q** of all rationals as a **Z**-module, prove that any non-zero element forms a **Z**-independent set, but any two distinct elements form a **Z**-dependent set; further, however, show that **Q** has no basis.

2. For any module M over a ring R and $x \in M$, the subset $\operatorname{Ann}_R(x) = \{r \in R \mid rx = 0\}$ is a left ideal of R, which is called the *order ideal* of x; and x is said to be *torsion-free* if $\operatorname{Ann}_R(x) = 0$. Further, $\operatorname{Ann}_R(M) = \bigcap_{x \in M} \operatorname{Ann}_R(x)$ is an ideal of R, called the *order ideal* of M (these are generalizations of the corresponding concept for groups).

   Prove: a non-zero element of a free module is torsion-free.

3. Complete the details of the proofs of Cor.4.5.6—4.5.8.

$^+$4. (*Projectivity of free modules*) Let F be a free module over a ring R. Prove:

   1) If M and N are R-modules and $h : M \to N$ is a surjective R-homomorphism and $f : F \to N$ is an R-homomorphism, then there is an R-homomorphism $g : F \to M$ such that $h \circ g = f$.

   2) Any short exact sequence $0 \to M' \to M \to F \to 0$ which is ended by F is split.

   3) For any short exact sequence $0 \to M' \xrightarrow{f} M \xrightarrow{g} M'' \to 0$ the following sequence of additive homomorphisms is exact:

   $$0 \longrightarrow \operatorname{Hom}(F, M') \xrightarrow{f_*} \operatorname{Hom}(F, M) \xrightarrow{g_*} \operatorname{Hom}(F, M'') \longrightarrow 0$$

   (Remark: this is a special case of a general kind of modules, see 4.10.1.)

5. Prove: a direct sum of some free modules is a free module.

6. Verify directly that

   1) $M_{n \times m}(R)$ is an additive group.

   2) $M_n(R)$ is a ring.

7. If R is a ring, and R' be a set equipped with two operations denoted by "+" and "·". If there is a bijection $f : R \to R'$ such that $f(r+s) = f(r) + f(s)$ and $f(r \cdot s) = f(r) \cdot f(s)$ for all $r, s \in R$; then R' with the two operations becomes a ring and f is a ring isomorphism.

   A similar result holds for groups.

   Apply the above fact to give another proof of Exer.6.

8. Prove: a ring which has an IBN quotient ring is also an IBN ring.

9. Let R be an IBN ring. Prove that two free R-modules are isomorphic if and only if their ranks are equal.

10. Let R be an IBN ring and F_1 and F_2 be two free R-modules. Prove that $\operatorname{rank}_R(F_1 \oplus F_2) = \operatorname{rank}_R(F_1) + \operatorname{rank}_R(F_2)$.

11. Complete the details of results (4.5.10*) — (4.5.15*).

$^+$*12. 1) For any R-independent subset Y of a module M over a ring R, there is a maximal R-independent subset X of M such that $Y \subset X$.

   (Hint: if X_i, $i \in I$, are independent subsets such that $Y \subset X_i$ and for any $i, j \in I$ either $X_i \subset X_j$ or $X_j \subset X_i$, then $\bigcup_{i \in I} X_i$ is also an independent subset; use Zorn's Lemma.)

2) If M is a free R-module and B is a basis, then B is a maximal R-independent subset. Show that a maximal R-independent subset of M is not necessarily a basis even if M is free.

13. Let R be a ring and V be a free R-module with a basis B, let $E = \text{End}^r_R(V)$ be the right R-endomorphism ring of V. Let U be a free R-module, and $\text{Hom}_R(V, U)$ be the additive group defined in Remark 4.2.13.

1) $\text{Hom}_R(V, U)$ is a left E-module, the scalar multiplication is:

$$E \times \text{Hom}_R(V, U) \longrightarrow \text{Hom}_R(V, U), \quad (\alpha, f) \mapsto \alpha \circ f.$$

2) If U has a basis X such that $|X| = |B|$, then the E-module $\text{Hom}_R(V, U)$ is isomorphic to the left regular E-module.

(Hint: a bijection $\varphi : X \to B$ induces an R-isomorphism $\widetilde{\varphi} : U \to V$, hence we have an E-module isomorphism:

$$\widetilde{\varphi}_* : \ \text{Hom}_R(V, U) \longrightarrow \text{Hom}_R(V, V), \quad f \longmapsto f \circ \widetilde{\varphi}.)$$

$^{+*}$14. Notations as above. Assume $B = \{b_1, b_2, b_3, \cdots\}$ is countably infinite.

1) Let $V_1 = \langle b_1, b_3, \cdots \rangle$ be the submodule of V generated by the b_i's with i odd, and $V_2 = \langle b_2, b_4, \cdots \rangle$ be the submodule of V generated by the b_i's with i even. Then both V_1 and V_2 are free and $V = V_1 \oplus V_2$.

2) Let $E_i = \{\alpha \in E \mid (V)\alpha \subset V_i\}$, $i = 1, 2$; then both E_1 and E_2 are left ideals of E and $E = E_1 \oplus E_2$.

3) As left E-modules, both E_1 and E_2 are isomorphic to the left regular module. Thus, let $1 = e_1 + e_2$ with $e_i \in E_i$, then both $\{1\}$ and $\{e_1, e_2\}$ are E-basis of the left regular module $_EE$.

$^{+*}$15. Let F be a free module over a ring R, and both B and B' be R-basis of F. If one of B and B' is infinite, then $|B'| = |B|$.

(Hint: For $b \in B$ there is a finite subset $B'_b \subset B'$ such that $b = \sum_{b' \in B'_b} r_{b'} b'$ with $r_{b'} \in R$; thus $\langle \bigcup_{b \in B} B'_b \rangle \supset \langle B \rangle = F$, hence $B' = \bigcup_{b \in B} B'_b$. So, if B is finite, then so is B'. Otherwise, $|B'| = |\bigcup_{b \in B} B'_b| \leq \sum_{b \in B} |B'_b| \leq |B| \cdot \aleph_0 = |B|$ because $|B| = \infty$, where $\aleph_0$ is the cardinality of $\mathbf{Z}$.)

4.6   Matrices over Division Rings

What about the rings R whose non-zero modules are all free?

If R is such a ring, then for any $t \in R$ the quotient module R/Rt is a free module; for $1 + Rt \in R/Rt$ it is clear that $Rt \subset \text{Ann}_R(1 + Rt)$, thus, by Exer.4.5.2, either $Rt = 0$ or $1 + Rt = 0$ in R/Rt; the latter implies $1 + Rt = Rt$, hence $1 \in Rt$, i.e. $Rt = R$; thus, by Exer.3.3.6, R is a division ring.

4.6.1 Proposition. *Let D be a ring. All non-zero D-modules are free if and only if D is a division ring.*

Proof. The necessity is proved as above, and the sufficiency follows from the following Lemma 4.6.3. □

For convenience, in the rest of this section we always assume (but some discussions in the following are general):

4.6.2 Assumption. D is a division ring (hence all D-modules are free) and V is a D-module.

4.6.3 Lemma. 1). *A subset B of V is a D-basis if and only if B is a maximal D-independent subset.*

2). *Any D-independent subset X of V can be extended to a D-basis; in particular, the D-basis of V exists.*

Proof. (1). The necessity follows from Exer.4.5.12(2). For the sufficiency, it is enough to show that $V = \langle B \rangle$. For any $v \in V$, if $v \in B$, then $v \in \langle B \rangle$. Otherwise, $B \cup \{v\}$ is D-dependent, and $v \in \langle B \rangle$ by Lemma 4.6.4(1) below.

(2). By Exer.4.5.12(1) any independent subset is contained in a maximal independent subset. □

4.6.4 Lemma. *Let X be a D-independent subset of V and $0 \neq y \in V - X$.*

1). *$X \cup \{y\}$ is dependent if and only if $y \in \langle X \rangle$.*

2). *If $y = d_1 x_1 + \cdots + d_n x_n \in \langle X \rangle$ with $x_i \in X$ and $d_i \in D$, then for any $d_j \neq 0$ we have that $X \cup \{y\} - \{x_j\}$ is independent and $\langle X \cup \{y\} - \{x_j\} \rangle = \langle X \rangle$.*

Proof. (1). If $y \in \langle X \rangle$, then $y = d_1 x_1 + \cdots + d_n x_n$ for some $x_1, \cdots, x_n \in X$ and $d_1, \cdots, d_n \in D$; thus $X \cup \{y\}$ is D-dependent. Next, assume that $X \cup \{y\}$ is dependent. Then $d_0 x_0 + d_1 x_1 + \cdots + d_n x_n = 0$ for some $x_0, x_1, \cdots, x_n \in X \cup \{y\}$ and non-zero $d_0, d_1, \cdots, d_n \in D$. Suppose $y \neq x_i$ for $i = 0, 1, \cdots, n$, then X is dependent, this is not the case. Hence for some i, say $i = 0$, we have that $x_0 = y$. Further, since $d_0 \neq 0$, $y = d_0^{-1} d_1 x_1 + \cdots + d_0^{-1} d_n x_n$, i.e. $y \in \langle X \rangle$.

(2). Since $d_j \neq 0$, we have $x_j = d_j^{-1}(y - \sum_{i \neq j} d_i x_i) \in \langle X \cup \{y\} - \{x_j\}\rangle$; hence $\langle X \cup \{y\} - \{x_j\}\rangle = \langle X \rangle$. Suppose $X \cup \{y\} - \{x_j\}$ is dependent, then the argument as above shows that $y = d_1' x_1' + \cdots + d_{n'}' x_{n'}'$ for $x_1', \cdots, x_{n'}' \in X - \{x_j\}$ and $d_1', \cdots, d_{n'}' \in D$. Hence $d_1' x_1' + \cdots + d_{n'}' x_{n'}' = d_1 x_1 + \cdots + d_n x_n$. Since $d_j x_j$ with $d_j \neq 0$ appears on the right-hand side but does not appear on the left-hand side, this turns out to be an expression of 0 by a non-trivial finite D-linear combination of X, which contradicts the independence of X. □

4.6.5 Lemma. *If both X and Y are finite D-independent subsets of V and $|Y| > |X|$, then there is a $y \in Y - X$ such that $X \cup \{y\}$ is D-independent.*

Proof. For convenience we can assume $X = \{x_1, \cdots, x_n\}$ and $Y = \{y_1, \cdots, y_n, y_{n+1}\}$. Suppose the assertion is false, then, by Lemma 4.6.4(1), $y_i \in \langle X \rangle$, $i = 1, \cdots, n + 1$; and by the above lemma again, there is an x_i, say x_1, such that $X_1 = \{y_1, x_2, \cdots, x_n\}$ is independent and $\langle X_1 \rangle = \langle X \rangle$. Since $y_2 \in \langle X_1 \rangle$, we have $y_2 = d_1 y_1 + d_2 x_2 + \cdots + d_n x_n$; if $d_2 = \cdots = d_n = 0$, then Y is dependent, which is impossible. So one of $d_2, \cdots, d_n$ is non-zero, say $d_2 \neq 0$; by Lemma 4.6.4(2), $X_2 = \{y_1, y_2, x_3, \cdots, x_n\}$ is independent and $\langle X_2 \rangle = \langle X_1 \rangle = \langle X \rangle$. Continuing the argument n times, we arrive at $X_n = \{y_1, y_2, \cdots, y_n\}$ is independent and $\langle X_n \rangle = \langle X \rangle$. Therefore, $y_{n+1} \in \langle X_n \rangle$ which implies, by Lemma 4.6.4(1), that Y is dependent; a contradiction. □

4.6.6 Corollary. *A division ring D is an IBN ring.*

Proof. Let B and B' be bases of a D-module V. If one of B and B' is infinite, then $|B| = |B'|$ by Prop.4.5.16. Otherwise, both B and B' are finite; by Lemma 4.6.3, both B and B' are maximal independent subsets; hence, by Lemma 4.6.5, it must be the case that $|B| = |B'|$. □

4.6.7 Definition. Since the D-modules for a division ring D are similar to the vector spaces, we call any D-module a *D-vector space*, or a *D-space* for short; and instead of "*D-rank*", we call the cardinality of a basis of a D-space V the *D-dimension* of V, denoted by $\dim_D V$. Note that we have left D-spaces and right D-spaces; as we agreed on, D-spaces without attributes means left D-spaces. As in Assumption (4.6.2), in the following V is always a D-space.

Many properties of vector spaces can be extended to D-spaces.

4.6.8 Proposition. **1)** *If W is a subspace of V, then there is a subspace W' such that $V = W \oplus W'$; in particular,*

$$\dim_D V = \dim_D W + \dim_D W' = \dim_D W + \dim_D(V/W);$$

and $W = V$ if and only if $\dim_D W = \dim_D V$.

2) *Two D-spaces are isomorphic if and only if their dimensions are equal.*

3) *If U and W are subspaces of V, then*

$$\dim_D(U + W) + \dim_D(U \cap W) = \dim_D U + \dim_D W.$$

4) *If $\alpha : V \to V'$ is a D-homomorphism of D-spaces, then*

$$\dim_D V = \dim_D(\mathrm{Ker}(\alpha)) + \dim_D(\mathrm{Im}(\alpha)).$$

5) *For two D-spaces V and V' of the same finite dimension, a D-homomorphism $\alpha : V \to V'$ is an isomorphism if and only if α is injective, if and only if α is surjective.*

Proof. Exer.1. $\qquad\qquad\qquad\qquad\qquad\qquad\qquad\qquad\qquad\qquad$ □

Let $\{v_1, \cdots, v_n\}$ and $\{u_1, \cdots, u_m\}$ resp. be basis of D-spaces V and U resp. Let $\alpha : V \to U$ be a D-homomorphism. By 4.5.10, α corresponds to a D-matrix $(\alpha_{ij})_{n \times m}$ such that

$$(v_i)\alpha = \sum_{j=1}^m \alpha_{ij} u_j \qquad\qquad i = 1, \cdots, n.$$

We consider the linear relation of the $(v_i)\alpha$'s, and the linear relation of the rows of the matrix; where the rows $\alpha_i = (\alpha_{i1} \cdots \alpha_{im})$, $i = 1, \cdots, n$, of the matrix $(\alpha_{ij})_{n \times m}$ are regarded as vectors in the left D-space $D^{(m)}$. For any subset I of $\{1, \cdots, n\}$ and $d_i \in D$ for $i \in I$, we have

$$\sum_{i \in I} d_i((v_i)\alpha) = \sum_{i \in I} \sum_{j=1}^m d_i \alpha_{ij} u_j = \sum_{j=1}^m \left(\sum_{i \in I} d_i \alpha_{ij} \right) u_j.$$

Thus, $\sum_{i \in I} d_i((v_i)\alpha) = 0$ if and only if $\sum_{i \in I} d_i \alpha_{ij} = 0$ for $j = 1, \cdots, m$. The above argument shows:

$$\sum_{i \in I} d_i((v_i)\alpha) = 0 \qquad \Longleftrightarrow \qquad \sum_{i \in I} d_i \alpha_{ij} = 0.$$

This implies that the maximal independent subsets of $(v_1)\alpha, \cdots, (v_n)\alpha$ and the maximal independent subsets of the rows of the matrix $(\alpha_{ij})_{n \times m}$ have the same cardinality.

4.6.9 Definition. The dimension of $\text{Im}(\alpha)$ is referred to as the *rank* of α. The dimension of the subspace of the *left* D-space $D^{(n)}$ generated by the rows of an $n \times m$ D-matrix $(\alpha_{ij})_{n \times m}$ is referred to as the *row rank* of the matrix $(\alpha_{ij})_{n \times m}$; while the dimension of the subspace of the *right* D-space $D^{(m)}$ generated by the columns of $(\alpha_{ij})_{n \times m}$ is referred to as the *column rank* of the matrix $(\alpha_{ij})_{n \times m}$. In fact, we will prove that the row rank and the column rank of a D-matrix are equal to each other, see 4.6.16 or 4.6.21; so we call that number the *rank* of the matrix.

Thus, for the D-homomorphism $\alpha : V \to U$ of D-spaces, by Exer.2, we get that *"the rank of α equals the row rank of its matrix."* On the other hand, if V and U are *right* D-spaces, then we apply the formula (4.5.10*) to get the matrix $(\alpha_{ij})_{m \times n}$ such that

$$\alpha(v_j) = \sum_{i=1}^{m} u_i \alpha_{ij} \qquad j = 1, \cdots, n;$$

and a similar conclusion can be reached. In short, we have

4.6.10 Proposition. *If V and U are left (right resp.) D-spaces and $\alpha : V \to U$ is a D-homomorphism, then the rank of α is equal to the row (column resp.) rank of its corresponding D-matrix.* $\square$

Now, we show two ways to compare the two ranks of a D-matrix.

The first way involves the equivalence of matrices, which is a general concept though we introduce it for D-matrices in the following.

Let $\alpha : V \to U$ be a homomorphism of left D-spaces; let both $\{v_1, \cdots, v_n\}$ and $\{v_1', \cdots, v_n'\}$ be bases of V, while both $\{u_1, \cdots, u_m\}$ and $\{u_1', \cdots, u_m'\}$ be bases of U. By 4.5.4 and 4.5.5, the map $v_i \mapsto v_i'$ ($u_i \mapsto u_i'$ resp.) determines an automorphism of V (of U resp.); hence, by Theorem 4.5.15, there is a matrix $P = (p_{ij}) \in M_n(D)$ ($Q \in M_m(D)$ resp.) which is invertible, i.e. $P^{-1} \in M_n(D)$ exists ($Q^{-1} \in M_m(D)$ exists resp.), such that (cf 4.5.10m and 4.5.11m)

$$\begin{pmatrix} v_1' \\ \vdots \\ v_n' \end{pmatrix} = P \begin{pmatrix} v_1 \\ \vdots \\ v_n \end{pmatrix} \qquad \text{and} \qquad \begin{pmatrix} u_1' \\ \vdots \\ u_m' \end{pmatrix} = Q \begin{pmatrix} u_1 \\ \vdots \\ u_m \end{pmatrix}.$$

4.6.11 Definition and Proposition. *The above invertible $n \times n$ matrix P is said to be the* transforming matrix *from a basis $\{v_1, \cdots, v_n\}$*

of V to another basis $\{v_1', \cdots, v_n'\}$ of V. Conversely, if $\{v_1, \cdots, v_n\}$ is a basis of V and $P \in M_n(R)$ is invertible, then $\begin{pmatrix} v_1' \\ \vdots \\ v_n' \end{pmatrix} = P \begin{pmatrix} v_1 \\ \vdots \\ v_n \end{pmatrix}$ *is also a basis of V.*

Proof. For the "conversely", by Theorem 4.5.15, the map $v_i \mapsto v_i'$ determines an automorphism of V, hence $\{v_1', \cdots, v_n'\}$ is a basis of V by Cor.4.5.5(3). □

4.6.12 Proposition. *Notations as above. Let $\alpha : V \to U$ be a D-homomorphism. If A is the matrix of α under the bases $\{v_1, \cdots, v_n\}$ and $\{u_1, \cdots, u_m\}$, then PAQ^{-1} is the matrix of α under the bases $\{v_1', \cdots, v_n'\}$ and $\{u_1', \cdots, u_m'\}$.*

Proof. It is derived from the following calculation (cf. (4.5.10m)):

$$\begin{pmatrix} v_1' \\ \vdots \\ v_n' \end{pmatrix} \alpha = P \begin{pmatrix} v_1 \\ \vdots \\ v_n \end{pmatrix} \alpha = PA \begin{pmatrix} u_1 \\ \vdots \\ u_m \end{pmatrix} = PAQ^{-1} \begin{pmatrix} u_1' \\ \vdots \\ u_m' \end{pmatrix}. \qquad \square$$

4.6.13 Definition. Two $n \times m$ D-matrix A and B are said to be *equivalent* if there are invertible $n \times n$ D-matrix P and invertible $m \times m$ D-matrix Q such that $B = PAQ$. It is clear that the "equivalence" of matrices are equivalence relation.

4.6.14 Corollary. *Two equivalent D-matrices have the same row rank and the same column rank.*

Proof. Assume A and B are $n \times m$ D-matrices such that there are invertible $P \in M_n(R)$ and invertible $Q \in M_m(R)$ such that $B = PAQ$. Take D-spaces V and U of dimension n and m resp. and take their bases $\{v_1, \cdots, v_n\}$ and $\{u_1, \cdots, u_m\}$ resp.; construct a D-homomorphism $\alpha : V \to U$ corresponding to the matrix A. Then, with transforming matrices P and Q^{-1} resp., we have another basis $\{v_1', \cdots, v_n'\}$ of V and another basis $\{u_1', \cdots, u_m'\}$ of U resp.; and by Prop.4.6.12, under the new bases of V and U, α corresponds to the matrix $B = PA(Q^{-1})^{-1} = PAQ$. Therefore the conclusion follows from Prop.4.6.10. □

Just like in the usual linear algebra, we have elementary operations on rows and on columns of a matrix.

4.6.15 Definition. Elementary operations on rows (columns resp.) of a D-matrix means:

1) interchange two rows (columns resp.);

2) multiply a row (a column resp.) by a non-zero $d \in D$ on left (on right resp.);

3) add left d times of a row to another row, where $d \in D$.

Then, the techniques of elementary operations on matrices in Linear Algebra can be applied to D-matrices, and to get, without any difficult (just somewhat long), the following theorem; see Exer.3–5.

4.6.16 Theorem. *For any D-matrix A the following are equivalent:*

i) *the row rank of A is k;*

ii) *the column rank A is k;*

iii) *A is equivalent to the matrix with 1 in (i,i)-entries for $i = 1, \cdots, k$ and 0 elsewhere* (called the *canonical equivalence form* of A). □

The second way to compare the two ranks of a D-matrix is through the so-called dual spaces and dual homomorphisms, which are in fact general concepts for rings.

Let R be a ring, and M a *left* R-module. Recall that $\mathrm{Hom}_R(M, R)$ is an additive group, see Remark 4.2.13; moreover, since R is in fact a left R- right R-bimodule, $\mathrm{Hom}_R(M, R)$ is in fact a *right* R-module which right scalar multiplication is defined as:

$$(x)(\varphi \cdot r) = ((x)\varphi)r \qquad \text{for} \quad \varphi \in \mathrm{Hom}_R(M, R), \ r \in R, \ x \in M;$$

we denote it by $M^* = \mathrm{Hom}_R(M, R)$. Further, if N is another left R-module and $\alpha : M \to N$ is an R-homomorphism, then by Exer.4.2.11(2) we have an additive group homomorphism

$$\alpha^* : \ N^* \longrightarrow M^*, \quad \varphi \longmapsto \alpha\varphi;$$

moreover, for $r \in R$ and $x \in M$ we have

$$\begin{aligned}
(x)(\alpha^*(\varphi \cdot r)) &= (x)(\alpha(\varphi \cdot r)) = ((x)\alpha)(\varphi \cdot r) \\
&= (((x)\alpha)\varphi)r = ((x)(\alpha\varphi))r = ((x)(\alpha^*(\varphi)))r \\
&= (x)(\alpha^*(\varphi) \cdot r);
\end{aligned}$$

that is, $\alpha^*(\varphi \cdot r) = \alpha^*(\varphi) \cdot r$. Thus α^* is an R-homomorphism.

4.6.17 Definition. The right R-module $M^* = \mathrm{Hom}_R(M, R)$ is called the *dual module* of the left R-module M; and the R-homomorphism α^* :

$N^* \to M^*$ is called the *dual homomorphism* of the R-homomorphism $\alpha : M \to N$ of left R-modules M and N. The dual modules and dual homomorphisms for right R-modules are defined similarly.

4.6.18 Lemma. *Let F and E be free left R-modules with finite bases.*

1) *If $\{b_1, \cdots, b_n\}$ is a basis of F, then for every i there is $b_i^* \in F^*$ such that $(b_j)b_i^* = \begin{cases} 1 & \text{if } i = j \\ 0 & \text{if } i \neq j \end{cases}$, and $\{b_1^*, \cdots, b_n^*\}$ is a basis of M^*, called the* dual basis; *in particular, $M^* \cong M$.*

2) *If $\alpha : F \to E$ is an R-homomorphism, then, under the dual basis, α and α^* have one and the same R-matrix.*

Proof. (1). By Theorem 4.5.4, $b_i^* \in F^*$ exists and is unique. For any $\varphi \in F^*$ it is easy to check that

$$(4.6.19) \qquad \varphi = \sum_{i=1}^{m} v_i^* \cdot \varphi(v_i).$$

The independence of $\{b_1^*, \cdots, b_n^*\}$ is clear.

(2). Further, let $\{a_1, \cdots, a_m\}$ be a basis of E. Then α corresponds to an R-matrix $(\alpha_{ij})_{n \times m}$ such that (see 4.5.10):

$$(b_i)\alpha = \sum_{j=1}^{m} \alpha_{ij}a_j, \qquad i = 1, \cdots, n.$$

On the other hand,

$$(b_i)(\alpha^*(a_j^*)) = ((b_i)\alpha)a_j^* = (\sum_{k=1}^{m} \alpha_{ik}a_k)a_j^* = \sum_{k=1}^{m} \alpha_{ik}((a_k)a_j^*) = \alpha_{ij};$$

so, by formula (4.6.19) we have the following formula which proves (2):

$$\alpha^*(a_j^*) = \sum_{i=1}^{n} b_i^* \cdot \alpha_{ij} \qquad j = 1, \cdots, m. \qquad \square$$

Returning to D-spaces, we have

4.6.20 Proposition. *If $\alpha : V \to U$ is a D-homomorphism of finite dimensional D-spaces V and U, then α and its dual homomorphism α^* have the same rank.*

Proof. Set $K = \text{Coker}(\alpha)$, and consider the exact sequence of D-homomorphisms (see Example 4.4.3):

$$V \xrightarrow{\alpha} U \to K \longrightarrow 0;$$

by Exer.4.4.8(2) we have an exact sequence:

$$0 \longrightarrow K^* \longrightarrow U^* \xrightarrow{\alpha^*} V^*.$$

Then, by Prop.4.6.8(1) and Prop.4.6.18(1), we have

$$\dim_D\Big(\mathrm{Im}(\alpha)\Big) \;=\; \dim_D U - \dim_D K$$
$$=\; \dim_D U^* - \dim_D K^* = \dim_D\Big(\mathrm{Im}(\alpha^*)\Big). \qquad \square$$

4.6.21 Theorem. *The row rank equals the column rank for any D-matrix A.*

Proof. Let A be of size $n \times m$. Then there are left D-spaces V and U of dimensions n and m resp. and a D-homomorphism $\alpha : V \to U$ such that, with given bases, α corresponds to the matrix A. By Lemma 4.6.18, with the dual bases, $\alpha^* : U^* \to V^*$ also corresponds to the matrix A. Thus the conclusion follows from Prop.4.6.10 and Prop.4.6.20. $\square$

Exercises 4.6

Note: in the following exercises D is a division ring and V is a D-space.

1. Prove Prop.4.6.8.

2. If $V = \langle X \rangle$, then
   1) a maximal independent subset of X is a basis of V.
   2) a minimal subset of X which generates V is a basis of V.

$^+$3. The matrix which is obtained by performing an elementary row (column) operation on the identity matrix is called the *elementary matrix* corresponding to the operation. Prove:
   1) Every elementary matrix is invertible.
   2) The matrix obtained by performing an elementary row (column resp.) operation on A equals the matrix obtained by left (right resp.) multiplying the corresponding elementary matrix to A.

$^+$4. For an $n \times n$ D-matrix A the following are equivalent:
   i) A is invertible;
   ii) After finitely many performances of elementary row operations, A is transformed to the identity matrix;
   iii) After finitely many performances of elementary column operations, A is transformed to the identity matrix;
   iv) A is a product of finitely many elementary matrices.
   v) The rank A equals n (such $n \times n$ matrix is said to be of *full-rank*).

$^{+}$5. Let A be an $n \times m$ D-matrix. If the n rows are D-independent, then after finitely many elementary row or column operations A can be transformed to a matrix which has 1 in (i,i)-entries for $i = 1, \cdots, n$ and 0 elsewhere. Show that a similar conclusion holds if the m columns of A are D-independent.

6. Prove Theorem 4.6.16.

7. Let A be an $n \times m$ D-matrix of rank k (hence $m \geq k$), and $X = \begin{pmatrix} x_1 \\ \vdots \\ x_m \end{pmatrix}$ be variables. Consider the system of homogeneous linear equations $AX = 0$ over D.

   1) The set of all the solutions of $AX = 0$ is a subspace (called the *solution subspace* of the system) of the right D-space $D^{(m)}$.

   2) Use the techniques of elementary operations to show that the dimension of the solution subspace of $AX = 0$ equals $m - k$.

8. Let V be a D-space and $w_1, \cdots, w_m \in V$.

   1) The set $\{(d_1, \cdots, d_m) \in D^{(m)} \mid \sum_{i=1}^{m} d_i w_i = 0\}$ is a subspace of the left D-space $D^{(m)}$.

   2) If $w_1, \cdots, w_k$ form a maximal independent set in all w_i's, then the dimension of the subspace in (1) is $m - k$. (Hint: $d_{k+1}, d_{k+2}, \cdots, d_m$ could take values arbitrarily, and then the other d_i's are determined.)

   3) Apply the above results to give another proof for the above Exer.7.

9. Notations as the above Exer.8. and $w \in V$.

   1) There are $(d_1, \cdots, d_m) \in D^{(m)}$ such that $\sum_{i=1}^{m} d_i w_i = w$ if and only if $\langle w, w_1, \cdots, w_m \rangle = \langle w_1, \cdots, w_m \rangle$.

   2) Assume $(t_1, \cdots, t_m) \in D^{(m)}$ satisfies $\sum_{i=1}^{m} t_i w_i = w$; then $(d_1, \cdots, d_m) \in D^{(m)}$ satisfies $\sum_{i=1}^{m} d_i w_i = w$ if and only if $\sum_{i=1}^{m} (d_i - t_i) w_i = 0$.

   3) Please apply the above results to give the theory on systems of non-homogeneous linear equations over D.

4.7 Matrices over Commutative Rings

4.7.0 Assumption. In this section, R is always a commutative ring, except for Lemma 4.7.12.

It is not difficult to extend the concept of the determinant and the fundamental properties on it in the usual linear algebra to the matrices over R. The details of the following 4.7.1–4.7.2 are left as Exer.1.

4.7.1 Definition. Let $A = (a_{ij})_{n \times n}$ be an R-matrix.

1) The *determinant* of A, denoted by $\det A$, is the sum

$$\det A = \sum_{\pi \in S_n} (\text{sign } \sigma) a_{1\pi(1)} a_{2\pi(2)} \cdots a_{n\pi(n)}$$

where S_n is the symmetric group (see Example 2.2.4) and sign σ is the sign of σ, i.e. it is 1 if σ is an even permutation, and -1 if it is odd, see Example 2.7.9.

2) The *transpose matrix* of A, denoted by A^T, is the matrix whose ith-row is the ith-column of A (hence whose jth-column is the jth-row of A).

3) For $1 \leq i, j \leq n$, the (i, j)-*cofactor* of the entry a_{ij}, denoted by A_{ij}, is $(-1)^{i+j}$ times the determinant of the $(n-1) \times (n-1)$ matrix obtained by eliminating the ith-row and the jth-column of A.

4) The *adjoint matrix* of A, denoted by adj A, is the matrix $(A_{ji})_{n \times n}$, whose (i, j)-entries are the (j, i)-cofactor A_{ji}.

4.7.2 Proposition. *Notations as above.*

1) *If the ith-row (ith-column) of A is a sum of two vectors v_1 and v_2 (recall that rows (columns) of A can be regarded as vectors of $R^{(n)}$), then $\det A = \det A^{(1)} + \det A^{(2)}$, where $A^{(i)}$, $i = 1, 2$, are the matrices obtained from A by replacing the ith-row by v_i.*

2) *The determinant of the matrix which is obtained by multiplying $r \in R$ to the ith-row (ith-column) of A is equal to r times the determinant of A.*

3) *If two distinct rows (columns) of A are the same, then $\det A = 0$.*

4) *The determinant of the identity matrix I is 1: $\det I = 1$.*

5) $\sum_{k=1}^{n} a_{ik} A_{jk} = \sum_{k=1}^{n} a_{ki} A_{kj} = \begin{cases} \det A & \text{if } i = j \\ 0 & \text{if } i \neq j \end{cases}.$

6) $A(\text{adj } A) = \text{diag}(\det A, \cdots, \det A) = (\text{adj } A)A$; *where* $\text{diag}(\cdots)$ *denotes the diagonal matrix.*

7) *If B is also an $n \times n$ R-matrix, then $\det(AB) = \det A \cdot \det B$.* $\square$

4.7.3 Corollary. *Notations as above. A is an invertible R-matrix if and only if $\det A$ is an invertible element of R.*

Proof. If there is a matrix B such that $AB = I$ (the identity matrix); then, by Prop.4.7.2(4) and (7), we have $\det A \cdot \det B = \det(AB) =$

$\det I = 1$; i.e. $\det A$ is invertible. Conversely, if $d = \det A$ is invertible in R, then

$$(d^{-1}(\text{adj } A))Ad^{-1}((\text{adj } A)A) = d^{-1}\text{diag}(d, \cdots, d) = \text{diag}(1, \cdots, 1)$$
$$= I = d^{-1}(A(\text{adj } A)) = A(d^{-1}(\text{adj } A));$$

that is, A is invertible and $A^{-1} = d^{-1}(\text{adj } A)$ is the inverse of A. $\qquad \square$

4.7.4 Proposition. *Let M be an R-module (hence M is both left and right R-module and a bimodule, see Prop.4.1.9). Then $\text{End}_R^l(M)$ is not only a ring, but also an R-module with the scalar multiplication $r\varphi$ for $r \in R$ and $\varphi \in \text{End}_R^l(M)$ as follows:*

$$(r\varphi)(x) = r((\varphi)(x)), \qquad \text{for all } \ x \in M;$$

and the following hold:

$$r(\alpha\beta) = (r\alpha)\beta = \alpha(r\beta), \qquad \text{for all } \ r \in R \ \text{and} \ \alpha, \beta \in \text{End}_R^l(M);$$

and the map $R \to \text{End}_R^l(M)$, $r \mapsto r \cdot \text{id}_M$, is a ring homomorphism.

The same is true for $\text{End}_R^r(M)$.

Proof. Check it directly, Exer.2. $\qquad \square$

4.7.5 Remark. In fact, the image $r \cdot \text{id}_M$ of $r \in R$ is just the image ρ_r of r in Prop.4.1.7: since R is commutative, ρ_r commutes with every $\rho_{r'}$ for $r' \in R$, i.e. ρ_r is an R-endomorphism of M. From this argument we also see why Prop.4.7.4 is false for general rings. In the following, for convenience, by r we also denote the image $r \cdot \text{id}_M$ in $\text{End}_R^l(M)$; in particular, $1 = 1_R = \text{id}_M$ if we speak of endomorphisms of M.

4.7.6 Proposition. *Let M be an R-module generated by n elements $x_1, \cdots, x_n$, let $\varphi \in \text{End}_R^l(M)$ and $R[\varphi]$ be the subring of $\text{End}_R^l(M)$ generated by φ over R (cf. (3.8.1)). Then M is an $R[\varphi]$-module and*

1) $R[\varphi]$ is a commutative subring of $\text{End}_R^l(M)$;

2) As an R-module, $R[\varphi]$ is generated by $\text{id}_M, \varphi, \cdots, \varphi^{n-1}$, i.e. $R[\varphi] = R \cdot \text{id}_M + R\varphi + \cdots + R\varphi^{n-1}$.

Proof. (2) follows from the next lemma (take $I = R$); the others are easy to check (M is in fact a left $\text{End}_R^l(M)$-module). $\qquad \square$

4.7.7 Lemma. *Notations as above. If $\varphi M \subset JM$ for an ideal J of R, then there are $a_1, a_2, \cdots, a_n \in J$ such that*

$$\varphi^n + a_1\varphi^{n-1} + a_2\varphi^{n-2} + \cdots + a_n = 0$$

Proof. Since $M = Rx_1 + \cdots + Rx_n$, we have $JM = \{l_1x_1 + \cdots + l_nx_n \mid l_i \in J\}$, cf. Exer.4.2.8. Then for every $\varphi(x_i)$ there are $a_{ij} \in J$ such that

$$\varphi(x_i) = \sum_{j=1}^{n} a_{ij}x_j \qquad i = 1, \cdots, n;$$

in other words, if we set

$$A = \begin{pmatrix} a_{11} & \cdots & a_{1n} \\ \cdots & \cdots & \cdots \\ a_{n1} & \cdots & a_{nn} \end{pmatrix}, \quad I = \begin{pmatrix} 1 & & \\ & \ddots & \\ & & 1 \end{pmatrix} \text{ and } X = \begin{pmatrix} x_1 \\ \vdots \\ x_n \end{pmatrix},$$

then $\varphi I - A$ is a matrix over the commutative ring $R[\varphi]$ (cf. 4.7.5) and

$$(\varphi I - A)X = 0.$$

Left multiplying the adjoint matrix $\text{adj}(\varphi I - A)$ to both sides of the above equation, by 4.7.2(6) we get

$$(\det(\varphi I - A) \cdot I) X = 0, \quad \text{or} \quad (\det(\varphi I - A)) x_i = 0 \text{ for } i = 1, \cdots, n;$$

hence $(\det(\varphi I - A) M = 0$, i.e. $\det(\varphi I - A) = 0$. Expanding the determinant and noting that all $a_{ij} \in J$, we get the desired equation. $\square$

4.7.8 Remark. In the above, if M is a free R-module with an R-basis $\{x_1, \cdots, x_n\}$, then A is just the matrix of φ under the basis; hence, in notation of linear algebra, $\det(\lambda I - A)$ is the *characteristic polynomial* of the endomorphism φ; and Lemma 4.7.7 is a generalization of the *Cayley-Hamilton Theorem* in linear algebra which says that the characteristic polynomial annihilates the endomorphism.

4.7.9 Corollary. *If M is a finitely generated R-module such that $M = JM$ for an ideal J of R, then there is an $a \in J$ such that $\text{id}_M = a \cdot \text{id}_M$. (i.e. $1 = a$ in $\text{End}_R^l(M)$ in the sense of Remark 4.7.5).*

Proof. In Lemma 4.7.7, taking $\varphi = \text{id}_M$, in $\text{End}_R^l(M)$ we have $a_i \in J$ such that $1 + a_1 + a_2 + \cdots + a_n = 0$. $\square$

4.7.10 Theorem. *If M is a finitely generated R-module, then any surjective R-endomorphism of M is an isomorphism.*

Proof. Let $\varphi \in \text{End}_R^l(M)$. By Prop.4.7.6, we consider M as an $R[\varphi]$-module; and consider the ideal $J = R[\varphi] \cdot \varphi$ of $R[\varphi]$; then $M = JM$ because φ is surjective. Thus there is a $\psi \cdot \varphi \in J = R[\varphi] \cdot \varphi$ such that $\text{id}_M = \psi \cdot \varphi$; i.e. φ is an isomorphism. $\square$

4.7.11 Corollary. *R (commutative ring, see (4.7.0)) is an IBN ring; in particular we can speak of the rank of a free R-module.*

Proof. This follows from Theorem 4.7.10 and the next lemma. □

4.7.12 Lemma. *Let R be an arbitrary ring (not necessarily commutative). If any surjective endomorphism of any finitely generated R-module is an isomorphism, then R is an IBN ring.*

Proof. By Prop.4.5.16, it is enough to show that, if a free R-module F has two bases $\{x_1, \cdots, x_n\}$ and $\{y_1, \cdots, y_m\}$, then $m = n$. If it is not the case, e.g. $m > n$, then we can have a surjective but not injective map ξ from $\{y_1, \cdots, y_m\}$ onto $\{x_1, \cdots, x_n\}$, and by Cor.4.5.5 we have a surjective but not injective endomorphism $\tilde{\xi}$ of F; this contradicts the hypothesis of the lemma. □

Exercises 4.7

In the following exercises, R is a commutative ring except for Exer.10.

1. Verify 4.7.1–4.7.3 in the way of the usual linear algebra.
2. Prove Prop.4.7.4, and complete the details of the proof of Prop.4.7.6.
3. Check that the **Z**-matrix $\begin{pmatrix} 1 & 1 & 1 \\ 1 & 2 & 3 \\ 1 & 3 & 6 \end{pmatrix}$ is invertible and find its inverse.
4. If $A, B \in M_n(R)$ and $AB = I$ (the identity matrix), then $BA = I$.
5. Prove: a diagonal matrix $\text{diag}(a_1, \cdots, a_n) \in M_n(R)$ is invertible if and only if every a_i is invertible in R.
6. Prove: $A \in C(M_n(R))$ (center of the ring) if and only if $A = aI$ for an $a \in R$.
7. If F is a free R-module of finite rank n and $x_1, \cdots, x_n$ generate F, then $x_1, \cdots, x_n$ form a basis of F. (Hint: Choose a basis $\{a_1, \cdots, a_n\}$ of F, the map $a_i \mapsto x_i$ induces a surjective endomorphism of F, hence an isomorphism.)
8. Let $A \in M_n(R)$ and $AX = 0$ is the system of homogeneous linear equations with coefficient matrix A. If $AX = 0$ has non-zero solutions, then $\det A$ is a zero divisor in R.
9. Apply Theorem 3.5.7 and Exer.4.5.8 to give another proof of Cor.4.7.11.
10. Recall that for any ring R and any positive integer n we have the matrix ring $M_n(R)$, see Theorem 4.5.15. If, for $A, B \in M_n(R)$, $BA = I$ (identity matrix) provided $AB = I$; then R is an IBN ring. Apply this and the above Exer.4 to give another proof of Cor.4.7.11.

(Hint: if F is a free R-module which has two bases $\{x_1, \cdots, x_n\}$ and $\{y_1, \cdots, y_m\}$ and $m > n$, then we have

$$
A = \begin{pmatrix} a_{11} & \cdots & a_{1n} & 0 & \cdots & 0 \\ a_{21} & \cdots & a_{2n} & 0 & \cdots & 0 \\ \cdots & \cdots & \cdots & 0 & \cdots & 0 \\ a_{m1} & \cdots & a_{mn} & 0 & \cdots & 0 \end{pmatrix}, \qquad Y = \begin{pmatrix} y_1 \\ \vdots \\ y_n \\ \vdots \\ y_m \end{pmatrix},
$$

$$
B = \begin{pmatrix} b_{11} & \cdots & b_{1m} \\ \cdots & \cdots & \cdots \\ b_{n1} & \cdots & b_{nm} \\ 0 & \cdots & 0 \\ \cdots & \cdots & \cdots \\ 0 & \cdots & 0 \end{pmatrix} \quad \text{and} \quad X = \begin{pmatrix} x_1 \\ \vdots \\ x_n \\ 0 \\ \vdots \\ 0 \end{pmatrix}
$$

such that $AX = Y$, and $BY = X$; thus $ABY = Y$ and $AB = 1$; Hence $BA = 1$, which is impossible.)

4.8   Algebras over Commutative Rings

4.8.0 Assumption. In this section R is *always a commutative ring*; but the other rings are *not* necessarily commutative.

Motivated by Prop.4.7.4, we introduce a kind of algebraic systems.

4.8.1 Definition. A is called an *algebra over R* (or an *R-algebra*) if A is both a ring and an R-module and the following two conditions hold:

1) the addition of the ring structure and the addition of the module structure coincide;

2) for any $r \in R$ and $a, b \in A$ we have $r(ab) = (ra)b = a(rb)$.

An R-algebra A is said to be *commutative* if the ring A is commutative; and, A is said to be *R-free (finitely R-generated* resp.) if A is R-free (finitely R-generated resp.) as an R-module; and the R-rank of the R-module A is also said to be the *R-rank of the algebra A* if A is R-free. In particular, if R is a field, then the dimension of the vector space of an R-algebra A is also said to be the *dimension of the algebra*.

A map $\varphi : A \to A'$ from an R-algebra A to an R-algebra A' is called an *R-algebra homomorphism* if it is both a ring homomorphism and an

R-module homomorphism. The *injective* (*surjective*) homomorphisms and the *isomorphisms* of R-algebras make clear sense.

Combining the criterions for ring homomorphisms and module homomorphisms, we can easily get a criterion for R-algebras homomorphisms; e.g. see Exer.1. Note that a ring homomorphism of two algebras is not necessarily an algebra homomorphism, see Exer.4(2).

From Theorem 4.5.15 (cf. Exer.4.5.6) and Prop.4.7.4, we have the following example at once.

4.8.2 Example. **1)** The R-matrix ring $M_n(R)$ of degree n, with the usual scalar multiplication, is an R-free R-algebra; in fact, the matrices E_{ij}, $1 \leq i, j \leq n$, with 1 in (i,j)-entry and 0 elsewhere, form an R-basis. $M_n(R)$ is called the R-*matrix algebra*.

2) If M is an R-module, then the left (right resp.) endomorphism ring $\mathrm{End}_R^l(M)$ ($\mathrm{End}_R^r(M)$ resp.) is an R-algebra. Further, if M is a free R-module of rank n, then $\mathrm{End}_R^l(M)$ is an R-free R-algebra of rank n^2. In fact, Prop.4.5.15 shows that $\mathrm{End}_R^l(M) \cong M_n(R)$ as rings in this case; and, it is easy to see that this isomorphism is also an R-module homomorphism; that is, $\mathrm{End}_R^l(M) \cong M_n(R)$ as R-algebras in this case.

4.8.3 Example. Any ring T is a **Z**-algebra, where **Z** is the integer ring. In fact, for $n \in \mathbf{Z}$ and $t \in T$, the scalar product $n \cdot t$ is just the n-multiple of t as defined in (2.2.13) (but in additive notation, cf. Remark 2.2.16 and §3.1).

Also, Prop.4.7.4 suggests another characterization of R-algebras.

4.8.4 Proposition. **1)** *If A is an R-algebra, then $\zeta : R \to C(A), r \mapsto r \cdot 1_A$ is a ring homomorphism, where $C(A)$ is the center of the ring A and 1_A is the identity element of A.*

2) *If A is a ring and $\zeta : R \to C(A)$ is a ring homomorphism, then A is an R-algebra with the scalar multiplication $r \cdot a = \zeta(r)a$ for $r \in R$ and $a \in A$.*

Proof. Exer.2. □

For example, if $A = M_n(R)$, the R-matrix algebra, then $r \cdot 1_A$ is just the scalar matrix by r, i.e. the diagonal matrix with r in diagonal entries; it is clear that $r \mapsto r \cdot 1_A$ is an injective ring homomorphism; in other words, we can assume that R is contained in the center of A as a subring. Before going into further discussions, we note that a subring of an algebra is not necessarily a subalgebra.

4.8.5 Definition. Let A be an R-algebra. We say that B is a *subalgebra* of A if B is both a subring of the ring A and a submodule of the R-module A. We say that I is an ideal of A if I is both an ideal of the ring A and a submodule of the R-module A.

4.8.6 Remark. The above is a natural definition, since an algebra has both a ring structure and a module structrue which are compatible with each other. Exer.4(1) shows a subring which is not a subalgebra. However, for an R-algebra A, by Prop.4.8.4, it is easy to see that an ideal of the ring A must also be a submodule of the module A; see Exer.6(1). Therefore, all the results on rings involving ideals are still valid for algebras. In particular, the fundamental theorem on homomorphisms holds for algebras; see Exer.6(2)(3).

Now we continue the discussion inspired by Prop.4.8.4. If R is a field, then the ring homomorphism ζ in Prop.4.8.4(1) is always injective, hence we can always assume that R is contained in the center $C(A)$ of the R-algebra A as a subring; see Exer.3.

However, the ring homomorphism ζ in Prop.4.8.4(1) is not necessarily injective if R is not a field. For example, the residue class ring $\mathbf{Z}_m$ is a $\mathbf{Z}$-algebra, see Example 4.8.3; but the ring homomorphism $r \mapsto r \cdot [1]$ is not injective provided $m > 0$. In fact, if we regard rings as $\mathbf{Z}$-algebras just like Example 4.8.3, then Prop.4.8.4 suggests the following result and definition.

4.8.7 Proposition. *Let T be a ring and $\mathbf{Z}$ be the integer ring. Then the image of the ring homomorphism $\zeta : \mathbf{Z} \to C(T), n \mapsto n \cdot 1_T$ is the smallest unitary subring, which is isomorphic to a residue class ring $\mathbf{Z}_m$ for a non-negative integer m.*

Proof. Any unitary subring of T must contain 1_T, hence contains all the elements $n \cdot 1_T$, $n \in \mathbf{Z}$. $\qquad\qquad\qquad\qquad\qquad\qquad \square$

4.8.8 Definition. The non-negative integer m in Prop.4.8.7 is called the *characteristic* of the ring T, and denoted by $\mathrm{char}(T) = m$. By Prop.4.8.7, it is obvious that $\mathrm{char}(T)$ is the order of 1_T in the additive group of T if $\mathrm{char}(T) > 0$.

The characteristic of a ring has some remarkable properties, see

Exer.7. For a given commutative ring R, in fact, we can get a similar result if we consider R-algebras; see Exer.8.

4.8.9 Definition. Let A be an R-algebra and $S \subset A$ be a subset. It is easy to see that the intersection of some subalgebras of A is also a subalgebra. Hence the intersection of all the subalgebras which contain S is the smallest subalgebra which contains S; this subalgebra is called the *subalgebra of A generated by S*, and denoted by $\langle S \rangle$. If $A = \langle S \rangle$, then we say that the algebra A is generated by S. If $A = \langle S \rangle$ for a finite subset S, then we say that A is *finitely algebraically generated* (note that this is different from the "finitely R-generated" defined in Def.4.8.1).

4.8.10 Example. The polynomial ring $R[x]$ in indeterminate x is an R-algebra generated by one element x; it is an R-free algebra of infinite R-rank, hence it is not finitely R-generated. Moreover, from Theorem 3.8.3 (and cf. Prop.4.8.4) it is easy to see that, if A is an R-algebra and $a \in A$, then there is a unique R-algebra homomorphism $\varphi : R[x] \to A$ such that $\varphi(x) = a$; thus the image $\text{Im}(\varphi) = R[a] = \{f(a) \mid f(x) \in R[x]\}$ is in fact the subalgebra $\langle a \rangle$ of A generated by a.

Exer.9 shows how to express the elements of the subalgebra $\langle S \rangle$ generated by an arbitrary subset $S \subset A$.

Now we consider the modules over an R-algebra A. An A-module M means a module over the ring A; i.e. M is an additive group and there is a ring homomorphism $\rho : A \to \text{End}^l(M)$. Combining this ring homomorphism with the ring homomorphism $\zeta : R \to C(A)$ in Prop.4.8.4, we have a ring homomorphism $\rho \circ \zeta : R \to \text{End}^l(M)$; in other words, M is an R-module. In fact, for $r \in R$ and $x \in M$ the scalar product of r and x is $rx = (r \cdot 1_A)x$. In this notation, furthermore, the module operation for M (see Def.4.1.1)

$$A \times M \longrightarrow M, \quad (a, x) \longmapsto ax$$

is not only a biadditive function, but also an R-bilinear function from the R-modules A and M to M. Moreover, we have the R-algebra $\text{End}_R^l(M)$ which is clearly contained in $\text{End}^l(M)$; and it is not difficult to see that the image $\text{Im}(\rho)$ of the structure map ρ is contatined in $\text{End}_R^l(M)$, and $\rho : A \to \text{End}_R^l(M)$ is an R-algebra homomorphism. Thus, an A-module M can be redefined as an R-module and an R-algebra homomorphism $\rho : A \to \text{End}_R^l(M)$. The details are left as Exer.10.

From a group, an R-algebra can be constructed in a natural way.

4.8.11 Example. Let G be a group. We can have the free R-module RG with the set G as a free basis, i.e. its elements are the finite R-linear combinations $\sum_{g \in G} a_g \cdot g$ where only finitely many a_g are non-zero. Since an associative multiplication has been defined on the basis G, by linear extension, we have a multiplication on RG:

$$\Big(\sum_{g \in G} a_g \cdot g \Big) \Big(\sum_{h \in G} a'_h \cdot h \Big) = \sum_{k \in G} \Big(\sum_{gh=k} a_g a'_h \Big) \cdot k \,;$$

where the right-hand side is clearly still a finite sum; and it is easy to see that this multiplication on RG is still associative. Of course we can identify that G is contained in RG since G is a basis of the free R-module RG; in fact $g \in G$ is the R-combination $\sum_{h \in G} a_g \cdot h$ with the coefficient $a_g = 1$ and all the other coefficients are zero. Then it is clear that the identity element 1_G is also the identity element of the ring RG. In short, RG is an R-algebra; called the *group algebra* of the group G over the commutative ring R. Recall that, for any R-algebra A, the multiplicative group A^* consists of all the invertible elements of A. It is obvious that $G \subset (RG)^*$; and for any R-algebra homomorphism $\varphi : RG \to A$ it is clear that the restriction map $\varphi|_G : G \to A$ takes its image in A^* and $\varphi|_G : G \to A^*$ is a group homomorphism. In fact this group homomorphism $\varphi|_G$ determines the algebra homomorphism φ.

4.8.12 Proposition. *Let RG be the group algebra of a group G and A be an R-algebra. Then for any group homomorphism $\gamma : G \to A^*$ there is a unique R-algebra homomorphism $\tilde{\gamma} : RG \to A$ such that $\tilde{\gamma}|_G = \gamma$ (sometimes $\tilde{\gamma}$ is said to be induced by γ).*

Proof. For any $\sum_{g \in G} a_g \cdot g \in RG$, since the coefficients a_g's are uniquely determined, a unique element $\sum_{g \in G} a_g \cdot \gamma(g) \in A$ is determined; and it is easy to check that

$$\tilde{\gamma} : \quad RG \longrightarrow A, \qquad \sum_{g \in G} a_g \cdot g \longmapsto \sum_{g \in G} a_g \cdot \gamma(g)$$

is an R-algebra homomorphism and $\tilde{\gamma}|_G = \gamma$. If $\varphi : RG \to A$ is an R-algebra homomorphism such that $\varphi|_G = \gamma$ then

$$\varphi \Big(\sum_{g \in G} a_g \cdot g \Big) = \sum_{g \in G} a_g \cdot \varphi(g) = \sum_{g \in G} a_g \cdot \gamma(g) = \tilde{\gamma} \Big(\sum_{g \in G} a_g \cdot g \Big);$$

that is, $\varphi = \tilde{\gamma}$ (compare with the map (4.5.5) and Theorem 4.5.4). $\square$

Recall that $M_n(R)$ is an R-algebra, see 4.8.2. Usually, the multiplicative group of $M_n(R)$ is called the *general linear group* over R of degree n, and denoted by $\mathrm{GL}_n(R)$. In general, a group homomorphism from a group G to $\mathrm{GL}_n(R)$ is said to be a *linear representation* of G over R of degree n. By Prop.4.8.12 above, such a linear representation of G is equivalent to an R-algebra homomorphism from the group algebra RG to $M_n(R)$. Recall, from 4.8.2 again, that $M_n(R) \cong \mathrm{End}_R^l(M)$ as R-algebras for a free R-module M of rank n. So a linear representation of the group G over R of degree n is equivalent to an R-algebra homomorphism from RG to $\mathrm{End}_R^l(M)$, in other words, is equivalent to an R-free RG-module M of R-rank n.

Exercises 4.8

Note: Assumption 4.8.0 is still valid in the following exercises.

$^+$1. Let A and A' be two R-algebras. A map $\varphi : A \to A'$ is an R-algebra homomorphism if and only if for any $a, b \in A$ and $r \in R$ we have

$$\varphi(a + b) = \varphi(a) + \varphi(b), \quad \varphi(ab) = \varphi(a)\varphi(b), \quad \text{and} \quad \varphi(ra) = r\varphi(a).$$

2. Prove Prop.4.8.4.

3. If R is a field and A is an R-algebra, then the ring homomorphism $r \mapsto r \cdot 1_A$ is injective. (Hint: note that all the ring homomorphisms in this chapter are assumed to be unitary.)

4. Take $R = \mathbf{C}$ to be the field of complexes and $\mathbf{R}$ to be the field of all reals. Then $\mathbf{C}$ is a $\mathbf{C}$-algebra.

   1) $\mathbf{R}$ is a subring of $\mathbf{C}$ but it is not a subalgebra of the $\mathbf{C}$-algebra $\mathbf{C}$.

   2) For any $a + bi \in \mathbf{C}$ where $a, b \in \mathbf{R}$ and i is the imaginary unit, the mapping $a + bi \mapsto a - bi$ is a ring homomorphism (in fact, a ring isomorphism) but not a $\mathbf{C}$-algebra homomorphism.

5. Show that, if $\varphi : A \to A'$ is an R-algebra homomorphism, then the image is a subalgebra of the algebra A' and the kernel of φ is an ideal of the algebra A.

6. Let A be an R-algebra and I be an ideal of the ring A. Then

   1) I must be a submodule of the R-module A, hence must be an ideal of the R-algebra A;

   2) the following is a well-defined scalar product on the quotient ring A/I:

$$r \cdot (a + I) = ra + I, \qquad \forall r \in R \text{ and } a \in A;$$

and A/I is an R-algebra too, and the natural homomorphism $A \to A/I, a \mapsto a + I$ is an R-algebra homomorphism.

3) Prove the Fundamental Theorem for Algebra Homomorphisms.

+7. Let T be a ring of characteristic m. Prove:

1) Any unitary subring of T has the same characteristic m as T.

2) For any $a \in T$ we have $m \cdot a = 0$; further, if $m = p$ is a prime and $a \neq 0$, then $na = 0 \iff p \mid n$.

3) If T is commutative and $m = p$ is a prime number, then we have a special binomial formula:

$$(a \pm b)^{p^n} = a^{p^n} \pm b^{p^n} \qquad \text{for all } a, b \in T.$$

(Hint: by induction on n; cf. Exer.3.1.13; and note that the binomial coefficient $\binom{p}{i}$, $0 < i < n$, is a multiple of p; and, if $\text{char}(T) = 2$, then $-b = b$ since $b + b = 2b = 0$.)

8. Let A be an R-algebra and $C(A)$ be the center of A.

1) $C(A)$ is a subalgebra of A.

2) The map $\zeta : R \to C(A), r \mapsto r \cdot 1_A$ is an R-algebra homomorphism, and the image $\text{Im}(\zeta) \cong R/I$ is the smallest unitary subalgebra of A, where $I = \text{Ker}(\zeta)$ is called the *characteristic ideal* of A; and $ra = 0$, $\forall r \in I, a \in A$.

3) If $\varphi : A \to A'$ is an R-algebra homomorphism, then φ maps the smallest subalgebra of A onto the smallest subalgebra of A'.

9. Let A be an R-algebra and $S \subset A$ be a subset. Let $\widehat{S}$ be the set of all finite products $s_1 \cdots s_n$ with $s_i \in S$. Prove that the subalgebra $\langle S \rangle = R\widehat{S}$, the R-submodule of the R-module A generated by $\widehat{S}$; i.e. $\langle S \rangle$ consists of all the finite R-linear combinations of $\widehat{S}$.

+10. Let A be an R-algebra.

1) If M is a module over the ring A with the structure map $\rho : A \to \text{End}^l(M)$, then M is an R-module with the scalar R-product $rx = (r \cdot 1_A)x$ for $r \in R$ and $x \in M$, and $\text{Im}(\rho) \subset \text{End}_R^l(M)$, and $\rho : A \to \text{End}_R^l(M)$ is an R-algebra homomorphism.

2) If M is an R-module and $\rho : A \to \text{End}_R^l(M)$ is an R-algebra homomorphism, then M is an A-module.

11. Let G be a group and RG be the group algebra.

1) The *trivial linear representation* $\tau : G \to R^*, g \mapsto 1_R$ (sending every $g \in G$ to 1_R) induces the *augmentation homomorphism*:

$$\text{aug} : \quad RG \longrightarrow R, \qquad \sum_{g \in G} a_g \cdot g \longmapsto \sum_{g \in G} a_g;$$

the kernel $\text{Ker}(\text{aug})$ is called the *augmentation ideal* of RG.

2) The augmentation ideal is an R-free ideal of RG which has an R-basis $\{1 - g \mid 1 \neq g \in G\}$.

12. Let H be a monoid (see Def.2.2.10). Let RH be the free R-module with free basis H.

1) With the following multiplication, RH is an R-algebra:

$$\Big(\sum_{h \in H} a_h \cdot h\Big)\Big(\sum_{h' \in H} a'_{h'} \cdot h'\Big) = \sum_{k \in H}\Big(\sum_{hh' = k} a_h a'_{h'}\Big) \cdot k$$

2) Let A be an R-algebra and $M(A)$ denote the multiplicative monoid of A. For any monoid homomorphism $\mu : H \to M(A)$ (i.e. $\mu(hh') = \mu(h)\mu(h')$, $\forall h, h' \in H$, and $\mu(1_H) = 1_A$) there is a unique R-algebra $\widetilde{\mu} : RH \to A$ such that the restriction map $\widetilde{\mu}|_H = \mu$.

4.9 Tensor Products

We begin with the tensor products of vector spaces over fields.

Let K be a field and V_1, V_2, W be K-vector spaces. Recall that a map $V_1 \times V_2 \to W$ is said to be *bilinear* if it is linear in both variables.

It is interesting to find a vector space T and a bilinear map $\tau : V_1 \times V_2 \to T$ such that any bilinear map $f : V_1 \times V_2 \to W$ can be "uniquely expressed" in a linear map $\widetilde{f} : T \to W$; precisely speaking, there is a unique linear map $\widetilde{f} : T \to W$ such that the following diagram is commutative:

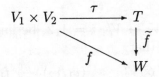

Such a vector space T and a bilinear map $\tau : V_1 \times V_2 \to T$ (if they exist) are called a *tensor product* of V_1 and V_2.

We know that any linear map is determined by its values on a basis of the domain vector space. And any bilinear map $V_1 \times V_2 \to W$ is determined by its values on the pairs $(u_{1i}, u_{2j}) \in B_1 \times B_2$, where B_1 and B_2 are K-bases of V_1 and V_2 resp. Thus, as follows, a tensor product of V_1 and V_2 really exists.

Let T be the K-vactor space with basis $\{u_{1i} \otimes u_{2j} \mid u_{1i} \in B_1, u_{2j} \in B_2\}$, where we denote the pairs (u_{1i}, u_{2j}) by $u_{1i} \otimes u_{2j}$ for later distinction, and let a bilinear map be defined as

(4.9.1) $\tau : \quad V_1 \times V_2 \longrightarrow T, \qquad (u_{1i}, u_{2j}) \longmapsto u_{1i} \otimes u_{2j}$.

Then, for any K-vector space W and any bilinear map $f : V_1 \times V_2 \to W$, it is easy to see that the linear map $\tilde{f} : T \to W$ determined by $\tilde{f}(u_{1i} \otimes u_{2j}) = f(u_{1i}, u_{2j})$ is the unique one such that $f = \tilde{f}\tau$. In other words, (4.9.1) is a tensor product of V_1 and V_2.

Moreover, if $\tau' : V_1 \times V_2 \to T'$ is also a tensor product of V_1 and V_2, then there are linear maps $\tilde{\tau} : T' \to T$ and $\tilde{\tau}' : T \to T'$ such that $\tau = \tilde{\tau}\tau'$ and $\tau' = \tilde{\tau}'\tau$. In particular, $\tau = \tilde{\tau}\tilde{\tau}'\tau$; that is, both $\tilde{\tau}\tilde{\tau}'$ and id_T make the following diagram commutative:

(4.9.2)

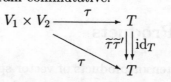

Then, by the uniqueness of the tensor product, we have $\tilde{\tau}\tilde{\tau}' = \mathrm{id}_T$. Similarly, we also have $\tilde{\tau}'\tilde{\tau} = \mathrm{id}_{T'}$. Thus both $\tilde{\tau}$ and $\tilde{\tau}'$ are K-isomorphisms.

4.9.3 Theorem.   *The tensor product of vector space V_1 and V_2 over a field K exists and is unique up to isomorphism.*   □

Notation. From the result we can denote the tensor product by $\tau : V_1 \times V_2 \to V_1 \otimes_K V_2$, and denote the specified bilinear map τ by $\otimes$, i.e. $\tau(v_1, v_2) = v_1 \otimes v_2$; in other words, $(v_1, v_2) \mapsto v_1 \otimes v_2$ is the bilinear map.

Moreover, let $f_i : V_i \to V_i'$, $i = 1, 2$, be linear maps of K-vector spaces. Then it is clear that

$$V_1 \times V_2 \longrightarrow V_1' \otimes V_2' , \qquad (v_1, v_2) \longmapsto f_1(v_1) \otimes f_2(v_2) ,$$

is a bilinear map; hence it determines a unique linear map

(4.9.4) $f_1 \otimes f_2 : \quad V_1 \otimes_K V_2 \longrightarrow V_1' \otimes_K V_2', \quad v_1 \otimes v_2 \longmapsto f_1(v_1) \otimes f_2(v_2)$;

which is also called the *tensor product* of f_1 and f_2.

4.9.5 Remark. It is not difficult to extend the tensor products of two vector spaces to several vector spaces. There are two ways to do it. One of them is to consider the so-called *poly-linear maps* $f : V_1 \times \cdots \times V_n \to W$. The other one is to prove the so-called associativity of tensor products: $(V_1 \otimes_K V_2) \otimes_K V_3 \cong V_1 \otimes_K (V_2 \otimes_K V_3)$. And, in fact, in the two ways we get the same space; see Exer.4.

Turn to the products of modules over rings. In the following we assume that R be a ring (not necessarily commutative).

Note that there are no longer bilinear maps for R-modules. And it is not reasonable to consider biadditive maps, because such maps are regardless of the R-scalar products on modules.

4.9.6 Definition. Let V be a (left) R-module, U be a right R-module and G be an additive group. A map $f : U \times V \to G$ is said to be *balanced* if it is biadditive and $f(ur, v) = f(u, rv)$, $\forall\, u \in U$, $v \in V$ and $r \in R$.

Modifying the tensor products of vector spaces, we introduce

4.9.7 Definition. Notations as above. An additive group T and a balanced map $\tau : U \times V \to T$ is called a *tensor product* of the module U and V if for any additive group G and any balanced map $f : U \times V \to G$ there is a unique group homomorphism $\tilde{f} : T \to G$ such that $f = \tilde{f}\tau$.

First we must show the existence of tensor products.

Notations as in Def.4.9.7 above. Let F be the free Abelian group with basis $U \times V$, i.e. F consists of all the finite sums $\sum_i n_i \cdot (u_i, v_i)$, $u_i \in U$, $v_i \in V$ and $n_i \in \mathbf{Z}$; then $U \times V \subset F$. Let H be the subgroup of F generated by the set of all the following three kinds of elements:

$$(4.9.8) \begin{cases} (u_1 + u_2, v) - (u_1, v) - (u_2, v), & u_1, u_2 \in U, \ v \in V\,; \\ (u, v_1 + v_2) - (u, v_1) - (u, v_2), & u \in U, \ v_1, v_2 \in V\,; \\ (ur, v) - (u, rv), & u \in U, \ v \in V, \ r \in R. \end{cases}$$

Let $T = F/H$; and $\tau : U \times V \to T$ be the composition map $U \times V \to F \to T$ where the former is the inclusion map and the latter is the natural map; in particular, we have that the image $\tau(U \times V)$ is a generator set of the Abelian group T. Then it is easy to check that τ is balanced; e.g.

$$\tau(ur, v) = (ur, v) + H = (u, rv) + \Big((ur, v) - (u, rv)\Big) + H$$
$$= (u, rv) + H = \tau(u, rv).$$

Next, assume that G is an additive group and $f : U \times V \to G$ is a balanced map. Since F is a free Abelian group with basis $U \times V$, there is a group homomorphism $\hat{f} : F \to G$ such that $\hat{f}|_{U \times V} = f$. Since f is a balanced map, the generators (4.9.8) of H are all mapped to zero by $\widehat{f}$, hence $H \subset \mathrm{Ker}(\widehat{f})$; and, by Exer.2.7.4, there is a homomorphism $\widetilde{f} : T = F/H \to G$ such that $\widetilde{f}(x + H) = \widehat{f}(x)$ for $x \in F$. Then it is clear that $f(u,v) = \widehat{f}(u,v) = \widetilde{f}((u,v) + H) = \widetilde{f}(\tau(u,v))$; i.e. $f = \widetilde{f}\tau$. If $g : T \to G$ is also a homomorphism such that $f = g\tau$, then $\widetilde{f}|_{\tau(U \times V)} = g|_{\tau(U \times V)}$ hence $g = \widetilde{f}$ because $\tau(U \times V)$ is a generator set of T, see Exer.2.4.9.

So far, we have proved the existence of the following theorem; and the uniqueness of the theorem is proved in the same way as argument (4.9.2).

4.9.9 Theorem. *The tensor product of a right R-module U and a left R-module V exists and is unique up to isomorphism.* □

4.9.10 Notation and Remark. The Abelian group T in the above tensor product is denoted by $U \otimes_R V$, and the specified balanced map τ is denoted by $\otimes$; i.e.

$$\otimes : \ U \times V \longrightarrow U \bigotimes_R V, \qquad (u,v) \longmapsto u \otimes v$$

is the tensor product of the right R-module U and the left R-module V. Moreover, the above arguments show that $\{u \times v \mid u \in U, \ v \in V\}$ is a generator set of the Abelian group $U \otimes_R V$; i.e. every $t \in U \otimes_R V$ is written as a finite sum $t = \sum_i u_i \otimes v_i$ for $u_i \in U$ and $v_i \in V$; but note that such expressions are not unique.

It is remarkable that the tensor products of general modules appear more complicated than the tensor products of vector spaces; Exer.7 exhibits this point.

The following result is proved by the same argument as for (4.9.4).

4.9.11 Proposition. *Let $f : U \to U'$ be an R-homomorphism of right R-modules, and $g : V \to V$ be an R-homomorphism of left R-modules. Then we have a group homomorphism*

$$f \otimes g : \ U \bigotimes_R V \longrightarrow U' \bigotimes_R V', \qquad u \otimes v \longmapsto f(u) \otimes g(v),$$

which is called the tensor product *of f and g.* □

4.9.12 Remark. Similarly to that for vector spaces, tensor products can be extended to several factors. Let R and S be rings, and U be a right R-module and V be a left R- right S-bimodule and W be a left S-module. We have two ways to define the tensor product $U \otimes_R V \otimes_S W$. One is through the *balanced maps* again: a map

$f : U \times V \times W \to G$ for Abelian group G is said to be balanced if it is triadditive and $f(ur, v, sw) = f(u, rvs, w)$. The other one is also related to the so-called associativity of tensor products; this way is not difficult but somewhat lengthy, because we have to make $U \otimes_R V$ into a right S-module and make $V \otimes_S W$ into a left R-module, and then prove $(U \otimes_R V) \otimes_S W \cong U \otimes_R (V \otimes_S W)$. For the details we refer to further readings.

Exercises 4.9

1. Let V_1 and V_2 be vector spaces over a field K. Show that
$$\dim_K(V_1 \otimes_K V_2) = \dim_K(V_1) \cdot \dim_K(V_2).$$

2. Let $f_i : V_i \to V_i'$, $i = 1, 2$, be linear maps of vector spaces over a field K, and $f_1 \otimes f_2 : V_1 \otimes_K V_2 \to V_1' \otimes_K V_2'$ be the tensor product of the linear maps. Prove: $f_1 \otimes f_2$ is surjective (injective, isomorphism) if both f_1 and f_2 are surjective (injective, isomorphism). (**Remark.** The latter two assertions are not valid for tensor products of modules.)

3. Notations as in Exer.2 above. Assume $B_i = (u_{i1}, \cdots, u_{in_i})$ are bases of V_i and $B_i' = (u_{i1}', \cdots, u_{in_i'}')$ are bases of V_i'; and assume $M_i = (m_{\alpha\beta}^i)$ are the matrices of f_i under the bases B_i and B_i'. Let $u_{1j} \otimes B_2 = (u_{1j} \otimes u_{21}, \cdots, u_{1j} \otimes u_{2n_2})$ etc. Prove: under the basis $(u_{11} \otimes B_2, \cdots, u_{1n_1} \otimes B_2)$ of $V_1 \otimes_K V_2$ and the basis $(u_{11}' \otimes B_2', \cdots, u_{1n_1'}' \otimes B_2')$ of $V_1' \otimes_K V_2'$, the matrix of $f_1 \otimes f_2$ is
$$\begin{pmatrix} m_{11}^1 M_2 & \cdots & m_{1n_1}^1 M_2 \\ \cdots & \cdots & \cdots \\ m_{n_1 1}^1 M_2 & \cdots & m_{n_1 n_1}^1 M_2 \end{pmatrix};$$
(which is sometimes called the *tensor product* of the matrices M_1 and M_2).

4. Let V_i, $i = 1, 2, 3$, be vector spaces over a field K. Show that
$$\tau : V_1 \times V_2 \times V_3 \longrightarrow (V_1 \otimes_K V_2) \otimes_K V_3, \quad (v_1, v_2, v_3) \mapsto (v_1 \otimes v_2) \otimes v_3$$
is a tri-linear map, and for any tri-linear map $f : V_1 \times V_2 \times V_3 \to W$ of vector spaces there is a unique linear map $\tilde{f} : (V_1 \otimes_K V_2) \otimes_K V_3 \to W$ such that $f = \tilde{f}\tau$.

5. If A, B and G are Abelian groups and $f : A \times B \to G$ be a biadditive map, then $f(na, b) = nf(a, b) = f(a, nb)$ for all $a \in A$, $b \in B$ and $n \in \mathbf{Z}$, in particular, $f(0, b) = 0 = f(a, 0)$ and $f(-u, v) = -f(u, v) = f(u, -v)$.

6. Let R be a ring and U be a right R-module and V be a left R-module. In $U \otimes_R V$, for $u \in U, v \in V$ and $m, n \in \mathbf{Z}$ we have $(mu) \otimes (nv) = (mn) \cdot (u \otimes v)$; in particular, $0 \otimes v = 0 = u \otimes 0$ and $(-u) \otimes v = -u \otimes v = u \otimes (-v)$.

(Hint: see Exer.5 above and the construction of $U \bigotimes_R V$ in the argument of Theorem 4.9.9.)

Remark. Recall that any **Z**-module is both left and right module, so the tensor product $A \bigotimes B$ (where the subscript **Z** is suppressed) of Abelian groups A and B make sense. This exercise shows that the **Z**-module structure of $A \bigotimes B$ is as: $n \cdot (a \otimes b) = (na) \otimes b = a \otimes (nb)$ for $a \in A, b \in B$ and $n \in \mathbf{Z}$.

In general, for commutative ring R, the tensor product $U \bigotimes_R V$ of R-modules U and V is not only an Abelian group, but also an R-module, where the scalar product is defined as: $r \cdot (u \otimes v) = (ru) \otimes v = u \otimes (rv)$; this is well-defined.

7. Let $A = \langle a \rangle$ and $B = \langle b \rangle$ be cyclic groups.

   1) $A \bigotimes B$ is a cyclic group generated by $a \otimes b$.

   2) If $|A| = m$ and $|B| = n$ and $d = (m, n)$ is the greatest common divisor of m and n, then $|A \bigotimes B| = d$.
   (Hint: $d = sm + tn$, hence $d \cdot (a \otimes b) = (sm) \cdot (a \otimes b) + tn \cdot (a \otimes b) = (sma) \otimes b + a \otimes (tnb) = 0$; if $|A \bigotimes B|$ is a proper divisor of d, then for the balanced map $f : A \times B \to \mathbf{Z}_d, (sa, tb) \mapsto [st]$ there is no homomorphism $\widetilde{f} : A \bigotimes B \to \mathbf{Z}_d$ such that $\widetilde{f}(a \times b) = [1]$.)

   3) $\mathbf{Z}_2 \bigotimes \mathbf{Z}_3 = 0$.

4.10   Projective Modules, Injective Modules

In this section R is an arbitrary ring (with unity of course).

4.10.1 Definition. An R-module P is said to be *projective* if for any surjective R-homomorphism $h : M \to N$ of R-modules and any R-homomorphism $f : P \to N$ there is an R-homomorphism $g : P \to M$ such that $h \circ g = f$.

Thus, by Exer.4.5.4, any free R-module is a projective module; in particular, we have

4.10.2 Corollary.   *Any R-module is the homomorphic image of a projective module.*   □

However, a projective module is not necessarily free. This can be seen from the following proposition which shows that projective modules are in fact direct summands of free modules. Exer.1 gives a direct summand of a free module which is no longer a free module.

4.10.3 Proposition.   *Let P be an R-module. The following three are equivalent:*

  i) *P is projective.*

  ii) *Every exact sequence $0 \to M' \to M \xrightarrow{g} P \to 0$ consisting of R-homomorphisms and ending in P is split.*

  iii) *P is isomorphic to a direct summand of a free R-module.*

  Proof. (i) $\implies$ (ii). By (i) there is an R-homomorphism $s : P \to M$ such that $g \circ s = \mathrm{id}_P$.

  (ii) $\implies$ (iii). By Cor.4.5.7 there is a free R-module F and a surjective R-homomorphism $g : F \to P$; hence by Example 4.4.3 we have an exact sequence $0 \to M' \to F \xrightarrow{g} P \to 0$. Thus, by Prop.4.4.7, (ii) shows that P is isomorphic to a direct summand of F.

  (iii) $\implies$ (i). It follows from the following result and the fact that any free module are projective. $\square$

4.10.4 Proposition.   *The direct sum $\bigoplus_{i \in I} P_i$ of R-modules P_i, $i \in I$, is projective if and only if every P_i, $i \in I$, is projective.*

  Proof. Let $P = \bigoplus_{i \in I} P_i$, let $\iota_i : P_i \to P$, $i \in I$, be the canonical injections while $\rho_i : P \to P_i$, $I \in I$, be the canonical projections; then $\rho_j \circ \iota_j = \mathrm{id}_{P_j}$ and $\sum_{i \in I} \iota_i \circ \rho_i = \mathrm{id}_P$, see 4.3.1, 4.3.3 and cf. 4.3.5.

  The necessity. Assume P is projective. Given $j \in I$. Let $h : M \to N$ be a surjective R-homomorphism and $f : P_j \to N$ be an R-homomorphism. Consider the composition homomorphism $f \circ \rho_j : P \to N$; then there is an R-homomorphism $k : P \to M$ such that $h \circ k = f \circ \rho_j$. Set $g = k \circ \iota_j : P_j \to M$; then $h \circ g = h \circ k \circ \iota_j = f \circ \rho_j \circ \iota_j = f \circ \mathrm{id}_{P_j} = f$.

  The sufficiency is proved in a dual way, which is left as Exer.2. $\square$

  The projective modules are characterized by notation $\mathrm{Hom}_R(-, -)$.

4.10.5 Proposition.   *An R-module P is projective if and only if for any exact sequence $0 \to M' \xrightarrow{f} M \xrightarrow{g} M'' \to 0$ of R-homomorphisms the following sequence of additive homomorphisms is exact:*

$$0 \longrightarrow \mathrm{Hom}(P, M') \xrightarrow{f_*} \mathrm{Hom}(P, M) \xrightarrow{g_*} \mathrm{Hom}(P, M'') \longrightarrow 0 \,.$$

  Proof. By Exer.4.4.8 it remains to prove that P is projective if and only if g^* is surjective for any surjective R-homomorphism $g : M \to M''$. But the surjectivity of g^* means that for any $f \in \mathrm{Hom}(P, M'')$ there is a

$k \in \text{Hom}(P, M)$ such that $f = g^*(k) = g \circ k$; this is just what Def.4.10.1 says. □

4.10.6 Definition. An R-homomorphism $f : M \to N$ of R-modules is said to be *projective* if there is a free R-module F and R-homomorphisms $f' : M \to F$ and $f'' : F \to N$ such that $f = f'' \circ f'$.

4.10.7 Proposition. *Let P be an R-module. The following four are equivalent:*

  i) *P is projective.*

  ii) *Any R-homomorphism $P \to M$ is projective.*

  iii) *Any R-homomorphism $M \to P$ is projective.*

  iv) *Any R-endomorphism of P is projective.*

Proof. (i) $\implies$ (ii). By Prop.4.10.3 from (i) we can assume that there is an R-module P' such that $P \oplus P'$ is a free R-module; let $\iota : P \to P \oplus P'$ be the injection map. For any R-homomorphism $f : P \to M$, we can define an R-homomorphism $g : P \oplus P' \to M, (x, x') \mapsto f(x)$; then it is obvious that $f = g \circ \iota$.

(ii) $\implies$ (iv). Trivial.

(iv) $\implies$ (i). By (iv) the identity endomorphism id_P is projective; thus there is a free R-module F and R-homomorphisms $s : P \to F$ and $g : F \to P$ such that $\text{id}_P = g \circ s$. By Prop.4.4.7, P is isomorphic to a direct summand of F; hence, by Prop.4.10.3, P is projective.

Similarly, we have (i) $\implies$ (iii) $\implies$ (iv). □

Recall that a free module has an R-independent generator set which is called a *free basis*. We show that a projective module has a special generator set which is called a *projective basis*.

4.10.8 Theorem (Projective Basis). *An R-module P is projective if and only if there are a subset $B \subset P$ and R-homomorphisms $\delta_b : P \to R$ indexed by $b \in B$ such that, for any δ_b only finitely many images $\delta_b(x)$ for $x \in P$ are non-zero, and $x = \sum_{b \in B} \varphi_b(x) \cdot b$ for all $x \in P$.*

Proof. Assume that P is projective, and assume that B is a generator set of P. Let F be a free R-module with a free basis A such that $|A| = |B|$. By Cor.4.5.7 there is a surjective R-homomorphism $g : F \to P$ such that $g|_A$ turns out a bijection from A onto B. Then, by Prop.4.10.3, there is an R-homomorphism $s : P \to F$ such that

$g \circ s = \mathrm{id}_P$; in particular, $s|_B : B \to A$ is just the inverse of $g|_A : A \to B$. On the other hand, $F = \bigoplus_{a \in A} Ra$ and $Ra \cong R$ for all $a \in A$. Let $\rho_a : F \to R$, $a \in A$, be the coordinate maps, i.e for $y \in F$ we have $y = \sum_{a \in A} \rho_a(y) \cdot a$; in particular, only finitely many images $\rho_a(y)$ for $a \in A$ are non-zero. For $b \in B$, set $\delta_b = \rho_{s(b)} \circ s : P \to R$; then only finitely many images $\delta_b(x)$ for $x \in P$ are non-zero, and for $x \in P$ we have

$$x = \mathrm{id}_P(x) = g \circ s(x) = g\Big(\sum_{a \in A} \rho_a(s(x)) \cdot a \Big) = \sum_{a \in A} \rho_a(s(x)) \cdot g(a);$$

putting $g(a) = b$ hence $a = s(b)$, we get

$$x = \sum_{b \in B} \rho_{s(b)}(s(x)) \cdot b = \sum_{b \in B} \delta_b(x) \cdot b.$$

Conversely, assume that there are $B \subset P$ and δ_b as in the theorem. Let $h : M \to N$ be a surjective R-homomorphism and $f : P \to N$ be an R-homomorphism. For every $b \in B$ choose a $b' \in M$ such that $h(b') = f(b)$. Define:

$$g : \quad P \longrightarrow M, \qquad x \longmapsto \sum_{b \in B} \delta_b(x) \cdot b';$$

then it is easy to check that g is an R-homomorphism and $h \circ g = f$. $\square$

The dual notion of the projectivity is the injectivity.

4.10.9 Definition. An R-module J is said to be *injective* if for any injective R-homomorphism $h : N \to M$ of R-modules and any R-homomorphism $f : N \to J$ there is an R-homomorphism $g : M \to J$ such that $g \circ h = f$.

And the dual of Prop.4.10.4 and Prop.4.10.5 are clearly true; their proofs are left as Exer.5.

4.10.10 Proposition. *The direct product $\prod_{i \in I} J_i$ of R-modules J_i, $i \in I$, is injective if and only if every J_i, $i \in I$, is injective.* $\square$

4.10.11 Proposition. *An R-module J is injective if and only if for any exact sequence $0 \to M' \xrightarrow{f} M \xrightarrow{g} M'' \to 0$ of R-homomorphisms the following sequence of additive homomorphisms is exact:*

$$0 \longrightarrow \mathrm{Hom}(M'', J) \xrightarrow{g^*} \mathrm{Hom}(M, J) \xrightarrow{f^*} \mathrm{Hom}(M', J) \longrightarrow 0 \,. \qquad \square$$

The duals of Cor.4.10.2 and Prop.4.10.3 are also true (in fact, nearly all the results for injective modules which are dual to the results for projective modules hold); but their arguments are somewhat complicated; in particular, some of them need citing the Zorn's Lemma (or its equivalent version). Another remarkable point is that the injective modules are related to the so-called divisible modules. We state a group of exercises (Exer.7–13) in the following.

Exercises 4.10

1. Let K be a field and n be a positive integer; let $R = M_n(K)$ be the matrix ring (in fact, algebra). Assume that e_{ij} is the matrix with 1 in the (i,j)-entry and 0 elsewhere. Then the R-module $R = Re_{11}$ is a projective R-module but not a free R-module. (Hint: $R = Re_{11} \oplus \cdots \oplus Re_{nn}$ by Exer.4.3.3.)

2. Complete the proof of Prop.4.10.4.

3. Let M be an R-module. Then the set consisting of all the projective R-endomorphisms forms an ideal of $\operatorname{End}_R^l(M)$.

4. Let P be a projective R-module, let I be an ideal of R. Show that P/IP is a projective R/I-module. (Hint: cf. Exer.4.2.8.)

5. Prove Props.4.10.10 and 4.10.11.

6. If R is a division ring, then any R-module is both projective and injective. (Hint: Prop.4.6.8 and Exer.4.4.9.)

*7. (*Baer Criterion*). A module J over a ring R is injective if and only if, for any left ideal L of R, every R-homomorphism $\varphi : L \to J$ can be extended to an R-homomorphism $\psi : R \to J$.

   (**Sketch** of a proof for the sufficiency. For a submodule N of an R-module M and an R-homomorphism $f : N \to J$, consider the set of all the pairs (N', f') where $N \subset N' \le M$ and $f' : N' \to J$ is an extended R-homomorphism of f; with a natural partial order relation, we can cite Zorn's Lemma to get a maximal pair (N_0, f_0). Suppose $N_0 \ne M$. Take $x \in M - N_0$, set $N_1 = N_0 + Rx$ and $I = \{r \in R \mid rx \in N_0\}$; then I is a left ideal of R and $Ix = (Rx) \cap N_0$; the composition map $I \to Ix \to J$, where the first one is $r \mapsto rx$ while the second one is $rx \mapsto f_0(rx)$, can be extended to an R-homomorphism $\psi : R \to J$; and ψ induces a well-defined R-homomorphism $h_0 : Rx \to J, rx \mapsto \psi(r)$. So h_0 and f_0 coincide in $(Rx) \cap N_0$; hence by Exer.4.2.10 we get an R-homomorphism $f_1 : N_1 \to J$ which extends f_0. This contradicts the maximality of (N_0, f_0).)

8. An R-module M is said to be *divisible* if for any non-zero divisor $s \in R$ and any $m \in M$ there is an $m' \in M$ such that $sm' = m$. Prove:

   1) An injective R-module is divisible.   (Hint: $f : Rs \to M, rs \mapsto rm$ is a

well-defined R-homomorphism, it is extended to an R-homomorphism $g :$ $R \to M$; hence $m = f(s) = g(s) = g(s \cdot 1_R) = s \cdot g(1_R)$.)

2) If R is a principal ideal domain, then an R-module is injective if and only if it is divisible. (Hint: Baer Criterion.)

9. An Abelian group is said to be divisible/injective if it is divisible/injective as $\mathbf{Z}$-module. Thus, by Exer.8 above, an Abelian group is injective if and only if it is divisible; and, by Prop.4.10.10, a direct product of Abelian groups is divisible if and only if every direct factor is divisible. Prove:

1) The rational additive group $\mathbf{Q}$ is divisible.

2) Any quotient group of a divisible Abelian group is also divisible. In particular, the quotient group $\mathbf{Q}/\mathbf{Z}$ is divisible.

3) For any Abelian group G and $0 \neq x \in G$, there is a group homomorphism $f_x : G \to \mathbf{Q}/\mathbf{Z}$ such that $f_x(x) \neq 0$. (Hint: for any positive integer n there is an element in $\mathbf{Q}/\mathbf{Z}$ of order n; hence there is a homomorphism $h_x : \langle x \rangle \to \mathbf{Q}/\mathbf{Z}$; by the injectivity of $\mathbf{Q}/\mathbf{Z}$, h_x can be extended to f_x.)

4) For any Abelian group G, there is an index set I and an injective homomorphism $f : G \to \prod_{i \in I} \mathbf{Q}/\mathbf{Z}$; the latter is the product of copies of $\mathbf{Q}/\mathbf{Z}$ indexed by I. In particular, G can be embedded into an injective Abelian group. (Hint: for $0 \neq x \in G$ set $f_x : G \to \mathbf{Q}/\mathbf{Z}$ be as above; let $I = G - \{0\}$; then quote Prop.4.3.2.)

*10. Consider a ring R as a left $\mathbf{Z}$-right R-bimodule. For any $\mathbf{Z}$-module G, recall from Exer.4.2.13 that $\mathrm{Hom}_{\mathbf{Z}}(R, G)$ is an R-module. Hence, for any R-module M, we have the Abelian group $\mathrm{Hom}_R(M, \mathrm{Hom}_{\mathbf{Z}}(R, G))$; on the other hand, regarding M as an Abelian group, we have the Abelian group $\mathrm{Hom}_{\mathbf{Z}}(M, G)$.

1) For $\gamma \in \mathrm{Hom}_{\mathbf{Z}}(M, G)$ define $\tilde{\gamma} : M \to \mathrm{Hom}_{\mathbf{Z}}(R, G)$ as: $(\tilde{\gamma}(m))(r) = \gamma(rm)$, $\forall m \in M$, $r \in R$. Prove that

$$\mathrm{Hom}_{\mathbf{Z}}(M, G) \longrightarrow \mathrm{Hom}_R(M, \mathrm{Hom}_{\mathbf{Z}}(R, G)), \qquad \gamma \longmapsto \tilde{\gamma}$$

is an R-isomorphism. (Hint: check that $\tilde{\gamma}(r'm) = r'\tilde{\gamma}(m)$; a map from $\mathrm{Hom}_R(M, \mathrm{Hom}_{\mathbf{Z}}(R, G))$ to $\mathrm{Hom}_{\mathbf{Z}}(M, G)$ is defined as follows: for $\mu : M \to \mathrm{Hom}_{\mathbf{Z}}(R, G)$, set $\mu' : M \to G$ to be $\mu'(m) = (\mu(m))(1_R)$.)

2) If $N \to M$ is an R-homomorphism of R-modules, then we have a commutative diagram where the vertical arrows represent isomorphisms:

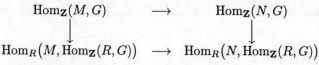

*11. If G is a divisible Abelian group, then $\mathrm{Hom}_{\mathbf{Z}}(R, G)$ is an injective R-module. (Hint: use Prop.4.10.11 and Exer.10 above.)

*12. Any R-module M can be embedded into an injective R-module.

(Hint: as an Abelian group M can be embedded into a divisible group G, see Exer.9(4) above; and $M \to \text{Hom}_{\mathbf{Z}}(R, G), m \mapsto \widehat{m}$, where $\widehat{m}(r) = rm \in G$, is an injective R-homomorphism.)

*13. An R-module J is injective if and only if every exact sequence $0 \to J \to M \to M'' \to 0$ of R-homomorphisms starting from J is split.

Chapter 5

Fields

5.1 Subfields and Extensions

The fields have been introduced in Def.3.2.6. Galois' ideas on solving polynomials by roots in the 1820s not only founded the base of the group theory, but also initiated the research on fields. One of the ideas is to consider how a field is generated based on a smaller field.

5.1.1 Definition. Let F be a field and $K \subset F$. If K is itself a field under the operations of F and $1_K = 1_F$, then we say that K is a *subfield* of F, and F is an *extension field* (or an *extension* for short) of K. A field with no proper subfields is called a *prime field*.

Example. The real number field $\mathbf{R}$ is an extension of the rational number field $\mathbf{Q}$; the complex number field $\mathbf{C}$ is an extension of the real number field $\mathbf{R}$. And, Exer.1 provides the examples of prime fields, which are in fact all the prime fields, see Cor.5.1.7 below.

For any unitary ring R, the following is a unitary homomorphism of rings (where $C(R)$ denotes the center of the ring R)

$$(5.1.2) \qquad \zeta: \ \mathbf{Z} \longrightarrow C(R), \qquad n \longmapsto n \cdot 1_R,$$

which induces, by Fundamental Theorem on Homomorphisms, an injective unitary ring homomorphism $\mathbf{Z}_m = \mathbf{Z}/\langle m \rangle \to C(R)$, where $\langle m \rangle = \text{Ker}(\zeta)$; and the non-negative integer m is called the characteristic of R and denoted $\text{char}(R) = m$; it is in fact the order of the identity element 1 in the additive group of R; cf. Prop.4.8.7 and Def.4.8.8.

In particular, the *characteristic* char(F) of a field F is defined.

5.1.3 Proposition.   *The characteristic of a field is a prime number if it is not zero (in fact, this holds for any unitary ring with no non-zero zero divisors).*

Proof. Let $m = \text{char}(F)$ for a field F. Assume $m = nk$; then $0 = (nk) \cdot 1 = nk \cdot (1 \cdot 1) = (n \cdot 1)(k \cdot 1)$. Thus either $n \cdot 1 = 0$ or $k \cdot 1 = 0$, hence either $n = m$ or $k = m$. That is, m is a prime number. $\square$

We list elementary properties on the characteristics and leave the proof as Exer.2. Note that, except the above fact, all the others of the following proposition are general facts on unitary rings; compare with Exer.4.8.7.

5.1.4 Proposition.   1) char$(K) = $ char(F) *for any subfield K of F.*

2) *If* char$(F) = 0$, *then* $na \neq 0$ *for any* $0 \neq a \in F$ *and* $0 \neq n \in \mathbf{Z}$.

3) *If* char$(F) = p > 0$, *then* $na = 0 \iff p|n$ *for all* $0 \neq a \in F$ *and* $n \in \mathbf{Z}$;   *and (binomial formula)*

$$(a \pm b)^{p^n} = a^{p^n} \pm b^{p^n}, \quad \text{for all } a, b \in F \text{ and } n > 0.\qquad \square$$

5.1.5 Definition. A unitary ring homomorphism $\varphi : F \to F'$ from a field F to a field F' is also called a *homomorphism of fields* (or *field homomorphism*). Such a map φ preserves the field structures, and it must be injective, see Exer.4; further, if φ is bijective, then φ is called an *isomorphism of fields* (or *field isomorphism*), and F is said to be isomorphic to F' and denote $F \cong F'$. The homomorphism from F to itself is called an *endomorphism* of F, while an isomorphism from F to itself is called an *automorphism* of F. Clearly, the set Aut(F) of all automorphisms of F is a group, called the *automorphism group* of F.

It is easy to see from Exer.3 that any field F has a unique minimal subfield (smallest subfield), i.e. the intersection of all subfields; and the smallest subfield has to be a prime field, hence it is the unique *prime subfield* of F. Since any subfield contains 1_F, we can recognize the smallest subfield easily.

5.1.6 Theorem.   *Let F be a field. If* char$(F) = p > 0$, *then F contains an isomorphic copy of $\mathbf{Z}_p$ (the residue class ring modulo p) as its smallest subfield; otherwise F contains an isomorphic copy of $\mathbf{Q}$ (the rational field) as its smallest subfield.*

Proof. Consider the homomorphism (5.1.2) again:

$$\zeta: \ \mathbf{Z} \longrightarrow F, \qquad n \longmapsto n \cdot 1.$$

If $\mathrm{char}(F) = p \neq 0$, then ζ induces an injective homomorphism $\overline{\zeta}$: $\mathbf{Z}_p \to F$; however, $\mathbf{Z}_p$ is a field, i.e. F contains the subfield $\mathrm{Im}(\overline{\zeta}) \cong \mathbf{Z}_p$; since any subfield contains 1, hence must contain $\mathrm{Im}(\overline{\zeta})$. Next assume $\mathrm{char}(F) = 0$, i.e. ζ is injective; in particular, $\zeta(n)$ for $0 \neq n \in \mathbf{Z}$ is invertible in F. Then, since the fraction field of $\mathbf{Z}$ is $\mathbf{Q}$, by Theorem 3.7.4 there is an injective homomorphism $\widetilde{\zeta}: \mathbf{Q} \to F$; hence, F contains the subfield $\mathrm{Im}(\widetilde{\zeta}) \cong \mathbf{Q}$. Similarly to the above, any subfield must contain $\mathrm{Im}(\widetilde{\zeta})$. $\qquad\square$

As a consequence, all prime fields are classified up to isomorphism.

5.1.7 Corollary. *Let K be a prime field. Then $K \cong \mathbf{Z}_p$ if $\mathrm{char}(K) = p > 0$; otherwise $K \cong \mathbf{Q}$.*

Proof. The smallest subfield of K is just K itself. $\qquad\square$

5.1.8 Remark. Let F be a field and φ be an endomorphism of F. The set $\{a \in F \mid \varphi(a) = a\}$ is a subfield (see Exer.5), called the *subfield of the φ-fixed points*. This subfield must contain the prime subfield of F. In other words, any endomorphism of F fixes every element of the prime subfield. In particular, a prime field has only the identity endomorphism.

5.1.9 Definition. Let F be an extension field of a field K and $S \subset F$. The smallest subfield containing both K and S (by Exer.3, it is in fact the intersection of all the subfields containing both K and S) is called the *extension generated by S over K*, and denoted by $K(S)$. If $S = \{s_1, \cdots, s_n\}$, then we also denote $K(S) = K(s_1, \cdots, s_n)$. In particular, the extension $K(u)$ over K generated by a single element u is called a *single extension*.

5.1.10 Remark. A related notation is $K[S]$, which denotes the subring of F generated by S over K, i.e. $K[S]$ consists of all the elements $f(s_1, \cdots, s_n)$, where $f(x_1, \cdots, x_n)$ runs over the K-polynomials in finitely many indeterminates x_i and $s_1, \cdots, s_n \in S$; see 3.8.1. Now, for $K(S)$ we have

5.1.11 Proposition. *Notations as above. Then $K(S)$ consists of the elements $f(s_1, \cdots, s_n)/g(s_1, \cdots, s_n)$, where $f(x_1, \cdots, x_n)/g(x_1 \cdots, x_n)$*

runs over the rational K-fractions (i.e. with K-coefficients) in finitely many indeterminates x_i and $s_1, \cdots, s_n \in S$ such that $g(s_1, \cdots, s_n) \neq 0$.

Proof. Let $K' = \{$all the elements $f(s_1, \cdots, s_n)/g(s_1, \cdots, s_n)\}$. It is easy to see that K' is a subfield of F and contains both K and S. Conversely, if a subfield F' of F contains both K and S, then F' contains all the elements $f(s_1, \cdots, s_n)/g(s_1, \cdots, s_n)$. Thus, K' is the smallest subfield of F containing both K and S; i.e. $K' = K(S)$.   □

5.1.12 Corollary. *Notations as above. For any $u \in K(S)$ there is a finite subset $S_0 \subset S$ such that $u \in K(S_0)$.*

Proof. By the above result, $u = f(s_1, \cdots, s_n)/g(s_1, \cdots, s_n)$; hence $u \in K(s_1, \cdots, s_n)$.   □

Exercises 5.1

+1. Prove that the rational field **Q** is a prime field of characteristic zero; while the residue ring $\mathbf{Z}_p$ of the integer ring **Z** modulo a prime integer p is a prime field of characteristic p.

   (Hint: any subfield must contain the identity element 1, hence contain all multiples $n \cdot 1$ of 1, and contain the inverse of $n \cdot 1$ if $n \cdot 1 \neq 0$.)

2. Prove Prop.5.1.4.

+3. Prove: the intersection of some subfields of a field is also a subfield.

4. Let F and F' be fields and $\varphi : F \to F'$ be a unitary ring homomorphism (i.e. a field homomorphism). Prove:

   1) $\text{Ker}(\varphi) = \{0\}$, i.e. φ is injective; and $\varphi(a)^{-1} = \varphi(a^{-1})$, $\forall \, 0 \neq a \in F$.

   2) $\text{Im}(\varphi)$ is a subfield of F'.

   3) φ maps the prime subfield of F bijectively onto the prime subfield of F'.

+5. If φ is an endomorphism of F, then $\{a \in F \mid \varphi(a) = a\}$ is a subfield of F.

+6. A field of characteristic zero is an infinite field; in particular, the characteristic of a finite field must be a prime.

7. Show that $\mathbf{Z}[i]/\langle 1 + i \rangle$ is a field; and find the characteristic of it.

+8. Let F be a field of characteristic $p > 0$. Prove:

   1) $(a_1 \pm \cdots \pm a_m)^{p^n} = a_1^{p^n} \pm \cdots \pm a_m^{p^n}$ for all $a_1, \cdots, a_m \in F$.

   2) The map $\rho : F \to F$, $a \mapsto a^p$ is an endomorphism; hence $\rho_t : F \to F$, $a \mapsto a^{p^t}$ for $t \geq 1$ is also an endomorphism and $\rho_t = \rho_1^t$.

   3) The subset $\{a \in F \mid a^{p^t} = a\}$ is a subfield of F. (cf. Exer.5 above.)

9. Assume F is a field and p is a prime number. If all the elements of F are roots of the polynomial $x^p - x$, then $F \cong \mathbf{Z}_p$.

10. Assume that p is a prime number and F is a field with $|F| \geq p$. If $(a+b)^p = a^p + b^p$ for all $a, b \in F$, then $\text{char}(F) = p$.

    (Hint: consider the equation $(x+1)^p = x^p + 1$.)

+11. Let F be a field. Prove: any finite subgroup of the multiplicative group F^* of F is a cyclic group. (Hint: use Prop.2.8.14 and Cor.3.10.9.)

+12. Let F be an extension field of K and $S, T \subset F$. Prove: $K(S)(T) = K(S \cup T)$.

5.2 Single Extensions

Let K be a field. We will follow the idea in §2.8 (where all cyclic groups are classified) to classify all single extensions over K. Our classification will be up to the following sense.

5.2.1 Definition. Let both F and F' be extensions of K. A homomorphism $\varphi : F \to F'$ is said to be a *K-homomorphism* if $\varphi(a) = a$ for all $a \in K$. And F is said to be *K-isomorphic* to F', denoted by $F \overset{K}{\cong} F'$, if there is a K-isomorphism $\varphi : F \overset{\sim}{\to} F'$. It is clear that the set $\text{Aut}_K(F)$ of all K-automorphisms of F is a group, called the *K-automorphism group* of the field F.

As a preparation (just like Lemma 2.8.1), for the polynomial ring $K[x]$ we list some properties which have been proved previously.

5.2.2 Lemma. 1) $K[x]$ *is a Euclidean ring, hence a principal ideal domain, hence a factorial domain. (Cor.3.9.15.)*

2) $p(x) \in K[x]$ *is irreducible* $\iff$ $p(x)$ *is prime* $\iff$ $\langle p(x) \rangle$ *is a prime ideal* $\iff$ $\langle p(x) \rangle$ *is a maximal ideal* $\iff$ $K[x]/\langle p(x) \rangle$ *is a field. (Cor.3.9.11 and Cor.3.5.10.)*

3) *The fraction field of $K[x]$ is the field $K(x)$ consisting of all the rational K-fractions $f(x)/g(x)$. (cf. Example 3.7.5.)* □

Let F be an extension of K and $u \in F$. Consider the single extension $K(u)$. Following Theorem 3.8.3, we have the following unitary ring homomorphism:

(5.2.3)
$$\xi : K[x] \longrightarrow K(u), \quad f(x) \longmapsto f(u);$$
$$\text{Ker}(\xi) = \{ f(x) \in K[x] \mid f(u) = 0 \} = \langle p(x) \rangle;$$

where Ker(ξ), by Lemma 5.2.2(1), is generated by a single $p(x)$. According to $p(x) \neq 0$ or $p(x) = 0$, there are two different cases far from each other (compare with Def.3.8.5).

5.2.4 Definition. Let F be an extension of a field K and $u \in F$ as above. $f(\lambda) \in K[\lambda]$ (λ denotes the indeterminate, because $K[x]$ has been considered) is said to be an *annihilating K-polynomial* of u if $f(u) = 0$. And u is called an *algebraic element* over K if u has non-zero annihilating K-polynomials; otherwise, u is called a *transcendental element* over K.

For algebraic elements we have a further description.

5.2.5 Proposition and Definition.   *Notations as above. If u is an algebraic element over a field K, then:*

1) *the set of all the annihilating K-polynomials of u is a non-zero ideal of $K[\lambda]$, called the* annihilating ideal *of u over K;*

2) *any non-zero annihilating K-polynomial $p(\lambda)$ of u with lowest degree is a generator of the annihilating ideal; in particular, a K-polynomial $f(\lambda)$ annihilates $u \iff p(\lambda) \mid f(\lambda)$; such a $p(\lambda)$ is called a* minimal polynomial *over K (or* minimal K-polynomial*) of u;*

3) *all minimal K-polynomials of u are associates (i.e. each is a non-zero constant multiple of the other), and they are irreducible.*

Proof. All the facts are clear from 5.2.2 and (5.2.3) except the last one. If $p(u) = p_1(u)p_2(u) = 0$ in $K(u)$, then one of the factors is zero; this proves the irreducibility of $p(\lambda)$ (the arguments for case 1 below contains another proof). Also, Exer.1 provides an elementary proof for all the facts.   □

Continuing to consider (5.2.3); according to the two cases, we will classify the single extensions respectively.

Case 1. $p(x) \neq 0$; i.e. u is an algebraic element over K with minimal K-polynomial $p(\lambda)$.

The Fundamental Theorem on Homomorphisms gives an injective homomorphism $\overline{\xi} : K[x]/\langle p(x) \rangle \to K(u)$. Note that $\operatorname{Im}(\overline{\xi})$ is an integral domain because $K(u)$ has no non-zero zero divisors, so $\langle p(x) \rangle$ is a prime ideal by Cor.3.5.4; hence $p(x)$ is an irreducible polynomial (this is just the last part of Prop.5.2.5), and consequently, $\operatorname{Im}(\xi) \cong K[x]/\langle p(x) \rangle$ is a field by Lemma 5.2.2(2). But $u \in \text{Im}(\xi)$, thus $K(u) = \text{Im}(\overline{\xi})$, and $\overline{\xi}$ is an isomorphism and the following diagram is commutative:

(5.2.6)

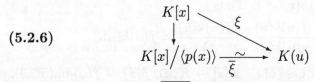

For convenience, denote $\overline{K[x]} = K[x]/\langle p(x)\rangle$. As usual, we can assume that $K \subset K[x]$; identifying K with its image in $\overline{K[x]}$, we can assume that $K \subset \overline{K[x]}$; and it is easy to verify that $\overline{\xi}$ is a K-isomorphism and $\overline{\xi}(v) = u$, where $v = \overline{x} = x + \langle p(x)\rangle \in \overline{K[x]}$.

On the other hand, consider the extension field $\overline{K[x]} = K[x]/\langle p(x)\rangle$ over K and $v = \overline{x}$ as above. Then every $\overline{f(x)} \in \overline{K[x]}$ for $f(x) = \sum_i a_i x^i \in K[x]$ is expressed as $\overline{f(x)} = \sum_i \overline{a_i}\overline{x} = \sum_i a_i v^i$ (remember that for $a_i \in K$ we have identified $\overline{a_i} \in \overline{K[x]}$ with a_i). That is, $\overline{K[x]} = K(v)$ is a single extension. Further, for $h(\lambda) = \sum_i a_i \lambda^i \in K[\lambda]$, similarly to the above we have $h(v) = \sum_i a_i v^i = \overline{\sum_i a_i x^i} = \overline{h(x)} \in \overline{K[x]}$; so $h(v) = 0 \iff h(x) \in \langle p(x)\rangle \iff p(\lambda) \mid h(\lambda)$. Therefore, in $\overline{K[x]}$, $v = \overline{x}$ is an algebraic element with minimal K-polynomial $p(\lambda)$.

Summarizing the above discussion, we get

5.2.7 Theorem. *Let K be a field and $p(\lambda)$ be an irreducible K-polynomial. Then there is a single extension $K(v)$ over K generated by an algebraic element v with minimal K-polynomial $p(\lambda)$; and for any single extension $K(u)$ generated by an algebraic element u with minimal K-polynomial $p(\lambda)$ there is a K-isomorphism $K(v) \overset{K}{\cong} K(u)$ mapping v to u.* □

5.2.8 Corollary. *Let F be an extension of K and $u \in F$. Then u is an algebraic element over K if and only if $K(u) = K[u]$.*

Proof. The "only if" is clear since $K(u) \cong K[x]/\langle p(x)\rangle$. For "if", since u has an inverse in $K(u) = K[u]$, we have a K-polynomial $f(\lambda)$ such that $u \cdot f(u) = 1$; in particular, $f(u) \neq 0$ hence $f(\lambda) \neq 0$; thus the non-zero K-polynomial $\lambda f(\lambda) - 1$ annihilates u. □

5.2.9 Remark. There is another way to show that $K(u) = K[u]$, i.e. the so-called "rationalizing denominator". Assume the algebraic element u over K has a minimal K-polynomial $p(\lambda)$. For any $f(u)/g(u) \in$

$K(u)$ we have $p(\lambda) \nmid g(\lambda)$ since $g(u) \neq 0$. Noting that $p(\lambda)$ is irreducible, we have $p(\lambda)h(\lambda) + g(\lambda)k(\lambda) = 1$ for some $h(\lambda), k(\lambda) \in K[\lambda]$; so $g(u)k(u) = 1 - p(u)h(u) = 1$. Thus

$$\frac{f(u)}{g(u)} = \frac{f(u)k(u)}{g(u)k(u)} = \frac{f(u)k(u)}{1} = f(u)k(u).$$

Consider the other case of $\xi : K[x] \to K(u)$, $f(x) \mapsto f(u)$ in (5.2.3).

Case 2. $p(x) = 0$; i.e. u is a transcendental element over K.

In this case, ξ is injective; so $\xi(f)$ is invertible in $K(u)$ for all $0 \neq f \in K[x]$. Recalling from Lemma 5.2.2(3) that $K(x)$ is the fraction field of $K[x]$, by Theorem 3.7.4 we have an injective homomorphism $\tilde{\xi} : K(x) \to K(u)$ which is also surjective since $u \in \mathrm{Im}(\tilde{\xi})$. Thus $\tilde{\xi}$ is an isomorphism and the following diagram is commutative

(5.2.10)

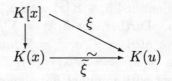

On the other hand, $K(x)$ is clearly a single extension of K generated by x which is a transcendental element over K. And, clearly, $\tilde{\xi}(a) = a$, $\forall a \in K$; and $\tilde{\xi}(x) = u$.

5.2.11 Theorem. *Let K be a field. Then there is a single extension $K(v)$ over K generated by a transcendental element v; and for any single extension $K(u)$ generated by a transcendental element u there is a K-isomorphism $K(v) \overset{K}{\cong} K(u)$ mapping v to u.* $\qquad\qquad\square$

5.2.12 Remark. The (5.2.6) and (5.2.10) resp. are the perfect complements to the existence of Theorem 5.2.7 and Theorem 5.2.11 resp., they provides the concrete single extensions $K[x]/\langle p(x)\rangle$ and $K(x)$ resp. On the other hand, the uniqueness of Theorem 5.2.7 and Theorem 5.2.11 can be extended slightly.

5.2.13 Lemma. *Let $\sigma : K \overset{\sim}{\to} K'$ be an isomorphism of fields K and K'.*

1) *σ can be extended to an isomorphism $\tilde{\sigma} : K[x] \overset{\sim}{\to} K'[x]$ such that $\tilde{\sigma}f(x) = \sum_i \sigma(a_i)x^i$ where $f(x) = \sum_i a_i x^i \in K[x]$.*

2) *For $p(x) \in K[x]$, $\widetilde{\sigma}$ induces an isomorphism $\overline{\widetilde{\sigma}} : K[x]/\langle p(x)\rangle \to$ $K'[x]/\langle\widetilde{\sigma}p(x)\rangle$ such that $\overline{\widetilde{\sigma}}(f(x) + \langle p(x)\rangle) = \widetilde{\sigma}f(x) + \langle\widetilde{\sigma}p(x)\rangle$; in particular, $p(x)$ is irreducible if and only if $\widetilde{\sigma}p(x)$ is irreducible.*

3) *$\widetilde{\sigma}$ can be extended to an isomorphism $\overline{\widetilde{\sigma}} : K(x) \overset{\sim}{\to} K'(x)$ such that $\overline{\widetilde{\sigma}}(f(x)/g(x)) = \widetilde{\sigma}f(x)/\widetilde{\sigma}g(x)$.*

Proof. (1) It follows from Theorem 3.8.3.

(2) The restriction map $\widetilde{\sigma}|_{\langle p(x)\rangle}$ turns out to be an isomorphism $\widetilde{\sigma}|_{\langle p(x)\rangle} : \langle p(x)\rangle \to \langle\widetilde{\sigma}p(x)\rangle$; hence the isomorphism $\overline{\widetilde{\sigma}}$ is induced by Theorem 3.4.1 (cf. 3.4.2).

(3). It follows from Theorem 3.7.4 at once. □

5.2.14 Theorem. *Let $\sigma : K \overset{\sim}{\to} K'$ be an isomorphism of fields K and K'.*

1) *If $K(u)$ and $K'(u')$ are extensions of K and K' resp. generated by algebraic elements u and u' resp., and if $p(x)$ is a minimal K-polynomial of u while $\widetilde{\sigma}p(x)$ is a minimal K'-polynomial of u', then σ can be uniquely extended to an isomorphism $\widehat{\sigma} : K(u) \overset{\sim}{\to} K'(u')$ such that $\widehat{\sigma}(u) = (u')$.*

2) *If $K(u)$ and $K'(u')$ are extensions of K and K' resp. generated by transcendental elements u and u' resp., then σ can be uniquely extended to an isomorphism $\widehat{\sigma} : K(u) \overset{\sim}{\to} K'(u')$ such that $\widehat{\sigma}(u) = (u')$.*

Proof. For the existence, (1) follows from (5.2.6) and Lemma 5.2.13(2); and (2) follows from (5.2.10) and Lemma 5.2.13(3). For the uniqueness, any element of $K(u)$ is expressed by u and elements of K, see Prop.5.1.11. □

Exercises 5.2

1. Let $F \supset K$ be fields.
   1) If $f(x) = g(x)h(x)$ be K-polynomials and $u \in F$ such that $f(u) = 0$, then either $g(u) = 0$ or $h(u) = 0$.
   2) Assume $f(x)$ and $p(x)$ are K-polynomials and $p(x)$ is irreducible. If $f(x)$ and $p(x)$ have a root in F in common, then $p(x)|f(x)$. (Hint: the g.c.d. of $f(x)$ and $p(x)$ is either 1 or $p(x)$.)
   3) Use the above results to prove Prop.5.2.5.
+2. Notations as in Exer.1. If $p(x)$ is an irreducible K-polynomial and $u \in F$ is a root of $p(x)$, then $p(x)$ is a minimal K-polynomial of u; show that,

however, $p(x)$ is not a minimal F-polynomial of u in general. (For example, $x^2 - 2$ is a minimal $\mathbf{Q}$-polynomial of $\sqrt{2}$, but not a minimal $\mathbf{R}$-polynomial of $\sqrt{2}$.)

3. Use the method in 5.2.9 to rationalize the denominator of $\frac{\sqrt[3]{2}+1}{\sqrt[3]{4}+\sqrt[3]{2}+1}$.

4. Find the minimal $\mathbf{Q}$-polynomial of $\frac{2i+1}{i-1}$, where i is the imaginary unit.

+5. Let $F \supset K$ be extension field. If u is an algebraic element over K, then u is an algebraic element over F. Show that an algebraic element over F is not necessarily an algebraic element over K. (Hint: π is an algebraic element over $\mathbf{R}$ but not an algebraic element over $\mathbf{Q}$.)

+6. Let $K(u)$ be an extension of a field K generated by an algebraic element u with a minimal K-polynomial $p(\lambda) = \sum_i a_i \lambda^i$; let F' be a field, and $\sigma : K \to F'$ a homomorphism and $K' = \sigma(K)$.

   1) If $\varphi : K(u) \to F$ is a homomorphism extending σ and $u' = \varphi(u)$, then $K'(u')$ is a single extension of K' and u' is algebraic over K' and has $\widetilde{\sigma}p(x) = \sum_i \sigma(a_i)\lambda^i$ as a minimal K'-polynomial.

   2) If $u' \in F'$ is an algebraic element over K' with $\widetilde{\sigma}p(\lambda)$ as a minimal K'-polynomial, then σ can be uniquely extended to a homomorphism $\varphi : K(u) \to F'$ such that $\varphi(u) = u'$.

7. State and prove the result for transcendental elements similar to Exer.6 above.

+8. Let $K(u)$ be a single extension over a field K and u have a K-polynomial $p(\lambda)$ of degree n. Prove: any K-automorphism of $K(u)$ maps u to a root of $p(\lambda)$; and $K(u)$ has at most n K-automorphisms.

9. Let $F = \mathbf{Z}_p(x)$ be the field of the rational fractions over the prime field $\mathbf{Z}_p$ (p is a prime number). Describe the single extension $F(u)$ by an algebraic element u with minimal F-polynomial $\lambda^p = x$ (in indeterminate λ), and factorize $\lambda^p - x$ in the extension field $F(u)$. (Hint: cf. Prop.5.1.4.)

10. Let $K(u)$ be a single extension of a field K by a transcendental element u, and $K \subsetneq F \subset K(u)$ be a subfield. Show that u is an algebraic element over F.

5.3   Algebraic Extensions

5.3.1 Definition. We say that F is an *algebraic extension* over a field K (or, the extension F over K is *algebraic*) if every $u \in F$ is an algebraic element over K; otherwise, we say that F is *transcendental* over K.

A remarkable fact is that *an extension generated by algebraic ele-*

ments is an algebraic extension. To prove it, first we show a criterion for algebraic elements in the following notation.

5.3.2 Definition. Let $F \supset K$ be fields. Multiplying the elements of K on F, by Examples 4.1.4 and 4.1.2, F is a vector space over K; the dimension of F over K, denoted by $|F : K|$, is called the *degree* of F over K. F is said to be a *finite extension* over K if $|F : K| < \infty$.

5.3.3 Lemma (Formula on Degrees). *Let $E \supset F \supset K$ be fields. Then $|E : K| = |E : F| \cdot |F : K|$.*

Remark. *The formula is clearly extended as: for fields $F_n \supset \cdots \supset F_1 \supset K$, we have $|F_n : K| = |F_n : F_{n-1}| \cdots |F_1 : K|$.*

Proof. Assume that U is a K-basis of F, and V is a F-basis of E. It is enough to show that uv, $u \in U$ and $v \in V$, form a K-basis of E.

First, if $u_1, \cdots, u_m \in U$ and $v_1, \cdots, v_n \in V$ and $a_{ij} \in K$ such that $\sum a_{ij} u_i v_j = 0$, then $0 = \sum_j (\sum_i a_{ij} u_i) v_j$, and by the F-independence of V we have $\sum_{i=1}^{m} a_{ij} u_i = 0$, $j = 1, \cdots, n$; then by the K-independence of U, we have $a_{ij} = 0$ for all $1 \le i \le m$ and $1 \le j \le n$. That is, uv, $u \in U$ and $v \in V$, are K-independent.

Second, for any $w \in E$ there are $v_1, \cdots, v_n \in V$ and $b_1, \cdots, b_n \in F$ such that $w = \sum_j b_j v_j$; next, every $b_j \in F$ is a finite K-combination of U; so we can find finitely many $u_1, \cdots, u_m \in U$ such that every $b_j = \sum_i a_{ij} u_i$ with $a_{ij} \in K$; thus $w = \sum_{i,j} a_{ij} u_i v_j$. $\square$

The following lemma shows why the dimension of an extension is called "degree".

5.3.4 Lemma. *Let K be a field and u be an algebraic element over K with minimal K-polynomial $p(\lambda)$. Then $|K(u) : K| = \deg(p(\lambda))$.*

Proof. Assume $p(\lambda) = \lambda^n + a_{n-1}\lambda^{n-1} + \cdots + a_0$; then $u^n = -a_{n-1}u^{n-1} - \cdots - a_0$. Note that $K(u) = K[u]$ by Cor.5.2.8; so $K[u] = K + Ku + \cdots + Ku^{n-1}$, i.e. $\{1, u, \cdots, u^{n-1}\}$ generates the K-space $K[u]$. Moreover, suppose $a_0 + a_1 u + \cdots + a_{n-1}u^{n-1} = 0$ but $a_0, a_1, \cdots, a_{-1} \in K$ were not all zero, then there would be a non-zero K-polynomial of degree $< n$ which annihilates u. Thus $1, u, \cdots, u^{n-1}$ form a K-basis of the K-vector space F. $\square$

5.3.5 Corollary. *An element u is algebraic over K if and only if $|K(u) : u| < \infty$.* $\square$

5.3.6 Corollary.

1) *A finite extension F of K is an algebraic extension of K.*

2) *F is a finite extension of K if and only if F is generated by finitely many algebraic elements over K.*

Proof. (1) For any $u \in F$, since $F \supset K(u) \supset K$, we clearly have $|K(u) : K| \leq |F : K| < \infty$; hence u is algebraic over K by Cor.5.3.5.

(2) The necessity is clear by (1). Conversely, let $F = K(s_1, \cdots, s_n)$ with every s_i algebraic over K. By Lemma 5.3.3 we have

$$|F : K| = |K(s_1, \cdots, s_{n-1}, s_n) : K(s_1, \cdots, s_{n-1})| \cdots |K(s_1) : K|.$$

Then, by Exer.5.2.5 and Cor.5.3.5, we have $|F : K| < \infty$.   □

Remark. In fact, any finite extension can be generated by a single algebraic element; but the proof is somewhat difficult.

5.3.7 Example. The complex field $\mathbf{C}$ is an extension of the real field $\mathbf{R}$ of degree 2; hence $\mathbf{C}$ is an algerbaic extension of $\mathbf{R}$.

5.3.8 Theorem. *Any extension of a field K generated by algebraic elements over K is an algebraic extension over K.*

Proof. Let $F = K(S)$ and every $s \in S$ is algebraic over K. If S is finite, then F is algebraic over K by Cor.5.3.6. In general case, for every $s \in F$, by Cor.5.1.12, there is a finite subset $S_0 \subset S$ such that $u \in K(S_0)$, and $K(S_0)$ is a algebraic over K since S_0 is finite; so u is an algebraic element over K.   □

5.3.9 Corollary. *In any extension of a field K, the subset of all algebraic elements over K is a subfield, hence it is an algebraic extension of K; in particular, the sum, difference, product and quotient (the denominator is non-zero) of two algebraic elements over K are also algebraic over K.*

Proof. Exer.2.   □

5.3.10 Theorem. *Let $E \supset F \supset K$ be fields. Then E is algebraic over K if and only if both E over F and F over K are algebraic.*

Proof. The necessity is clear, cf. Exer.5.2.5.

Conversely, if $u \in E$ is algebraic over F, then there are $a_0, a_1, \cdots, a_n \in F$ with $a_n \neq 0$ such that $a_n u^n + \cdots + a_1 u + a_0 = 0$; in particular,

$$|K(a_0, a_1, \cdots, a_n)(u) : K(a_0, a_1, \cdots, a_n)| < \infty.$$

Further, since F over K is algebraic, by Cor.5.3.6(2) we have

$$|K(a_0, a_1, \cdots, a_n) : K| < \infty.$$

Thus $|K(a_0, a_1, \cdots, a_n, u) : K| < \infty$ by lemma 5.3.3. Then u is algebraic over K by Cor.5.3.6(1). □

5.3.11 Example. The complex field $\mathbf{C}$ is a transcendental extension over the rational field $\mathbf{Q}$. A complex number a is said to be an *algebraic number* if a is an algebraic element over $\mathbf{Q}$. Thus we have that the sum, the difference, the product and the quotient of algebraic numbers are also algebraic numbers. An *algebraic number field* means a subfield of $\mathbf{C}$ which is algebraic over $\mathbf{Q}$, i.e. consists of algebraic numbers. In particular, by Cor.5.3.9 and Theorem 5.3.10 (cf. the proof of 5.3.13 below), we have

"*All the algebraic numbers form a field E, called the* algebraic number field; *and E has no proper algebraic extension.* "

That is, in the following notation, E is an algebraic closure of $\mathbf{Q}$.

5.3.12 Definition. A field F is called an *algebraically closed field* if it has no proper algebraic extension. A field F is called an *algebraic closure* of a field K if F is an algebraic extenson of K and F is algebraically closed.

Similarly to Example 5.3.11, we have

5.3.13 Proposition. *Let F be an algebraic extension over a field K. If every K-polynomial of degree > 0 has a root in F, then F is algebraically closed, hence F is an algebraic closure of K.*

Proof. By Prop.5.3.14 below, every K-polynomial of degree > 0 is split in F. Assume that E is an algebraic extension of F; then E is algebraic over K by Theorem 5.3.10, hence any $u \in E$ is algebraic over K; so the minimal K-polynomial of u is split in F, consequently, $u \in F$; that is, $E = F$. Thus F is algebraically closed. □

5.3.14 Proposition. *Let F be a field. The following four are equivalent:*

i) *F is algebraically closed.*

ii) *Any F-polynomial of degree > 0 has at least one root in F.*

iii) *Any F-polynomial of degree > 0 can be factorized as a product of F-polynomials of degree 1 (in that case we say the polynomial is split in F).*

iv) *Any irreducible F-polynomial has degree* 1.

    Proof.   Exer.10.   □

Exercises 5.3

1. Prove Cor.5.3.5.
2. Prove Cor.5.3.9.
3. Let $F \supset K$ be fields and $u, v \in F$. If $|K(u) : K| = m$ and $|K(v) : K| = n$ and $(m, n) = 1$ (i.e. m and n are coprime), then $|K(u, v) : K| = mn$.
4. If $|K(u) : K|$ is an odd number, then $K(u^2) = K(u)$. (Hint: $|K(u) : K(u^2)| =$ either 1 or 2.)
5. Prove: $\mathbf{Q}(\sqrt{2}, \sqrt[3]{2}) = \mathbf{Q}(\sqrt{2} + \sqrt[3]{2})$.
6. If $|F : K|$ is a prime integer, then for any $u \in F - K$ we have $F = K(u)$.
7. Let $F \supset K$ be field, and both L, M be fields between F and K. By LM we denote the smallest subfield of F which contains both L and M.
   1) If $L = K(S)$, then $LM = M(S)$.
   2) If $L = K(S)$ and $M = K(T)$, then $LM = K(S \cup T)$.
   3) Find a condition such that $LM = L \cup M$.
8. Let F, K, L and M be as in Exer.7 above. Prove:
   1) $|LM : K|$ is finite if and only if both $|L : K|$ and $|M : K|$ are finite.
   2) If $|LM : K|$ is finite, then $|L : K|\big||LM : K|$ and $|M : K|\big||LM : K|$ and $|LM : K| \leq |L : K| \cdot |M : K|$.
   3) If $|L : K|$ and $|M : K|$ are finite and coprime, then $|LM : K| = |L : K| \cdot |M : K|$.
9. Let F, K, L and M be as in Exer.7 above. Assume $|F : K| < \infty$. Prove:
   1) If $|LM : K| = |L : K| \cdot |M : K|$, then $L \cap M = K$.
   2) The converse of (1) holds if one of $|L : K|$ and $|M : K|$ equals 2.
10. Prove Prop.5.3.14.
11. Any finite field is not algebraically closed.

5.4   Splitting Fields, Normal Extensions

In order to solve a polynomial equation over a field K, it is reasonable to consider the extensions which contain *exactly* all the roots of the polynomial; such extensions exist and are unique up to isomorphism.

5.4.1 Definition. Let K be a field and $f(x) \in K[x]$ with $\deg f(x) > 0$. An extension F of K is called a *splitting field* of $f(x)$ over K if the following two hold:

    1) $f(x) = a(x - u_1)(x - u_2) \cdots (x - u_n)$ with all $u_i \in F$ (for convenience we say $f(x)$ is *split* in F if this is the case);

    2) $F = K(u_1, u_2, \cdots, u_n)$.

Example. If a field F is algebraically closed, then F is a splitting field of any F-polynomial of degree > 0. The converse also holds; see Prop.5.3.13.

We show an enlightening example. Consider the rational field $\mathbf{Q}$ and $f(x) = (x^2 - 2)(x^2 - 3)$. In the complex field $\mathbf{C}$, $\mathbf{Q}(\sqrt{2})$ is a splitting field of $x^2 - 2$. But, $\mathbf{Q}(\sqrt{2})$ is not a splitting field of $f(x)$; for, otherwise we have $(a + b\sqrt{2})^2 - 3 = 0$ with $a, b \in \mathbf{Q}$, i.e. $a^2 + 2b^2 - 3 = 0$ and $2ab = 0$ which are impossible for $a, b \in \mathbf{Q}$. Obviously, $\mathbf{Q}(\sqrt{2}, \sqrt{3})$ is a splitting field of $f(x)$.

Putting the complex field aside, on the other hand, the theory on single extensions in §5.2 suggests us to construct a splitting field for $f(x)$ as follows. Firstly, since $x^2 - 2$ is an irreducible $\mathbf{Q}$-polynomial, by Theorem 5.2.7, there is an extension $F_1 = \mathbf{Q}(u)$ generated by u with minimal $\mathbf{Q}$-polynomial $x^2 - 2$; hence $u^2 - 2 = 0$ and $x^2 - 2 = (x - u)(x + u)$. Secondly, similarly to the above, $x^2 - 3$ has no root in F_1, hence $x^2 - 3$ is an irreducible F_1-polynomial; by Theorem 5.2.7 again, we have an extension $F = F_1(v)$ such that v has $x^2 - 3$ as a minimal F_1-polynomial; hence $x^2 - 3 = ((x - v)(x + v))$. Thus $F = \mathbf{Q}(u, v) = \mathbf{Q}(u, -u, v, -v)$ and $f(x) = (x - u)(x + u)(x - v)(x + v)$ in F. That is, F is a splitting field of $f(x)$.

There is no reason to say that $F = \mathbf{Q}(u, v) \subset \mathbf{C}$. But, from the theory in §5.2, we can show $F \cong \mathbf{Q}(\sqrt{2}, \sqrt{3})$ easily. Firstly, since u and $\sqrt{2}$ have the same minimal $\mathbf{Q}$-polynomial $x^2 - 2$, by Theorem 5.2.14 there is a $\mathbf{Q}$-isomorphism $\sigma : F_1 \to \mathbf{Q}(\sqrt{2})$ such that $\sigma(u) = \sqrt{2}$. Secondly, $x^2 - 3$ is a minimal $\mathbf{Q}(u)$-polynomial of v, and it is also a minimal $\mathbf{Q}(\sqrt{2})$-polynomial of $\sqrt{3}$; by Theorem 5.2.14 again, we have an isomorphism $\hat{\sigma} : F \to \mathbf{Q}(\sqrt{2}, \sqrt{3})$ such that $\hat{\sigma}(v) = \sqrt{3}$ and $\hat{\sigma}|_{F_1} = \sigma$; in particular, $\hat{\sigma}$ is a $\mathbf{Q}$-isomorphism.

Reviewing the above argument, we see that the argument can proceed by induction; it suggests the following main theorem on spltting fields.

5.4.2 Theorem.   *Let K be a field. The splitting fields over K of any K-polynomial $f(x)$ of degree > 0 exist and are unique up to K-isomorphism.*

Proof. It is trivial if $\deg f(x) = 1$, since K is the unique splitting field of $f(x)$ in this case. Assume $\deg f(x) > 1$.

Let $f(x) = p(x)h(x)$ in $K[x]$ and $p(x)$ is irreducible in $K[x]$. By Theorem 5.2.7 there is an extension $F_1 = K(u)$ such that u has $p(x)$ as a minimal K-polynomial. In $F_1[x]$ we have $f(x) = (x - u)g(x)$ and $\deg g(x) > 0$. By induction, there is a splitting field F over F_1 of $g(x)$. Then $f(x) = (x - u)g(x) = a(x - u)(x - v) \cdots (x - t)$ with $u \in F_1 \subset F$ and $v, \cdots, t \in F$ and $F = F_1(v, \cdots, t) = K(u, v, \cdots, t)$; that is, F is a splitting field over K of the K-polynomial $f(x)$.

The uniqueness follows by taking $\sigma = \mathrm{id}_K : K \to K$ in Theorem 5.4.3 below. $\square$

5.4.3 Theorem.   *Let $\sigma : K \overset{\sim}{\to} K'$ be an isomorphism of fields K and K'. Let $f(x) \in K[x]$ and $\deg f(x) > 0$, and $\tilde{\sigma} f(x) \in K'[x]$ be as in Lemma 5.2.13(1). If F and F' are splitting fields of $f(x)$ and $\tilde{\sigma} f(x)$ resp., then σ can be extended to an isomorphism $\sigma' : F \overset{\sim}{\to} F'$.*

Proof. It is trivial if $\deg f(x) = 1$. Assume $\deg f(x) > 1$. Let $f(x) = p(x)h(x)$ in $K[x]$ and $p(x)$ is irreducible, then $\tilde{\sigma} f(x) = \tilde{\sigma} p(x) \cdot \tilde{\sigma} h(x)$ in $K'[x]$ and $\tilde{\sigma} p(x)$ is irreducible. By the splittingness of F and F' resp., we have $u \in F$ and $u' \in F'$ such that $p(u) = 0$ and $\tilde{\sigma} p(u') = 0$; hence, by Exer.5.2.2, u has $p(x)$ as a minimal K-polynomial, while u' has $\tilde{\sigma} p(x)$ as a minimal K'-polynomial. By Theorem 5.2.14 there is an isomorphism $\hat{\sigma} : K(u) \to K'(u')$ such that $\hat{\sigma}(u) = u'$ and $\hat{\sigma}|_K = \sigma$. In $K(u)[x]$, we have $f(x) = (x - u)g(x)$ and $\deg g(x) > 0$; hence in $K'(u')[x]$ we have $\tilde{\sigma} f(x) = (x - u') \cdot \tilde{\sigma} g(x)$. By Exer.2, F is a splitting field of the $K(u)$-polynomial $g(x)$, while F' is a splitting field of the $K'(u')$-polynomial $\tilde{\sigma} g(x)$. Since $\deg g(x) < \deg f(x)$, by induction, $\hat{\sigma} : K(u) \to K'(u')$ can be extended to an isomorphism $\sigma' : F \to F'$ which is also an extension of $\sigma : K \to K'$. $\square$

Remark. The extended isomorphism σ' is not unique though σ' must map the roots of an irreducible factor $p(x)$ of $f(x)$ onto the roots of $\tilde{\sigma}p(x)$. It is delicate which root can be mapped to which root by σ'; this is close to the Galois' idea for solving polynomials.

The splitting field is closely related to the so-called normality of extensions.

5.4.4 Definition. An algebraic extension F of a field K is said to be *normal over K* if any irreducible K-polynomial $p(x)$ is split in F provided $p(x)$ has a root in F.

5.4.5 Theorem. *A finite extension F of a field K is normal over K if and only if F is a splitting field of a K-polynomial.*

Proof. Assume F is normal over K. Since F is a finite extension, by Cor.5.3.6 we can assume $F = K(u_1, \cdots, u_n)$ and every u_i is algebraic over K. Let $p_i(x)$ be a minimal K-polynomial of u_i, $i = 1, \cdots, n$; and let $f(x) = p_1(x) \cdots p_n(x)$. By the normality of F over K, $f(x)$ is split in F; and, by our assumption, F is generated over K by all the roots in F of $f(x)$. That is, F is a splitting field of $f(x)$ over K.

Conversely, assume $F = K(r_1, \cdots, r_n)$ and $f(x) = (x - r_1) \cdots (x - r_n)$ is a K-polynomial. Assume $p(x)$ is an irreducible K-polynomial and $p(u) = 0$ for a $u \in F$. Suppose $p(x)$ is not split in F, i.e. in $F[x]$ there is an irreducible factor $q(x)$ of $p(x)$ and $\deg q(x) > 1$. By Theorem 5.2.7 we have an extension $F(v)$ where v has $q(x)$ as a minimal F-polynomial. Hence $p(v) = 0$ and, by Lemma 5.3.4, $|F(v) : F| > 1$. By Exer.5.2.2, both u and v has $p(x)$ as a minimal K-polynomial (not minimal F-polynomial!); and by Theorem 5.2.14, there is a K-isomorphism $\sigma : K(u) \to K(v)$ such that $\sigma(u) = v$. By definition and cf. Exer.1, $F = K(r_1, \cdots, r_n) = K(u)(r_1, \cdots, r_n)$ is a splitting field of $f(x)$ over $K(u)$, while $F(v) = K(r_1, \cdots, r_n)(v) = K(v)(r_1, \cdots, r_n)$ is a splitting field of $f(x)$ over $K(v)$. Therefore, by Theorem 5.4.3, σ can be extended to an isomorphism $\sigma' : F \to F(v)$; in particular, σ' is a K-isomorphism. Hence, as K-vector spaces F and $F(v)$ have the same dimension; however, by Lemma 5.3.3 we have $|F(v) : K| = |F(v) : F| \cdot |F : K| > |F : K|$; a contradiction. So $p(x)$ has to be split in F. $\square$

From Exer.1 we immediately have the following corollary.

5.4.6 Corollary. *Let $E \supset F \supset K$ be algebraic extensions of fields. If E is normal over K, then E is normal over F.* $\square$

If an algebraic extension is not normal, it is reasonable to find a minimal normal extension containing it.

5.4.7 Definition. Let F be an algebraic extension of a field K. An algebraic extension E over F is called a *normal closure* of F over K if E is normal over K and, for any subfield L such that $E \supset L \supset F$ and L is normal over K, we have $E = L$.

5.4.8 Lemma.   *Let $F \supset K$ be as above. Assume N is a normal extension of K and $N \supset F$. Then*

1) *the intersection E of all the subfields of N which are normal over K and contain F is a normal closure of F over K;*

2) *for any K-homomorphism $\varphi : F \to N$, the image $\varphi(F) \subset E$.*

Proof. (1) follows from Exer.9 at once.

(2) For $u \in F$, let $p(x)$ be a minimal K-polynomial of u; then $p(x)$ has a root $u \in E$, hence $p(x)$ is split in E by the normality of E over K; on the other hand, $\varphi(u)$ must be a root of $p(x)$ by Exer.5.2.6; hence $\varphi(u) \in E$.   □

Even if there is no such N as in the lemma beforehand, the normal closure can be constructed and is in fact unique.

5.4.9 Theorem.   *Let $F \supset K$ be fields and $|F : K| < \infty$. The normal closures of F over K exist and are unique up to K-isomorphism.*

Proof. By Cor.5.3.6 we assume that $F = K(u_1, \cdots, u_n)$ and $p_i(x)$ is a minimal K-polynomial of u_i, $i = 1, \cdots, n$; let $f(x) = p_1(x) \cdots p_n(x)$. The theorem follows from Theorem 5.4.2 and the following conclusion (which is similar to Theorem 5.4.5).

(5.4.10).   *E is a normal closure of F over K if and only if E is a splitting field over K of $f(x)$.*

Indeed, if E is a splitting field of $f(x)$, then E is normal over K; and for $E \supset L \supset F$ with L normal over K, we have $u_i \in L$ hence $p_i(x)$ is split in L for $i = 1, \cdots, n$; so L contains all the roots of $f(x)$, hence $L = E$. Conversely, if E is a normal closure of F over K, then, by Def.5.4.7, $f(x)$ is split in E, hence E contains a splitting field E' of $f(x)$ over K (cf. Exer.3); in particular, $E' \supset F$ and, by Theorem 5.4.5, E' is normal over K; then, by Def.5.4.7, $E = E'$.   □

5.4.11 Example. Consider the rational field $\mathbf{Q}$. Let p be a prime number, Consider $x^p - 1$. We have $x^p - 1 = (x-1)(x^{p-1} + x^{p-2} + \cdots + 1)$ and $x^{p-1} + x^{p-2} + \cdots + 1$ is irreducible in $\mathbf{Q}[x]$, see Exer.3.10.8. Let ζ be a primitive pth root of unity. Then $\mathbf{Q}(\zeta) = \mathbf{Q}(1, \zeta, \cdots, \zeta^{p-1})$ be a splitting field of $x^p - 1$, hence a normal extension of $\mathbf{Q}$.

Now consider $x^p - 2$ and assume p is an odd prime number. All the roots of $x^p - 2$ are $\sqrt{2}, \sqrt{2}\zeta, \cdots, \sqrt{2}\zeta^{p-1}$. Thus $\mathbf{Q}(\sqrt{2})$ is not a normal extension of $\mathbf{Q}$; it is easy to see that the normal closure of $\mathbf{Q}(\sqrt{2})$ in $\mathbf{C}$ is $\mathbf{Q}(\sqrt{2}, \zeta)$.

Exercises 5.4

+1. Let $E \supset F \supset K$ be fields, and $f(x)$ be a K-polynomial of degree > 0; hence $f(x)$ is also an F-polynomial. If E is a splitting field of $f(x)$ over K, then E is a splitting field of $f(x)$ over F.

+2. Let K be a field and $f(x)$ be a K-polynomial of degree > 1 and F be a splitting field of $f(x)$. If $u \in F$ such that $f(u) = 0$ and $f(x) = (x-u)g(x)$ in $K(u)[x]$, then F is a splitting field of the $K(u)$-polynomial $g(x)$.

+3. If a K-polynomial $f(x)$ of degree > 0 is split in an extension field F over K, then F contains a (unique) splitting field over K of $f(x)$.

4. 1) Find a splitting field of $x^4 - 2$ over $\mathbf{Q}$;
   2) Find a splitting field of $x^{p^n} - 1$ over the finite field $\mathbf{Z}_p$, where p is a prime number.

5. Assume F is a splitting field over a field K of a K-polynomial $f(x)$ and $\deg f(x) = n$. Show that $|F : K| \le n!$. Moreover, prove that $|F : K| \,|\, n!$.

6. Prove that any extension of a field K of degree 2 is normal over K, and is generated over K by an element with a minimal K-polynomial $x^2 - a$ where $a \in K$ (the root of $x^2 - a$ is denoted by $\sqrt{a}$ and called a *square root of* a).

7. Let K be a field and $\mathcal{S}$ be a set of K-polynomials. Let F be the extension over K generated by all the roots of all $f(x) \in \mathcal{S}$. Then F is normal over K. (Hint: if $u \in F$ such that $p(u) = 0$, then there are $f_1, \cdots, f_n \in \mathcal{S}$ such that $u \in F'$ where F' is generated over K by the roots of $f_i(x)$, $i = 1, \cdots, n$.)

8. Assume E is a normal extension over F and F is a normal extension over K. Is E normal over K? (Hint: $\mathbf{Q}(2^{1/4}) \supset \mathbf{Q}(2^{1/2}) \supset \mathbf{Q}$.)

+9. Let E be an extension of K and both F and L be fields between E and K. If both F and L are normal over K, then both $F \cap L$ and FL are normal over K. (cf. Exer.5.3.7.)

10. Let F be a splitting field over K of a K-polynomial $f(x)$, let $F \supset L \supset K$. Prove: any K-homomorphism $\sigma : L \to F$ can be extended to a K-automorphism $\tilde{\sigma}$ of F. (Hint: F is also a splitting field of $f(x)$ over $\sigma(L)$.)

11. If F is an algebraically closed field, then F is normal over any subfield K.

5.5   Two Applications

In this section we exhibit how to apply the field theory we discussed earlier to the Fundamental Theorem of Algebra, and to the Ruler and Compass Constructions.

5.5.1 Fundamental Theorem of Algebra.   *Any complex polynomial of degree $n > 0$ has a complex root.*

Proof. This is equivalent to the statement that *the complex field* **C** *is algebraically closed.* Since **C** is an algebraic extension of the real field **R**, by Prop.5.3.13 it is enough to show that any real polynomial of degree > 0 has a complex root.

Let $f(x)$ be a real polynomial of degree n; decomposing $n = 2^k m$ with m being odd, we apply induction on k.

First assume $k = 0$. Then $f(x)$ is a real polynomial of odd degree, hence, we have a real number x_0 such that $f(x_0) > 0$ and $f(-x_0) < 0$. Thus, as a continuous real function, $f(x)$ takes zero at a point between x_0 and $-x_0$. That is $f(x)$ has a real root.

Now assume $k > 0$. By Theorem 5.4.2 there is a splitting field E of $f(x)$ over **C**; let $e_1, \cdots, e_n$ be all roots in E of $f(x)$.

For any $r \in \mathbf{R}$, set

$$d_{ij} = e_i e_j + r(e_i + e_j) \qquad 1 \le i < j \le n$$

and set

$$g(x) = \prod_{1 \le i < j \le n} (x - d_{ij}).$$

Then $\deg g(x) = \frac{1}{2}n(n-1) = 2^{k-1}(m(2^k m - 1))$ where $m(2^k m - 1)$ is odd. Thus, every coefficient of $g(x)$ is written as an elementary symmetric polynomial of d_{ij}'s, hence a symmetric polynomial of e_i's; by Theorem 3.8.12, it is expressed as a polynomial of the elementary symmetric polynomials of e_i's; however, the elementary symmetric polynomials of

e_i's are the coefficients of $f(x)$, hence belong to $\mathbf{R}$; in other words, $g(x)$ is also a real polynomial. By induction, one of d_{ij}'s belongs to $\mathbf{C}$.

Note that for every $r \in \mathbf{R}$ we have a pair (i, j) such that $d_{ij} = e_i e_j + r(e_i + e_j) \in \mathbf{C}$. There are only finitely many such pairs, but $\mathbf{R}$ is infinite; so there must be a pair (i, j) and $r \neq r' \in \mathbf{R}$ such that

$$c = e_i e_j + r(e_i + e_j) \in \mathbf{C} \qquad \text{and} \qquad c' = e_i e_j + r'(e_i + e_j) \in \mathbf{C}.$$

Hence

$$e_i + e_j = \frac{c - c'}{r - r'} \in \mathbf{C} \qquad \text{and} \qquad e_i e_j = \frac{rc' - r'c}{r - r'} \in \mathbf{C}.$$

So, e_i and e_j are the roots of the complex polynomial $x^2 - (e_i + e_j)x + e_i e_j$ of degree 2; thus $e_i, e_j \in \mathbf{C}$. That is, $f(x)$ has a complex root. $\square$

Now consider the question "ruler and compass construction". There are several ways to formulate it precisely. We describe it as follows.

5.5.2 The regulations of the ruler and compass constructions.

Some points $P_1, \cdots, P_m$ are given as known; and the following operations are allowed:

(O1) If two distinct points P and P' are given, then the line through P and P' is obtained;

(O2) If two distinct points P and P' are given, then the circle having center P and containing P' (i.e. having radius the segment PP') is obtained;

(O3) The intersection points of a pair of lines by (O1), or a pair of circles by (O2), or a pair of a line by (O1) and a circle by (O2), are obtained as new known points and for further operations.

Can a required geometric object be constructed after a finite sequence of the above operations? And how to construct it?

As examples, two ancient famous questions are described below.

Trisection of a given angle: Given points O, U, P; find a point Q such that the angle$QOU = \frac{1}{3}$(anglePOU) as illustrated below:

(Figure T)

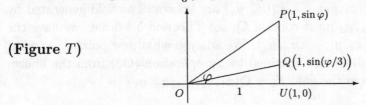

Duplication of a given cube: Given points O, A; find a point D such that $(\text{segment}OD)^3 = 2(\text{segment}OA)^3$ as illustrated below:

Review the regulation 5.5.2 again. The operations (O1) and (O2) are determined by known points; they do not produce any new "known points"; for example, the points on the line obtained by (O1) cannot be used for the next operations, except P and P' (which are the known points determining this line) and the points on this line obtained by an operation (O3).

Thus, essentially speaking, only (O3) turns out new data. Now we translate it into algebraic versions; also, there are slightly different ways to do it. We take the following.

Taking a known point as the origin and a known segment as the unity length (e.g. a segment determined by two known points), we establish an orthogonal coordinate system in the plane (or space). Then any point is characterized by its coordinates which are real numbers. In particular, we can assume that the points $O(0,0)$ and $U(1,0)$ are given known. For example, see the above Figure T for the question on trisection of an angle.

On the other hand, the required geometric object is determined by certain points, i.e. by certain real numbers.

5.5.3 Definition. Assume that, in a Cartesian plane, points $O(0,0)$, $U(1,0)$ and $P_1(x_1,y_1), \cdots, P_m(x_m, y_m)$ are given as known data. A point $Q(x_0, y_0)$ is said to be *constructible* (by ruler and compass from the known points) if it can be obtained by a finite sequence of the allowed operations in 5.5.2; a real number r is said to be *constructible* (by ruler and compass from the known numbers $x_1, y_1, \cdots, x_m, y_m$) if it is a co-ordinate of a constructible point.

Further, assume that, after a finite sequence of allowed operations, the points $P_1(x_1, y_1), \cdots, P_n(x_n, y_n)$ are known. The field generated by $x_1, y_1, \cdots, x_n, y_n$ must contain $\mathbf{Q}$, see Theorem 5.1.6; i.e. we have the field $F_k = \mathbf{Q}(x_1, y_1, \cdots, x_n, y_n)$. We analyze what new points, i.e. what new real numbers, are obtained by an operation (O3) from the known points, i.e. from the field $F_k = \mathbf{Q}(x_1, y_1, \cdots, x_n, y_n)$.

The line obtained by (O1) is a linear equation with coefficients in F_k; while the circle obtained by (O2) is an equation like this: $(x-a)^2 + (y-b)^2 = c$ with $a, b, c \in F_k$. Let $P_{n+1}(x_{n+1}, y_{n+1})$ be a point produced by (O3). Then x_{n+1} is a root of an F_k-polynomial of degree 1 or 2 or like this $z^4 + dz^2 + e$. In the first case, $F_k(x_{n+1}) = F_k$; in the second case, $|F_k(x_{n+1}) : F_k| \le 2$; and in the third case, $|F_k(x_{n+1}^2) : F_k| \le 2$, and $|F_k(x_{n+1}) : F_k(x_{n+1}^2)| \le 2$. It is the same for y_{n+1}. In other words, there is a series $F_k \subset F_{k+1} \subset \cdots \subset F_{k'}$ of fields such that $F_{k'} = F_k(x_{n+1}, y_{n+1})$ and $|F_i : F_{i-1}| \le 2$ for $i = k+1, \cdots, k'$.

Since only finite number of operations can be done, we have proved the necessity of the following criterion.

5.5.4 Theorem. *A real number r is constructible (by ruler and compass) from real numbers $s_1, \cdots, s_m$ if and only if there is a series $F_0 \subset F_1 \subset \cdots \subset F_n$ of fields such that $F_0 = \mathbf{Q}(s_1, \cdots, s_m)$ and $r \in F_n$ and $|F_i : F_{i-1}| = 2$ for $i = 1,, \cdots, n$.*

Proof. It remains to prove the sufficiency. Using induction on n, we show that every $r \in F_i$ is constructible. First the unity length 1 is constructible by Definition 5.5.3; hence, applying the intersections by parallel lines, we see that any $q \in \mathbf{Q}$ is constructible. In a similar way, $r \in F_0 = \mathbf{Q}(s_1, \cdots, s_m)$ is constructible. Assume every $r_k \in F_k$ is constructible and $r \in F_{k+1}$. Since $|F_{k+1} : F_k| = 2$, a F_k-polynomial of degree 2 annihilates r; so $r = a + \sqrt{b}$ for $a, b \in F_k$. But $\sqrt{b}$ can be constructed from 1 and b as follows.

Hence r is constructible. □

5.5.5 Corollary. *If a real number r is constructible from real numbers $s_1, \cdots, s_m$, then $|\mathbf{Q}(s_1, \cdots, s_m, r) : \mathbf{Q}(s_1, \cdots, s_m)| = 2^k$ for $k \ge 0$.*

Proof. It follows from Theorem 5.5.4 and Lemma 5.3.3. □

5.5.6 Remark. By Galois Theory, it is easy to extend the above to get a necessary and sufficient condition for the constructbility:

A real number r is constructible from real numbers $s_1, \cdots, s_m$ if and only if r is algebraic over the field $F = \mathbf{Q}(s_1, \cdots, s_m)$ and the normal closure of $F(r)$ over F has degree 2^n over F.

5.5.7 Remark. As consequences we see at once that both the *Duplication of a cube* and the *Squaring a circle* are *unconstructible*. The former is equivalent to the unconstructibility of $\sqrt[3]{2}$ from $\mathbf{Q}$ (taking the edge of the given cube as the unity length); while the latter is equivalent to the unconstructibility of $\sqrt{\pi}$ from $\mathbf{Q}$ (taking the radius of the given circle as the unity length).

5.5.8 Proposition. *An angle φ can be trisected if and only if the polynomial $4x^3 - 3x + \sin\varphi$ is reducible in $\mathbf{Q}(\sin\varphi)$.*

Proof. By the condition, $\sin\varphi$ is known; and φ can be trisected if and only if $\sin(\varphi/3)$ is constructible from $\sin\varphi$. By the trigonometric formula, $\sin(\varphi/3)$ is a root of the polynomial $4x^3 - 3x + \sin\varphi$; hence a minimal $\mathbf{Q}(\sin\varphi)$-polynomial of $\sin(\varphi/3)$ has degree ≤ 3; i.e. $|\mathbf{Q}(\sin\varphi)\big(\sin(\varphi/3)\big) : \mathbf{Q}(\sin\varphi)| \leq 3$. Therefore, by Cor.5.5.5, $\sin(\varphi/3)$ is constructible if and only if this degree ≤ 2; in other words, if and only if $4x^3 - 3x + \sin\varphi$ is reducible in $\mathbf{Q}(\sin\varphi)$.   □

Example. Let $\varphi = 90^o$. Then $\sin\varphi = 1$ and $4x^3 - 3x + 1 = (2x-1)(2x^2+x-1)$; so $\varphi = 90^o$ can be trisected by ruler and compass.

Next, let $\varphi = 30^o$. Then $\sin\varphi = 1/2$ and we claim the irreducibility of $4x^3 - 3x + \frac{1}{2}$ over $\mathbf{Q}$, equivalently, the irreducibility of $f(x) = 8x^3 - 6x + 1$ over $\mathbf{Q}$. Suppose $f(x)$ is reducible, then it has a rational root n/m with $n = \pm 1$ and $m = 1, 2, 4, 8$; but it is clear that all of them are impossible. Thus $\varphi = 30^o$ cannot be trisected by ruler and compass.

Now we consider the question on construction of regular n-gon.

5.5.9 Proposition. *Let p be a prime. If the regular p-gon can be constructed by ruler and compass, then $p = 2^{2^m} + 1$ is a Fermat prime.*

Proof. In that case, $\cos\frac{2\pi}{p}$, $\sin\frac{2\pi}{p}$ are constructible. By Cor.5.5.5, we have $|\mathbf{Q}(\cos\frac{2\pi}{p}, \sin\frac{2\pi}{p}) : \mathbf{Q}| = 2^k$.

Let $\zeta = \cos\frac{2\pi}{p} + i\sin\frac{2\pi}{p}$; then $\mathbf{Q}(\zeta) \subset \mathbf{Q}(\cos\frac{2\pi}{p}, \sin\frac{2\pi}{p}, i)$ and the degree $|\mathbf{Q}(\cos\frac{2\pi}{p}, \sin\frac{2\pi}{p}, i) : \mathbf{Q}| = 2^{k+1}$. Thus $|\mathbf{Q}(\zeta) : \mathbf{Q}| = 2^n$ where $n \leq k+1$. On the other hand, ζ is a root of $x^{p-1} + \cdots + x + 1$ which is irreducible over $\mathbf{Q}$ by Exer.3.10.8; So $|\mathbf{Q}(\zeta) : \mathbf{Q}| = p - 1$.

Therefore $p - 1 = 2^n$, i.e. $p = 2^n + 1$. If $n = st$ with $t > 1$ odd, then $2^n + 1 = (2^s + 1)\big((2^s)^{t-1} - (2^s)^{t-2} + \cdots + (-1)^{t-1}\big)$ is not a prime number. That is, $n = 2^m$ and $p = 2^{2^m} + 1$ is a Fermat prime.   □

Example. From Prop.5.5.9 we see that the regular 7-gon, 11-gon and 13-gon cannot be constructed by ruler and compass. On the other hand, Gauss showed a construction of the regular 17-gon.

Exercises 5.5

1. Show that $72°$ can be trisected by ruler and compass; but $60°$ cannot be trisected by ruler and compass.
2. Show that the regular 9-gon cannot be constructed by ruler and compass.

5.6 Separability, Multiple Roots

Different from the previous sections, the separabilities is related to the characteristic of fields. We begin with the multiple roots of polynomials.

Let K be a field.

5.6.1 Definition. Any K-polynomial $f(x)$ is uniquely expressed, in its splitting field F, into $f(x) = a(x - r_1)^{m_1} \cdots (x - r_t)^{m_t}$ with $r_i \neq r_j$ for $1 \leq i \neq j \leq t$ and $m_i \geq 1$ for $1 \leq i \leq t$. The positive integer m_i is called the *multiplicity* of the root r_i. A root r_i is said to be a *multiple root* of $f(x)$ if $m_i > 1$; otherwise, r_i is said to be a *single root* of $f(x)$.

The following notation for polynomials and criterion for multiple roots are well-known in the complex field; but they are in fact valid generally.

5.6.2 On Derivatives. Let $f(x) = \sum_{i=0}^{n} a_i x^i$ be a K-polynomial in indeterminate x. $f'(x) = \sum_{i=1}^{n-1} i a_i x^{i-1}$ is called the *derivative* of $f(x)$. The following formulas on derivatives (which are familar in annalysis) are easy to verify:

$$\Big(f(x) + g(x)\Big)' = f'(x) + g'(x); \qquad \Big(af(x)\Big)' = af'(x);$$

$$\Big(f(x)g(x)\Big)' = f'(x)g(x) + f(x)g'(x); \qquad \Big(f(x)^m\Big)' = mf(x)^{m-1}f'(x).$$

Let $d(x) = (f(x), f'(x))$ be the greatest common divisor. Assume $f(x) = (x-r)^m g(x)$ with $m > 0$ and $(x-r) \nmid g(x)$; i.e. r is a root of $f(x)$ of multiplicity $m > 0$. Then $f'(x) = (x - r)^{m-1}\Big((x - r)g'(x) + mg(x)\Big)$; hence we have

5.6.3 Lemma. *Notations as above. Then* $(x - r)^{m-1} \mid d(x)$. *Further,*
$(x - r)^m \nmid d(x)$ *if* $\mathrm{char}K \nmid m$ *(in particular, if* $m = 1$ *or* $\mathrm{char}K = 0$*).* □

5.6.4 Corollary. *Notations as above.*

1) r *is a multiple root of* $f(x) \iff (x - r) \mid d(x)$.

2) $f(x)$ *has no multiple roots* $\iff d(x) = (f(x), f'(x)) = 1$.

3) *Let* $f(x)$ *be irreducible.* $f(x)$ *has multiple roots* $\iff f'(x) = 0$.

Proof. (1) follows from Lemma 5.6.3 directly. (2) is a consequence of (1). For (3), $f(x)$ has multiple roots if and only if $d(x) =$
$(p(x), p'(x))$ has degree > 0; but both $f(x)$ and $f'(x)$ are K-polynomials
and $\deg f'(x) < \deg f(x)$; hence, by the irreducibility of $f(x)$, $\deg d(x) >$
0 if and only if $p'(x) = 0$.   □

5.6.5 Corollary. *Let* $p(x)$ *be an irreducible* K-*polynimial.*

1) *When* $\mathrm{char}K = 0$, *then* $p(x)$ *has no multiple roots.*

2) *When* $\mathrm{char}K = p > 0$, *then* $p(x)$ *has multiple roots if and only
if the exponent of any nonzero term of* $p(x)$ *is divided by* p, *i.e.* $p(x) =$
$q(x^p)$ *for a* K-*polynomial* $q(y)$.

Proof. Let $p(x) = \sum_{i=0}^{n} a_i x^i$. Then, $p(x)$ has multiple roots, if
and only if $p'(x) = 0$, if and only if $\sum_{i=1}^{n} i d_i x^{i-1} = 0$, if and only if
either $a_i = 0$ or $p \mid i$ for $i = 1, \cdots, n$.   □

5.6.6 Notation. Let K be a field with $\mathrm{char}K = p > 0$. Recall
from Exer.5.1.8 that map $\rho : K \to K$, $a \mapsto a^p$, is an endomorphism of
K and the ρ-fixed point set $\{a \in K \mid a^p = a\}$ is the prime subfield $\mathbf{Z}_p$ of
K; set $K^p = \mathrm{Im}(\rho) = \{a^p \mid a \in K\}$.

With the above information, we can show a typical example. Let
$K = \mathbf{Z}_p(t)$ be the field of fractions in an indeterminate t over the prime
field $\mathbf{Z}_p$ where p is a prime number. We claim that:

(5.6.7) $x^p - t$ *is irreducible and has a unique root of multiplicity* p.

The latter conclusion is obvious: in a splitting field, $x^p - t$ has a
root r, so $r^p = t$ and $x^p - t = x^p - r^p = (x - r)^p$.

In order to show the former, we first show that $x^p - t$ has no roots
in K: otherwise, suppose $a(t)/b(t) \in K = \mathbf{Z}_p[t]$ is a root of $x^p - t$;
remembering $f(t)^p = f(t^p)$ for any $f(t) \in K$ by (5.6.6), we have $t =$
$a(t)^p/b(t)^p = a(t^p)/b(t^p)$; hence $tb(t^p) = a(t^p) \neq 0$, this is impossible
since $p \mid \deg a(t^p)$ but $p \nmid \deg(tb(t^p))$. Suppose $x^p - a$ is reducible in

$K[x]$; then, since $x^p - t = (x - r)^p$ in a splitting field, there is a $1 \leq k \leq p - 1$ such that $(x - r)^k \in K[x]$; in particular, the constant coefficient of $(x - r)^k$ belongs to K, i.e. $r^k \in K$; but $(k, p) = 1$ implies that $mp + lk = 1$ for some integers k, l; hence $r = r^{pm+kl} = r^{pm} r^{kl} \in K$ as $r^{pm} = t^m$, $r^{kl} = (r^k)^l \in K$. Thus $x^p - t$ has the root r in K, a contradiction.   □

To some extent, the above example has its general interest.

Assume $\operatorname{char} K = p > 0$ and $p(x)$ is an irreducible K-polynomial. If $p(x)$ has multiple roots, then, by Cor.5.6.5.2, $p(x) = p_1(x^p)$ for $p_1(x) \in K[x]$. If $p_1(x)$ has multiple roots too, then $p_1(x) = p_2(x^p)$ for $p_2(x) \in K[x]$, hence $p(x) = p_1(x^p) = p_2(x^{p^2})$; and so on. After finite steps, we get a $p_k(x) \in K[x]$ such that

$$p(x) = p_k(x^{p^k}) \quad \text{and } p_k(x) \text{ has no multiple roots.}$$

Every $p_i(x)$ is irreducible in $K[x]$, because otherwise $p(x) = p_i(x^{p^i})$ would be reducible. Assume that, in its splitting field, $p_k(x) = a(x - s_1) \cdots (x - s_n)$; then $s_i \neq s_j$ for $1 \leq i \neq j \leq n$. And assume, in a large enough field (so that it contains a splitting field of $x^{p^k} - s_i$, $i = 1, \cdots, n$), $x^{p^k} - s_i$ has a root r_i, hence $x^{p^k} - s_i = x^{p^k} - r_i^{p^k} = (x - r_i)^{p^k}$. Therefore

$$p(x) = a(x^{p^k} - s_1) \cdots (x^{p^k} - s_n) = a(x - r_1)^{p^k} \cdots (x - r_n)^{p^k}.$$

Summarizing the above observations, we have

5.6.8 Theorem. *Let K be a field with $\operatorname{char} K = p > 0$. Let $p(x) \in K[x]$ be irreducible. Then there are a unique $k \geq 0$ and a unique irreducible $q(x) \in K[x]$ which has no multiple roots such that $p(x) = q(x^{p^k})$ and any two distinct roots of $p(x)$ has the same multiplicity p^k.*   □

5.6.9 Definition. Notations as above theorem.

1) If $k = 0$ (i.e. $p(x)$ has no multiple roots), then $p(x)$ is said to be *separable*; otherwise $p(x)$ is said to be *inseparable*. A K-polynomial $f(x)$ is said to be *separable* if every irreducible K-factor is separable (i.e. $f(x)$ has no multiple roots); otherwise $f(x)$ is said to be *inseparable*.

2) p^k is called the *inseparability degree* of $p(x)$, denoted $\deg_i p(x)$; and k is called the *inseparability index* of $p(x)$.

3) The K-polynomial $q(x)$ is called the *separability form* of $p(x)$; and $\deg q(x)$ is called the *separability degree* of $p(x)$, denoted $\deg_s p(x)$. (Hence $\deg p(x) = \deg_i p(x) \deg_s p(x)$.)

4) An algebraic element u over K is said to be *separable* if its minimal K-polynomial is separable; otherwise, u is said to be *inseparable*. In particular, the irreducible K-polynomial $x^{p^k} - a$ is said to be *purely inseparable*, and its root is called a *purely inseparable element* over K.

5) An algebraic extension F over K is said to be *separable* if every $u \in F$ is separable over K; otherwise F is said to be *inseparable* over K. In particular, if every $u \in F$ is purely inseparable over K, then F is said to be *purely inseparable* over K.

5.6.10 Remark. If $\operatorname{char} K = 0$, then any irreducible K-polynomial $p(x)$ is said to be *separable* and $\deg_i p(x) = 1$; of course, every algebraic extension is separable.

We listed a lot of notations without a break. The essential point is just the fact that an irreducible polynomial may have multiple roots, which implies that for a single extension of degree n the number of the embeddings of it into its normal closure may properly be less than n.

5.6.11 Lemma. *Let $\sigma : K \to K'$ be an isomorphism of fields and $E' \supset K'$ be a normal extension; let $K(u)$ be an extension and u have minimal K-polynomial $p(x)$; let $\widetilde{\sigma} p(x) \in K'[x]$ be as in Lemma 5.2.13(1).*

1) *The number h of the homomorphisms $K(u) \to E'$ which extend σ is non-zero if and only if $\widetilde{\sigma} p(x)$ has a root in E';*

2) *If $h > 0$, then $h = \deg_s p(x)$, i.e. the number of distinct roots of $p(x)$.*

Proof. It is clear that, for any homomorphisms $\varphi : K(u) \to E'$ extending σ, the image $\varphi(u)$ is a root of $\widetilde{\sigma} p(x)$, and φ is determined by $\varphi(u)$; conversely, if $u' \in E'$ is a root of $\widetilde{\sigma} p(x)$, then $\sum_i a_i u^i \mapsto \sum_i \sigma(a_i) u'^i$ is a homomorphism from $K(u)$ to E' which extends σ; cf. Exer.5.2.6. Then the lemma is obvious.   □

The above fact suggests the following notation, which appears in a more general case, see Lemma 5.6.13 below.

5.6.12 Definition. Let F be a finite extension of a field K, and N be a normal extension of K containing F. The number of the K-homomorphisms from F into N is called the *separability degree* of F over K, and denoted by $|F : K|_s$.

Note that, this definition make sense. For: firstly, such N exists, (e.g. the normal closure of F over K). Secondly, it is independent of the choice of N, because, by Lemma 5.4.8, N contains a normal closure E of F over K and any K-homomorphism from F into N takes its image in E, i.e. is in fact a K-homomorphism from F into E; and by Theorem 5.4.9, the normal closure of F over K is unique up to K-isomorphism.

5.6.13 Lemma. *Let $\sigma : K \to K'$ be an isomorphism of fields. Let F be a finite extension over K, and N' be a normal extension of K' such that there is a homomorphism $\psi : F \to N'$ which extends σ. Then there are exactly $|F : K|_s$ homomorphisms from F into N' which extend σ.*

Proof. Let $F' = \varphi(F)$. The isomorphism $\varphi_0 = \psi|_F : F \overset{\sim}{\to} F'$ extends the isomorphism $\sigma : K \overset{\sim}{\to} K'$. Hence $|F' : K'|_s = |F : K|_s$. Any homomorphism $\varphi : F \to N'$ which extends σ corresponds to a K'-homomorphism $\varphi \varphi_0^{-1} : F' \to N'$. Conversely, any K'-homomorphism $\varphi' : F' \to N'$ corresponds to a homomorphism $\varphi' \varphi_0 : F \to N'$ which extends σ. Thus the number of the homomorphisms from F into N' which extend σ is equal to $|F' : K'|_s = |F : K|_s$. □

The following treatments are similar to §5.3. The results 5.6.14–5.6.21 are similar to results 5.3.3–5.3.14.

5.6.14 Lemma. *Let $L \supset F \supset K$ be finite extensions of fields. Then $|L : K|_s = |L : F|_s \cdot |F : K|_s$.*

Proof. Let N be a normal extension of K such that $N \supset L$. For a K-homomorphism $\varphi : F \to N$, by Lemma 5.6.13 there are exactly $|L : F|_s$ homomorphisms from L into N which extend φ. In this way we can get $|L : F|_s \cdot |F : K|_s$ K-homomorphisms from L into N altogether. On the other hand, every K-homomorphism $\tau : L \to N$ can be obtained in this way, because $\tau|_F : F \to N$ is a K-homomorphism. Thus the formula follows. □

In Lemma 5.6.11, taking $\sigma = \mathrm{id}_K : K \to K$ and $E' = E$ to be the normal closure of $K(u)$ over K, we have

5.6.15 Corollary. *Let $K(u)$ be an extension with u having minimal K-polynomial $p(x)$.*

   1) $|K(u) : K|_s = \deg_s p(x)$.

   2) u *is serapable over K* $\iff$ $|K(u) : K|_s = |K(u) : K|$.

   3) u *is purely inserapable over K* $\iff$ $|K(u) : K|_s = 1$. □

5.6.16 Theorem. *Let $F \supset K$ be finite extensions of fields. Then $|F : K|_s \leq |F : K|$; and*

1) $|F:K|_s = |F:K|$ *if and only if F is separable over K;*

2) $|F:K|_s = 1$ *if and only if F is purely inseparable over K.*

Proof. $F = K(u_1, \cdots, u_n)$ since $|F:K| < \infty$. If $n = 1$, it is just Lemma 5.6.11. Let $F_1 = K(u_1)$. By induction on n we have

$$|F:K|_s = |F:F_1|_s \cdot |F_1:K|_s \leq |F:F_1| \cdot |F_1:K| = |F:K|.$$

(1). If F is separable over K, then u_1 is separable over K, hence $|F_1:K|_s = |F_1:K|$ by Cor.5.6.15; on the other hand, since $u_2, \cdots, u_n$ are separable over F_1 by Exer.4, by induction we have

$$|F:K|_s = |F:F_1|_s \cdot |F_1:K|_s = |F:F_1| \cdot |F_1:K| = |F:K|.$$

Conversely, if F is inseparable over K, then there is a $u_1 \in F - K$ inseparable over K; hence $|F_1:K|_s < |F_1:K|$ by Cor.5.6.15 again; so, by the above argument we have

$$|F:K|_s = |F:F_1|_s \cdot |F_1:K|_s < |F:F_1| \cdot |F_1:K| = |F:K|.$$

(2) is proved similarly.   □

The following is an immediate consequence.

5.6.17 Corollary. *Let F be a finite normal extension of K. Let $\mathrm{Aut}_K(F)$ be the group of all K-automorphisms of F. Then $|\mathrm{Aut}_K(F)| \leq |F:K|$; and the equality holds if and only if F is separable over K.* □

5.6.18 Theorem. *Let F be an algebraic extension of a field K.*

1) *F is separable over K if and only if F is generated over K by separable elements over K.*

2) *If F is separable over K and E is a separable extension over F, then E is separable over K.*

Proof. (1). The necessity is obvious. Conversely, assume $F = K(S)$ and every $s \in S$ is separable over K. For any $u \in F$, we have $u_1, \cdots, u_n \in S$ such that $u \in K(u_1, \cdots, u_n)$ by Cor.5.1.12. Since every u_i is separable over K, by Cor.5.6.15 and Exer.4 and Lemma 5.6.14 we have $|K(u_1, \cdots, u_n) : K|_s = |K(u_1, \cdots, u_n) : K|$. Thus, by Theorem 5.6.16, $K(u_1, \cdots, u_n)$ is separable over K, hence u is separable over K.

(2). Similar to the above and the argument for Theorem 5.3.10. □

5.6.19 Corollary. *Let F ba an algebraic extension of a field K. The subset $F_s \subset F$ of all separable elements over K is a subfield containing K and is separable over K. In particular, the sum, difference, product* *and quotient (the denominator is non-zero) of two separable elements over K are separable over K too. Moreover, if $|F : K| < \infty$, then $|F : K|_s = |F_s : K|$.*

Proof. Exer.9.   □

5.6.20 Definition. A field K is said to be *perfect* if K has no proper inseparable (algebraic) extension.

Example. Of course, any field of characteristic 0 is perfect, see Remark 5.6.10; while the example (5.6.7) shows an imperfect field $\mathbf{Z}_p(t)$ of the rational fractions over $\mathbf{Z}_p$.

5.6.21 Proposition. *Let K be a field. The following are equivalent:*

   i) *K is perfect;*

   ii) *Every irreducible K-polynomial is separable;*

   iii) *Every algebraic extension of K is separable;*

   iv) *Either* $\operatorname{char} K = 0$, *or* $K = K^p$ *where* $p = \operatorname{char} K > 0$ *and* $K^p = \{a^p \mid a \in K\}$ *as in (5.6.6).*

Proof. Exer.12   □

Now we prove the so-called "Theorem on Primitive Elements".

5.6.22 Theorem. *If F is a finite separable extension of a field K, then there is a $w \in F$ such that $F = K(w)$.*

Proof. This is true if K is finite, see Prop. 5.7.3 below. In the following we assume K is infinite. Since $F = K(u, v, \cdots, t)$ for finitely many $u, v, \cdots, t$, it is enough to prove $K(u, v) = K(w)$.

Let $f(x)$ and $g(x)$ be the minimal K-polynomials of u and v resp.; let E be a splitting field of $f(x)g(x)$ over F. Assume $u_1 = u, \cdots, u_m$ are all the roots of $f(x)$ in E, while $v_1 = v, \cdots, v_n$ are all the roots of $g(x)$ in E. Consider the following equations in a variable y:

$$u_i + y v_j = u_k + y v_1 \qquad 1 < j \leq n, \ 1 \leq i, k \leq m.$$

Each of them has at most one solution as $v_j \neq v_1$; but K is infinite; so there is a $c \in K$ such that c satisfies none of the equations. Set

$$w = u_1 + c v_1, \qquad \text{i.e. } w = u + cv.$$

Then the polynomial $f(w - cx)$ (in variable x) and the polynomial $g(x)$ have exactly one common root v. Since $g(x)$ has no multiple roots, $x - v$ is the greatest common divisor of the two polynomials which have coefficients in $K(w)$. Thus, by Lemma 5.2.2(1), there are $s(x), t(x) \in K(w)[x]$ such that

$$x - v = s(x)f(w - cx) + t(x)g(x).$$

Comparing the constant coefficient on both sides, we have $v \in K(w)$; so $u = w - cv \in K(w)$. Thus $K(u,v) \subset K(w)$, i.e. $K(u,v) = K(w)$.   □

Remark. This theorem can be extended slightly to get a necessary and sufficient condition for the existence of primitive elements; but it involves some more delicate notations. We just exhibit, in Exer.13 and Exer.14, some features of this question.

Exercises 5.6

1. Let K be a field and $\operatorname{char}K = p > 0$. Let $p(x) \in K[x]$ be irreducible with separability form $q(x)$ and $\deg_i p(x) = p^k$. If, in an extension, u is a root of $p(x)$, then u^{p^k} is a root of $q(x)$, hence is a separable element over K.

2. Let K be a field and $\operatorname{char}K = p > 0$.
   1) The K-polynomial $x^n - 1$ has multiple roots if and only if $p \mid n$. Hence, the number of the nth roots of unity is less than n when $p \mid n$.
   2) The K-polynomial $x^{p^m} - x$ has no multiple roots.

3. If $\operatorname{char}K = p > 0$, then $x^{p^k} - a$ is irreducible in K iff $a \notin K^p$. (Hint: cf. the proof of (5.6.7).)

$+$4. Let $F \supset M \supset K$ be fields, and $u \in F$.
   1) If u is separable over K, then u is separable over M.
   2) If u is purely inseparable over K, then u is purely separable over M.
   3) If u is both separable and purely inseparable over K, then $u \in K$.

5. Assume that $\operatorname{char}K = p > 0$, F is an extension of K and $f(x) \in K[x]$.
   1) If $f(x)$ is irreducible in $K[x]$ and $\deg f(x)$ is prime to p, then $f(x)$ is separable over K.
   2) If $|F : K|$ is prime to p, then F is separable over K.

6. Assume that $\operatorname{char}K = p > 0$ and u is an algebraic element over K. Prove: u is separable over $K \iff K(u) = K(u^{p^n})$ for every positive integer n.

7. Let $F \supset K$ be fields and $u, v \in F$. If u is separable over K while v is purely inseparable over K, then $K(u,v) = K(u + v)$.

8. Complete the proof of Theorem 5.6.18(2).

9. Prove Cor.5.6.19. (Hint: for the last conclusion, cf. Exer.10 below.)

10. Let $F \supset K$ and F_s be as in Cor 5.6.19. Show that F is purely inseparable over F_s. (Hint: cf. Exer.1 above.)

11. If char$K = p > 0$, then $K^p = \{a^p \mid a \in K\}$ is a subfield of K.

12. Prove Prop.5.6.21. (Hint: For (iv), if $K^p \neq K$, find $a \in K - K^p$; $x^p - a$ is irreducible by Exer.3 above, and is inseparable by 5.6.5(2). Suppose $K^p = K$; for irreducible $p(x) \in K[x]$, if it is inseparable, then $p(x) = \sum_i a_i (x^p)^i$ by 5.6.5(2); for each a_i find $b_i \in K$ such that $a_i = b_i^p$; hence $p(x) = \sum_i b_i^p (x^p)^i = (\sum_i b_i x^i)^p$, contradicts the irreducibility of $p(x)$.)

13. Let $F = \mathbf{Z}_p(x, y)$, $K = F^p$. Prove: 1) $K = \mathbf{Z}_p(x^p, y^p)$.
    2) F is not a single extension over K.

14. Let $F = K(u, v)$ be an algebraic extension over a field K. If one of u and v is separable over K, then F is a single exension over K. (Hint: cf. the proof of Theorem 5.6.22.)

15. If F is a purely inseparable extension over K and E is a purely inseparable extension over F, prove: E is a purely inseparable over K.

5.7 Finite Fields

The finite fields are also called *Galois fields*. Their structures are clear. And they are applied extensively to the computer science, the theory of communications and codings, combinatorics, etc.

5.7.1 Proposition. *Any finite field F has a prime characteristic p, and $|F| = p^m$ for a positive integer m.*

Proof. char$F = p > 0$ by Exer.5.1.6; so $F \supset \mathbf{Z}_p$ by Theorem 5.1.6, and $m = |F : \mathbf{Z}| < \infty$ as $|F| < \infty$. Let $\{u_1, \cdots, u_m\}$ be a $\mathbf{Z}_p$-basis of F; then each $v \in F$ is uniquely written as $v = a_1 u_1 + \cdots + a_n u_n$ with coefficients $a_i \in \mathbf{Z}_p$. □

We collect several results as follows which are known from §5.1.

5.7.2 Proposition. *Let F be a finite field of order p^m (hence char$F = p$ is a prime number).*

  1) *The multiplicative group F^* is a cyclic group of order $p^m - 1$.*

  2) *If $F^* = \langle u \rangle$ is generated by u, then $F = K(u)$ is a single extension over K for any subfield $K \subset F$.*

  3) *For any power p^t of p with $t \geq 0$, the map $\rho_t : F \to F, a \mapsto a^{p^t}$ is an automorphism of F ($\rho_0 = \mathrm{id}_F$).*

  4) *F is perfect; in particular, any irreducible polynomial over F is separable (i.e. has no multiple roots).*

Proof. (1). F^* is cyclic by Exer.5.1.11 and $|F^*| = |F| - 1 = p^m - 1$.

(2). This is clear from (1). However, we remark that a generator (i.e. a primitive element) of F over K is not necessarily a generator of the cyclic group F^*; see Example 5.7.15 below.

3). ρ_t is an endomorphism of F by Exer.5.1.8(2); it must be injective, see (5.1.5); hence it has to be bijective since F is finite.

4). Taking $t = 1$ in (3), we have $K^p = K$; hence, by Prop.5.6.21, K is perfect. $\qquad\square$

Continuing the above proposition, we consider the subfield of the ρ_t-fixed points, see (5.1.8). $\rho_t(a) = a$ if and only if $a^{p^t} = a$; in other words, the roots in F of the polynomial $x^{p^t} - x$ form a subfield F'. By Prop.5.7.1, $|F'| = p^k$ for some k. Thus, by Prop.5.7.2(1), $|F'^*| = p^k - 1$ and F'^* is a subgroup of the cyclic group F^* of order $p^m - 1$. We can determine k as follows. Assume $a \in F'^*$ has order $p^k - 1$. Since $a^{p^t} - a = 0$, we see that $a^{p^t-1} = 1$, i.e. $p^k - 1$ (the order of a) divides both $p^m - 1$ and $p^t - 1$. We have a number-theoretic fact

(5.7.3). $p^n - 1 \mid p^m - 1 \iff n \mid m$.

To see it, let $m = nh + r$ with $0 \le r < n$, then we have $p^m - 1 = p^{nh}p^r - p^r + p^r - 1 = (p^n - 1)\big((p^n)^{h-1} + \cdots + p^n + 1\big)p^r + (p^r - 1)$.

Continuing the above discussion and letting $d = (m, t)$ be the g.c.d. of m and t, we see that $k \mid d$. Then, as $p^d - 1 \mid p^m - 1$, by Theorem 2.8.7, F^* has a subgroup H of order $p^d - 1$; and, as $p^k - 1 \mid p^d - 1$, by Theorem 2.8.7 again, F'^* is a subgroup of H. On the other hand, since $p^d - 1 \mid p^t - 1$, every $a \in H$ satisfies $a^{p^t-1} = 1$; in other words, $H \subset F'$. Therefore, $H = F'^*$ and $k = d$. Summarizing the discussion, we get

5.7.4 Lemma. *If F is a field of order p^m, then all the roots in F of the polynomial $x^{p^t} - x$ form a subfield of order p^d where $d = (m, t)$ is the g.c.d. of m and t. In particular, the elements of F are exactly all the roots of $x^{p^m} - x$.* $\qquad\square$

Then most of the information on finite fields can be deduced from this lemma and the ideas of its proof.

5.7.5 Theorem. *For any prime p and positive integer m, the fields of order p^m exist and are unique up to isomorphism, which consists of exactly the p^m roots of the polynomial $x^{p^m} - x$.*

Proof. By Theorem 5.1.6 and Lemma 5.7.5, such fields are exactly the splitting fields of $x^{p^m} - x$ over $\mathbf{Z}_p$. Thus the theorem follows from the existence and uniqueness of splitting fields (Theorem 5.4.2). $\square$

5.7.6 Notation. Identifying isomorphic fields with each other, one can denote the field of order p^m by a fixed notation. In many literatures, it is denoted by $GF(p^m)$ which means "Galois field of order p^m". In short, we use $\mathbf{F}_{p^m}$ to represent it.

5.7.7 Corollary. $\mathbf{F}_{p^n} \subset \mathbf{F}_{p^m} \iff n \mid m \iff p^n - 1 \mid p^m - 1.$

Proof. The former follows from Lemma 5.7.4 by considering the roots of $x^{p^n} - x$ in $\mathbf{F}_{p^m}$; and the latter follows from (5.7.3). $\square$

5.7.8 Corollary. *If $\mathbf{F}_{p^n} \subset \mathbf{F}_{p^m}$ with $m = nd$, then*

1) $\mathbf{F}_{p^m}$ *is normal and separable over $\mathbf{F}_{p^n}$;*

2) *The $\mathbf{F}_{p^n}$-automorphism group $\mathrm{Aut}_{\mathbf{F}_{p^n}}(\mathbf{F}_{p^m}) = \langle \rho_n \rangle$ is a cyclic group of order d; where $\rho_n : \mathbf{F}_{p^m} \to \mathbf{F}_{p^m}$, $a \mapsto a^{p^n}$ as in Prop.5.7.2(3).*

Proof. (1). The normality follows from Theorem 5.7.5 and Theorem 5.4.5. The separability follows from Prop.5.7.2(4).

(2). It is easy to see from Lemma 5.7.4 that $\rho_0, \rho_n, \cdots, \rho_{n(d-1)}$ are distinct automorphisms of $\mathbf{F}_{p^m}$ which fixes the elements of $\mathbf{F}_{p^n}$. On the other hand, $|\mathrm{Aut}_{\mathbf{F}_{p^n}}(\mathbf{F}_{p^m})| = d$ by the above (1) and Cor.5.6.17. $\square$

We mentioned in the proof of Prop.5.7.2(2) that a primitive element w of $\mathbf{F}_{p^{nd}}$ over $\mathbf{F}_{p^n}$ is not necessarily a generator of the cyclic group $(\mathbf{F}_{p^{nd}})^*$, i.e. not necessarily a primitive p^{nd}-th root of unity. Now we show a condition for primitive elements.

5.7.9 Lemma. $\mathbf{F}_{p^m}$ *contains a primitive k-th root of unity if and only if $k \mid (p^m - 1)$.*

Proof. A k-th root of unity is an element of order k in the multiplicative group $\mathbf{F}_{p^m}^*$ which is a cyclic group of order $p^m - 1$. $\square$

5.7.10 Corollary. $\mathbf{F}_{p^{nd}} = \mathbf{F}_{p^n}(w)$ *if and only if w is a primitive k-th root of unity with k satisfying the following*

$(5.7.11) \quad k \mid p^{nd} - 1$ *but* $k \nmid p^{nl} - 1$ *for any proper divisor l of d.*

Proof. By Cor.5.7.7, $\mathbf{F}_{p^n} \subset \mathbf{F}_{p^{nl}} \subset \mathbf{F}_{p^{nd}}$ if and only if $l \mid d$. Hence the result follows from Lemma 5.7.9 at once. $\square$

Remark. Conditon (5.7.11) is equivalent to the following one:

(5.7.11') *In the multiplicative group* $\mathbf{Z}_k^*$, *the order of* $[p^n]$ *is* d.

Turn to the minimal polynomials of the element w, we can describe the irreducible polynomials over finite fields as follows.

5.7.12 Theorem. **1)** *The irreducible* $\mathbf{F}_{p^n}$*-polynomials* $p_{n,d}(x)$ *of degree* d *exist, they are factors of* $x^{p^{m-1}} - 1$ *provided* $(nd) \mid m$.

2) *Any root* w *of* $p_{n,d}(x)$ *is a primitive* k-*th root of unity with* k *satisfying condition (5.7.11), and*

$$p_{n,d}(x) = (x - w)(x - w^{p^n}) \cdots (x - w^{p^{n(d-1)}}).$$

3) *If* $n \mid m$, *then in* $\mathbf{F}_{p^n}[x]$ *the prime factorization of* $x^{p^{m-1}} - 1$ *is the product of all the distinct* $p_{n,d}(x)$ *with* $nd \mid m$.

Proof. (1). $\mathbf{F}_{p^{nd}} = \mathbf{F}_{p^n}(w)$ by Prop.5.7.2(2); then the minimal $\mathbf{F}_{p^n}$-polynomial $p_{n,d}(x)$ of w is irreducible and of degree d. Conversely, such a $p_{n,d}(x)$ determines a single extension over $\mathbf{F}_{p^n}$ of degree d, i.e. determines $\mathbf{F}_{p^{nd}}$. If $nd \mid m$, then $\mathbf{F}_{p^{nd}} \subset \mathbf{F}_{p^m}$ by Cor.5.7.7; so $x^{p^m} - x$ annihilates w by Theorem 5.7.5, hence $p_{n,d}(x) \mid x^{p^{m-1}} - 1$.

(2). Since $\mathbf{F}_{p^{nd}} = \mathbf{F}_{p^n}(w)$ as shown in the above (1), the first conclusion follows from Cor.5.7.10 directly. For the second one, by Exer.5.2.8 and Cor 5.7.8(2) we see that all $w, w^{p^n}, \cdots, w^{p^{n(d-1)}}$ are distinct roots of $p_{n,d}(x)$; but $\deg p_{n,d}(x) = d$; thus the factorization of $p_{n,d}(x)$ follows.

(3). If $nd \mid m$, then $p_{n,d}(x) \big| x^{p^{m-1}} - 1$ by the above (1). Conversely, considering $\mathbf{F}_{p^m}$, we see from Theorem 5.7.5 that every prime factor $p(x)$ of degree d of $x^{p^{m-1}} - 1$ has a root $w \in \mathbf{F}_{p^m}$; hence $p(x)$ determines a single extension $\mathbf{F}_{p^n}(w) \subset \mathbf{F}_{p^m}$ and $|\mathbf{F}_{p^n}(w) : \mathbf{F}_{p^n}| = d$; i.e. $\mathbf{F}_{p^{nd}} \subset \mathbf{F}_{p^m}$; thus $(nd)\big|m$. Hence $p(x)$ must be one of $p_{n,d}(x)$'s. Further, $x^{p^{m-1}} - 1$ has no multiple roots by Cor.5.6.4(3) (cf.Exer.5.6.2), so $p_{n,d}(x)$ cannot be repeated in the factorization of $x^{p^{m-1}} - 1$. $\square$

5.7.13 Remark. Applying the arguments to the so-called *cyclotomic polynomials*, one can get their explicit factorizations; see Exer.11.

5.7.14 Example. Considering $\mathbf{F}_2[x] = \mathbf{Z}_2[x]$. By Theorem 5.7.12, $x^7 - 1 = x^{2^3 - 1} - 1$ has irreducible factors of degrees 1 and 3; so its prime factorization is as follows:

$$x^7 - 1 = (x - 1)(x^3 + x^2 + 1)(x^3 + x + 1).$$

Let w be a root of $x^3 + x + 1$ (hence w is a primitive 7-th root of unity). Then $w^3 = w + 1$; and $\mathbf{F}_8 = \mathbf{F}_2(w)$ whose elements are expressed as:

$$a_0 + a_1 w + a_2 w^2, \qquad a_t \in \mathbf{F}_2.$$

5.7.15 Example. Consider $\mathbf{F}_3[x] = \mathbf{Z}_3[x]$. By Theorem 5.7.12, $x^8 - 1 = x^{3^2 - 1} - 1$ has irreducible factors of degrees 1 and 2; so its prime factorization is as follows:

$$x^8 - 1 = (x - 1)(x + 1)(x^2 + 1)(x^2 - x - 1)(x^2 + x - 1).$$

Let i be a root of $x^2 + 1$ (hence i is a primitive 4-th root of unity). Then $i^2 = -1$; and $\mathbf{F}_9 = \mathbf{F}_3(i)$ whose elements are expressed as:

$$a_0 + a_1 i, \qquad a_t \in \mathbf{F}_3.$$

If, instead of the above i, choose a root u of $x^2 - x - 1$, then u is a primitive 8-th root of unity; the expressions of the elements of $\mathbf{F}_9 = \mathbf{F}_3(u)$ are the same as above but, instead of $i^2 = -1$, we have $u^2 = u + 1$.

Exercises 5.7

Note: p always denotes a prime number in the following exercises.

1. If K is a finite field of $\mathrm{char}K = p > 0$, then every element of K has a unique p-th root in K.
2. Let $p(x) = x^p - x - 1 \in \mathbf{F}_p[x]$.
   1) Show that $p(x)$ has no roots in $\mathbf{F}_p$.
   2) If u is a root of $p(x)$ in an extension, then $u + 1$ is also a root of $p(x)$.
   3) Show that $p(x)$ is irreducible over $\mathbf{F}_p$. (Hint: otherwise, by (2) we have $(x - u - a_1) \cdots (x - u - a_k) \in F_p[x]$ with $a_i \in \mathbf{F}_p$ and $1 \leq k < p$; expanding the product and considering the coefficient of x^{k-1}, we have $ku \in \mathbf{F}_p$ hence $u \in \mathbf{F}_p$, contradicts (1).)
   4) Find the degree of $\mathbf{F}_p(u)$ over $\mathbf{F}_p$
3. Let $q = p^m$ and $f(x) \in \mathbf{F}_q[x]$ be irreducible. Then $f(x) \mid x^{q^n} - x \iff \deg f(x) \mid n$.
4. Construct the field of order 4 and write out its addition table and its multiplicaation table.
5. Construct the field of order 64; and find all the subfields of it.
6. Find all the primitive elements of $\mathbf{F}_9$ over $\mathbf{F}_3$ (i.e. such u that $\mathbf{F}_9 = \mathbf{F}_3(u)$).
7. 1) If $(r, p^n - 1) = 1$. then every element of $\mathbf{F}_{p^n}$ has an r-th root in $\mathbf{F}_{p^n}$.
   2) Let $r \mid (p^n - 1)$; then $a \in \mathbf{F}_{p^n}$ has an r-th root in $\mathbf{F}_{p^n} \iff a^{(p^n - 1)/r} = 1$.

8. Every element in a finite field can be written as a sum of two squares of elements in the field. (Hint: If $\text{char}F = 2$, then $F^2 = F$; ok. If $\text{char}F \neq 2$ and $|F| = p^m$, then $|(F^*)^2| = (p^m - 1)/2$ hence $|F^2| = (p^m + 1)/2$; so $|a - F^2| = (p^m + 1)/2$; thus $F^2 \cap (a - F^2) \neq \emptyset$; i.e. there are $u^2 \in F^2$ and $a - v^2 \in a - F^2$ such that $u^2 = a - v^2$.)

9. Prove: there are irreducible polynomials over $\mathbf{F}_{p^n}$ of arbitrary large degree.

10. Find the conditions such that $\mathbf{F}_{p^m} \cap \mathbf{F}_{p^n} = \mathbf{F}_p$.

+11. Let F be *any field* and E be a splitting field of $x^k - 1$ over F. Further, assume that k and $\text{char}F$ are coprime. Obviously, all the roots in E of $x^k - 1$ form a cyclic (multiplicative) group of order k. Thus, by Cor 2.8.10, there are $\varphi(k)$ primitive k-th roots of unity altogether, where $\varphi(k)$ is the number-theoretic Euler function. Let $w_1, \cdots, w_{\varphi(k)}$ be all the primitive k-th roots of unity. The polynomial $\Phi_k(x) = (x - w_1) \cdots (x - w_{\varphi(k)})$ is called the *k-th cyclotomic polynomial*. Prove:

  1) $x^k - 1 = \prod_{d|k} \Phi_d(x)$;

     (Hint. Any k-th root of unity has order $d|k$; conversely for any $d|k$ there are $\varphi(d)$ primitive d-th roots of unity; compare with Cor.2.8.11.)

  2) Every coefficient of $\Phi_k(x)$ is a multiple of 1_F, and the leading one is 1_F while the constant one is $\pm 1_F$. (Hint: by induction on k.)

  3) If $F = \mathbf{F}_{p^m}$, then the cyclotomic polynomial $\Phi_k(x)$ is factorized into a product of irreducible F-polynomials of degree d where d and k satisfy the condition $(5.7.11')$.

     (*Remark.* For example, $\Phi_8(x) = (x^2 - x - 1)(x^2 + x - 1)$ in $\mathbf{F}_3$. As a comparison, it is well known that $\Phi_k(x)$ is always irreducible if $\text{char}F = 0$. The proof is not straightforward, so we omit it.)

12. Let $\mathbf{F} = \mathbf{F}_q$ where $q = p^m$, and n be a positive integer.

  1) Prove that the number of the distinct 1-dimensional subspaces of $\mathbf{F}^n$ is equal to $(q^n - 1)/(q - 1)$.

  2) Find the number of the distinct k-dimensional subspaces of $\mathbf{F}^n$.

5.8   Coding

Generally speaking, a code is a set of words spelt according to certain rules with alphabets. For example, the *ASCII code* in computer science is the set of the words over the alphabet of 16 letters; e.g. "4F 4B" are two words standing for "OK". We are interested in the so-called linear codes which are defined as subspaces of linear spaces over finite fields.

Differently from the usual linear algebra, however, codes emphasize the coordinates of vectors; e.g. the two codewords $(01001111), (01001011)$ are two vectors of dimension 8 over $\mathbf{F}_2$, they are just the binary form of the ASCII codewords for "OK". With the help of suitable mathematical structures, the linear codes can possess strong capabilities of error-corrections; precisely speaking, they are called *error-correcting codes*.

The following notations are preserved throughout this section.

5.8.1 Notation. $\mathbf{F} = \mathbf{F}_q$ stands for a finite field of order $q = p^l$, hence $\operatorname{char}\mathbf{F} = p$; and $\mathbf{F}^n = \{(a_1, \cdots, a_n) \mid a_i \in \mathbf{F}\}$ denotes the $\mathbf{F}$-vector space consisting of the n-tuples over $\mathbf{F}$. Any k-dimensional linear subspace C of $\mathbf{F}^n$ is said to be a *q-ary code* of *length* n and *dimension* k, or a *q-ary* $[n, k]$ *code* briefly. Any $v \in \mathbf{F}^n$ is called a *word* (of length n over $\mathbf{F}$). On the other hand, once the code C is given, any $c \in C$ is called a *codeword*.

5.8.2 Definition. Let $\mathbf{Z}^+$ denote the set of all non-negative integers. The map $w : \mathbf{F}^n \to \mathbf{Z}^+$, $(a_1, \cdots, a_n) \mapsto |\{i \mid a_i \neq 0\}|$ is called the *weight function*; and $w(a)$ is called the *weight* of $a = (a_1, \cdots, a_n) \in \mathbf{F}$; by definition it is the number of the non-zero components of a.

For any pair $(a, b) \in \mathbf{F}^n \times \mathbf{F}^n$, we define $d(a, b) = w(a - b)$ and call it the *Hamming distance* between a and b; it is the number of the components of a which are different from the corresponding components of b. We also call $d(a, b)$ the distance in short.

Moreover, if $C \leq \mathbf{F}^n$ is a code, the minimal value of the weights $w(c)$ for $0 \neq c \in C$ is called the *minimal weight* of the code C, while the minimal value of the distances $d(c, c')$ for $c \neq c' \in C$ is called the *minimal distance* of the code C.

It is easy to see that $d(a, b)$ is indeed a distance function on $\mathbf{F}^n$, see Exer.2. On the other hand, the above two minimal values for a code are in fact the same quantity.

5.8.3 Proposition and Definition. *For a given code C, the minimal weight and the minimal distance equal each other; hence we denote it by $t(C)$, and an $[n, k]$ code of minimal weight t is called a $[n, k, t]$ code.*

Proof. By definition we have that $\min\{d(a, b) \mid a \neq b \in \mathbf{F}^n\} = \min\{w(a - b) \mid a \neq b \in \mathbf{F}^n\} = \min\{w(a) \mid 0 \neq a \in \mathbf{F}^n\}$. $\square$

In the practice of communications, a code C is usually given; and a message is expressed by the codewords in C, and then transmitted by some way, e.g. by radio. However, because of some disturbances, errors may occur when a codeword c is transmitted, so that the received word r maybe different from c, and $s = c - r$ is just the "error part" and $w(s) = d(c, r)$ is the number of errors in the transmission of the word c. A very natural treatment for receivers is as follows.

5.8.4 Definition. **1)** The following way is called *maximal likelihood decoding*: If the received r is not a codeword, i.e. if $r \notin C$, then decode r to be a codeword $c' \in C$ such that $d(c', r)$ is as small as possible.

2) A code C is said to be *e-error-correcting code*, if any received word r from c with errors $\leq e$ (i.e. $d(r, c) \leq e$) can be uniquely decoded to c itself by maximal likelihood decoding.

The following result shows that a code of minimal distance $2e + 1$ is e-error-correcting but not $(e + 1)$-error-correcting.

5.8.5 Proposition. *Let C be an $[n, k, 2e + 1]$ code over $\mathbf{F}$.*

1) *If $c \in C$ and $r \in \mathbf{F}^n$ such that $d(c, r) \leq e$, then $d(c', r) > e \geq d(c, r)$ for any $c \neq c' \in C$.*

2) *There are $c, c' \in C$ and $r \in \mathbf{F}^n$ such that $d(c, r) > e$ and $d(c', r) \leq d(c, r)$.*

Proof. (1). By Exer.2, $2e + 1 \leq d(c', c) \leq d(c', r) + d(c, r) \leq d(c', r) + e$; i.e. $d(c', r) \geq e + 1 > e$.

(2). By the definition of $t(C)$, there are $c, c' \in C$ such that $d(c, c') = 2e + 1$, i.e. without lose of generality we can assume

$$c = (c_1, \cdots, c_e, c_{e+1}, \cdots, c_{2e+1}, c_{2e+2}, \cdots, c_n),$$
$$c' = (c'_1, \cdots, c'_e, c'_{e+1}, \cdots, c'_{2e+1}, c_{2e+2}, \cdots, c_n),$$

with $c_i \neq c'_i$ for $1 \leq i \leq 2e + 1$ and the other coordinates are equal. Set

$$r = (c_1, \cdots, c_e, c'_{e+1}, \cdots, c'_{2e+1}, c_{2e+2}, \cdots, c_n).$$

Then $d(c, r) = e + 1$ while $d(c', r) = e$; (2) holds.   □

We turn to discuss how to determine the minimal distance of a code, and how to construct a code of large enough minimal distance.

Recall from linear algebra that there are two ways to characterize a subspace of a linear space: one is to give a basis of a subspace, and the other is to describe it as a solution space of a system of homogeneous linear equations. Thus we have two matrices to characterize a code C.

5.8.6 Proposition and Definition. *Let $C \leq \mathbf{F}^n$ be an $[n, k]$ code.*

1) *There is a $k \times n$ matrix G over $\mathbf{F}$ such that, for $v \in \mathbf{F}^n$, $v \in C$ iff there is a $w = (w_1, \cdots, w_k) \in \mathbf{F}^k$ such that $v = wG$. Such a matrix G is called a* generator matrix *of the code C.*

2) *There is an $(n - k) \times n$ matrix H over $\mathbf{F}$ such that, for $v \in \mathbf{F}^n$, $v \in C$ iff $vH^T = 0$, where H^T denotes the transpose matrix of H. Such a matrix H is called a* parity check matrix *of the code C.*

Proof. (1). Take an $\mathbf{F}$-basis $\{g_1, \cdots, g_k\}$ of C, then the $k \times n$-matrix G with rows $g_1, \cdots, g_k$ clearly satisfies (1) by linear algebra.

(2). Let G be as in the above (1), and let $x = (x_1, \cdots, x_n)$ be the vector in the indeterminates $x_1, \cdots, x_n$. Consider the system of homogeneous linear equations $Gx^T = 0$. Since the rank of G is k, by linear algebra again, the solutions of $Gx^T = 0$ form a subspace of $\mathbf{F}^n$ of dimension $n - k$. Take a basis $\{h_1, \cdots, h_{n-k}\} \subset \mathbf{F}^n$, and set H to be the $(n - k) \times n$ matrix with rows $h_1, \cdots, h_{n-k}$. Then $GH^T = 0$ which shows that the k rows of G form a basis of the solution subspace of the system of equations $xH^T = 0$. Therefore, $v \in C$ if and only if $v = wG$ for w as in (1), if and only if $vH^T = 0$. □

The above arguments also show how to find a generator matrix and how to find a parity check matrix of a code C.

Conversely, even if there is no code beforehand, we can construct a code C by a suitable matrix G or H as above.

5.8.7 Corollary. 1) *If G is a $k \times n$ matrix over $\mathbf{F}$ of rank k, then the following is an $[n, k]$ code over $\mathbf{F}$ with generator matrix G:*

$$C = \{c \in \mathbf{F}^n \mid c = wG \quad \text{for a } w \in \mathbf{F}^k\}.$$

2) *If H is an $(n - k) \times n$ matrix over $\mathbf{F}$ of rank $n - k$, then the following is an $[n, k]$ code over $\mathbf{F}$ with parity check matrix H:*

$$C = \{c \in \mathbf{F}^n \mid cH^T = 0\}.$$

Proof. Exer.9. □

In some cases the parity check matrices are more convenient; for example, we can determine the minimal distance of a code.

5.8.8 Theorem. *Let C be an $[n, k]$ code over $\mathbf{F}$, and $(n - k) \times n$ matrix H be a parity check matrix of C.*

1) *If H has t linearly dependent columns, then $t(C) \leq t$.*

2) *If H has no $t - 1$ linearly dependent columns, then $t(C) \geq t$.*

Proof. Let $H = (H_1, \cdots, H_n)$ with i-th column H_i. Then $c = (c_1, \cdots, c_n) \in C$ if and only if $c_1 H_1 + \cdots + c_n H_n = 0$ (Prop.5.8.6(2)). So both (1) and (2) are easy to verify. The detail is left as Exer.10. $\square$

5.8.9 Example. Take $\mathbf{F} = \mathbf{F}_2$, $n = 7$, $k = 4$. Take

$$H = \begin{pmatrix} 1 & 0 & 1 & 1 & 1 & 0 & 0 \\ 0 & 1 & 0 & 1 & 1 & 1 & 0 \\ 0 & 0 & 1 & 0 & 1 & 1 & 1 \end{pmatrix}$$

as a parity check matrix, then, by Cor.5.8.7 and Theorem 5.8.8, a $[7, 4, 3]$ binary code C is determined; by Prop.5.8.5, C is a 1-error-correcting but not a 2-error-correcting code.

Moreover, using the parity check matrices of codes, a practical way to perform the maximal likelihood decoding can be described; due to the space constraint, we omit it.

Now we show a well-known kind of codes.

5.8.10 Remark. For a technical reason, we extend $\mathbf{F}$ to a large field $\overline{\mathbf{F}}$; and allow a $k' \times n$ matrix $\overline{H}$ over $\overline{\mathbf{F}}$ as a parity check matrix of a code $C \leq \mathbf{F}^n$ (*not* $\leq \overline{\mathbf{F}}^n$); i.e. C is the subspace of $\mathbf{F}^n$ consisting of the all solutions *in* $\mathbf{F}$ (*not* in $\overline{\mathbf{F}}$) of the system of homogeneous linear equations $(x_1, \cdots, x_n)\overline{H}^T = 0$; But note that, in this case, the $\mathbf{F}$-dimension of C maybe less than $n - k'$.

5.8.11 Definition. Assume $l \geq 0$, $d \geq 2$ and $n \geq l + d - 2$. Let w be a primitive n-th root of unity in a splitting field of $x^n - 1$ over $\mathbf{F}$. Then the code C over $\mathbf{F}$ with the following parity check matrix $\overline{H}$ is called a *BCH-code* of *designed distance* d:

$$\overline{H} = \begin{pmatrix} 1 & w^l & w^{2l} & \cdots & w^{(n-1)l} \\ 1 & w^{l+1} & w^{2(l+1)} & \cdots & w^{(n-1)(l+1)} \\ \cdots \cdots & \cdots & \cdots \cdots \\ 1 & w^{l+d-2} & w^{2(l+d-2)} & \cdots & w^{(n-1)(l+d-2)} \end{pmatrix}.$$

And it is referred to as a *narrow-sense* BCH code if $l = 1$. While it is referred to as a *primitive* BCH code if $n = q^m - 1$, i.e. w is a generator of the multiplicative group of $\mathbf{F}_{q^m}$ which is an extension of $\mathbf{F} = \mathbf{F}_q$.

The name "BCH" was due to R.C. Bose and D.K. Ray-Chaudhuri (1960) and A. Hocquenghem (1959). And the "designed distance" is explained in the following theorem.

5.8.12 Theorem. *Notations are as in Def.5.8.11. If $C \neq 0$, then $t(C) \geq d$.*

Proof. For arbitrary distinct $d-1$ indexes $0 \leq j_0, \cdots, j_{d-2} \leq n-1$, the $d-1$ columns of $\overline{H}$ form a $(d-1) \times (d-1)$ matrix with determinant:

$$\det \begin{pmatrix} w^{j_0 l} & w^{j_1 l} & \cdots & w^{j_{d-2} l} \\ w^{j_0 (l+1)} & w^{j_1 (l+1)} & \cdots & w^{j_{d-2}(l+1)} \\ \cdots & \cdots & \cdots \cdots \\ w^{j_0(l+d-2)} & w^{j_1(l+d-2)} & \cdots & w^{j_{d-2}(l+d-2)} \end{pmatrix}$$

$$= (w^{j_0 l} w^{j_1 l} \cdots w^{j_{d-2} l}) \cdot \det \begin{pmatrix} 1 & 1 & \cdots & 1 \\ w^{j_0} & w^{j_1} & \cdots & w^{j_{d-2}} \\ \cdots & \cdots & \cdots \cdots \\ w^{j_0(d-2)} & w^{j_1(d-2)} & \cdots & w^{j_{d-2}(d-2)} \end{pmatrix}$$

$$= w^{(j_0 + j_1 \cdots + j_{d-2})l} \prod_{0 \leq r < s \leq d-2} (w^{j_r} - w^{j_s}) \neq 0,$$

where the last inequality follows from the assumption that w is a primitive n-th root of unity and $0 \leq j_0, \cdots, j_{d-2} \leq n - 1$ are distinct from each other. Therefore $t(C) \geq d$ by Theorem 5.8.8(2). $\square$

5.8.13 Remark. Here are three remarks about the result.

1) If $l > 0$ and $l + d - 2 < n$, then $C > 0$; i.e. the condition of Theorem 5.8.12 is satisfied; see Cor.5.8.17 below.

2) $t(C) \geq d$ in Theorem 5.8.12 may be a strict inequality; e.g. in Example 5.8.18 below, $n = 7$, $l = 1$, $d = 2$ but $t(C) = 3$. In general, it is difficult to determine the minimal distance of a BCH code.

3) A common way to understand and to deal with the BCH codes is through the so-called cyclic codes.

5.8.14 Definition. Let $X = \{\xi_0 = 1, \xi_1, \cdots, \xi_{n-1}\}$ be a finite group of order n, and let $\mathbf{F}X$ denote the group algebra of X over $\mathbf{F}$, see Example 4.8.11. Then $\mathbf{F}X$ is an n-dimensional $\mathbf{F}$-space with basis $\{1, \xi_1, \cdots, \xi_{n-1}\}$; so any element $a_0 + a_1\xi_1 + \cdots + a_{n-1}\xi_{n-1} \in \mathbf{F}X$ is just a word $(a_0, a_1, \cdots, a_{n-1})$ over $\mathbf{F}$ of length n. Thus any ideal C of $\mathbf{F}X$ is a code of length n over $\mathbf{F}$; such a code is called a *code of the group algebra*. In particular, if X is a cyclic group, then a code of the group algebra $\mathbf{F}X$ is called a *cyclic code*.

In the following, let $X = \{\xi^0 = 1, \xi^1, \cdots, \xi^{n-1}\}$ be a cyclic group of order n generated by ξ; then any $a \in \mathbf{F}X$ is written as $a(\xi) = a_0 + a_1\xi + \cdots + a_{n-1}\xi^{n-1}$, regarded as an $\mathbf{F}$-polynomial in ξ. But, notice that $\xi^n - 1 = 0$; hence, obviously, a polynomial $f(\xi) = 0$ if and only if $x^n - 1 \mid f(x)$ for polynomials in indeterminate x. The following is a precise formulation for this observation.

5.8.15 Theorem and Notation (on cyclic codes). *Let* $X = \langle \xi \rangle$ *be a cyclic group of order* n. *Then*

$$\varphi : \mathbf{F}[x] \longrightarrow \mathbf{F}X, \qquad f(x) \longmapsto f(\xi)$$

is a surjective ring homomorphism with kernel $\langle x^n - 1 \rangle$. *And, if* C *is a code (i.e. an ideal) of* $\mathbf{F}X$ *of dimension* k, *then there is an* $\mathbf{F}$-*decomposition* $x^n - 1 = g(x)h(x)$ *such that* $\deg h(x) = k$ *and the following two hold:*

1) $C = \langle g(\xi) \rangle = g(\xi) \cdot \mathbf{F}X$; *so* $g(x)$ *is called a* generator polynomial *of the cyclic code* C;

2) $c(\xi) \in C$ *if and only if* $c(\xi)h(\xi) = 0$; *so* $h(x)$ *is called a* parity check polynomial *of the cyclic code* C.

(**Remark.** As a complement, see Exer.14.)

Proof. The result on φ follows from Theorem 3.8.3 and Cor.3.8.4 straightforwardly. The inverse image $\varphi^{-1}(C)$ is an ideal of $\mathbf{F}[x]$.

By Cor.3.9.15 (or Lemma 5.5.2(1)), $\mathbf{F}[x]$ is a principal ideal domain; hence there is a $g(x)$ such that $\varphi^{-1}(C) = \langle g(x) \rangle$. Thus $C = \langle \varphi(g(x)) \rangle = \langle g(\xi) \rangle$; (1) holds.

Furthermore, since $x^n - 1 \in \varphi^{-1}(C)$, we have $x^n - 1 = g(x)h(x)$.

If $c(\xi) \in C$, then $c(\xi) = f(\xi)g(\xi)$ by (1), hence

$$c(\xi)h(\xi) = f(\xi)g(\xi)h(\xi) = f(\xi)(\xi^n - 1) = 0.$$

Conversely, if $c(\xi)h(\xi) = 0$, then $x^n - 1 \big| c(x)h(x)$, i.e. $g(x)h(x) \big| c(x)h(x)$, hence $g(x) \big| c(x)$; thus $c(x) \in \varphi^{-1}(C)$, so that $c(\xi) \in C$. (2) is proved.

At last, assume $\deg g(x) = k'$, hence $\deg h(x) = n - k'$. Noting that $X = \{1, \xi, \cdots, \xi^{n-1}\}$ is an $\mathbf{F}$-basis of $\mathbf{F}X$ and any $c(\xi) \in C$ can be written as $c(\xi) = f(\xi)g(\xi)$ with $\deg f(\xi) < n - k'$, we see that

$$g(\xi), \quad \xi g(\xi), \quad \cdots, \quad \xi^{n-k'-1}g(\xi)$$

form an $\mathbf{F}$-basis of C. Thus $n - k' = \dim(C) = k$; i.e. $\deg h(x) = k$. $\square$

As a corollary, we can describe the BCH codes as follows.

5.8.16 Corollary. *Notation as in Def.5.8.11. Let $p_{l+i}(x)$ be a minimal $\mathbf{F}$-polynomial of w^{l+i} for $i = 0, \cdots, d-2$; and let $g(x)$ be the least common multiple of $p_l(x), \cdots, p_{l+d-2}(x)$. Then $g(x) \mid (x^n - 1)$; and the BCH code C is a cyclic code of length n with generator polynomial $g(x)$.*

Proof. $x^n - 1$ annihilates every w^{l+i} because they are n-th roots of unity; so every $p_{l+i}(x) \mid (x^n - 1)$, and hence $g(x) \mid (x^n - 1)$. By 5.8.11, $(c_0, \cdots, c_{n-1}) \in C$ if and only if

$$c_0 + c_1 w^{l+i} + c_2 w^{2(l+i)} + \cdots + c_{n-1} w^{(n-1)(l+i)} = 0, \quad i = 0, \cdots, d-2;$$

if and only if the $\mathbf{F}$-polynomial $c(x) = c_0 + c_1 x + \cdots + c_{n-1} x^{n-1}$ annihilates w^{l+i} for $i = 0, \cdots, d-2$; if and only if $p_{l+i}(x) \mid c(x)$ for $i = 0, \cdots, d-2$; if and only if $g(x) \mid c(x)$; if and only if $c(\xi) \in \langle g(\xi) \rangle$. $\square$

5.8.17 Corollary. *Notations as above. If $l > 0$ and $l + d - 2 < n$, then $C \neq 0$.*

Proof. In this case every $w^{l+i} \neq 1$ for $i = 0, \cdots, d-2$; thus every $p_{l+i}(x) \mid \frac{x^n - 1}{x - 1}$ for $i = 0, \cdots, d-2$; hence $g(x) \mid \frac{x^n - 1}{x - 1}$ where $g(x)$ is as in Cor.5.8.16 above. In particular, $\deg g(x) < n$; and $\dim_{\mathbf{F}}(C) = n - \deg g(x) \geq 1$ by Theorem 5.8.15. $\square$

5.8.18 Example. Take $\mathbf{F} = \mathbf{F}_2$, $n = 7$, w to be a root of $x^3 + x + 1$. From Example 5.7.14 we know that

$$x^7 - 1 = (x - 1)(x^3 + x^2 + 1)(x^3 + x + 1).$$

is an irreducible decomposition in $\mathbf{F}[x]$ and w is a primitive 7-th root of unity. In Def.5.8.11, take $l = 1$ and $d = 2$ and $\overline{H} = (1\, w\, w^2\, \cdots\, w^6)$, a 1×7 matrix over $\mathbf{F}(w) = \mathbf{F}_8$; then, with $\overline{H}$ as a parity check matrix, we get a primitive narrow-sense BCH code C over $\mathbf{F}$. Furthermore, by Cor.5.8.16, C is a cyclic code with generator polynomial $g(x) = x^3 + x + 1$, hence with parity check polynomial $h(x) = (x - 1)(x^3 + x^2 + 1) = x^4 + x^2 + x + 1$. Thus, by Exer.14, the following

$$H = \begin{pmatrix} 1\,0\,1\,1\,1\,0\,0 \\ 0\,1\,0\,1\,1\,1\,0 \\ 0\,0\,1\,0\,1\,1\,1 \end{pmatrix}$$

is the parity check $\mathbf{F}$-matrix of C. Therefore, it is a $[7, 4, 3]$ binary code, and we again obtain Example 5.8.9.

Exercises 5.8

1. Prove that the weight function $w(a)$ on $\mathbf{F}^n$ satisfies:
   a) $w(a) \geq 0$, and $w(a) = 0$ only if $a = 0$;
   b) $w(a + b) \leq w(a) + w(b)$.

+2. Prove that the distance function $d(a, b)$ on $\mathbf{F}^n$ satisfies:
   a) $d(a, b) \geq 0$, and $d(a, b) = 0 \iff a = b$;
   b) $d(a, b) = d(b, a)$);
   c) $d(a, b) \leq d(a, c) + d(c, b)$.

3. Assume C is a binary code with generator matrix $\begin{pmatrix} 1101 \\ 0111 \end{pmatrix}$. Apply the maximal likelihood decoding to decode the received word $r = (0110)$; is the code C 1-error-correcting?

4. Is a $[6, 3]$ code 2-error-correcting? why?

5. Let C be a binary $[n, k]$ code. If there is a $c \in C$ such that $w(c)$ is odd, then all the codewords of even weight form a $[n, k - 1]$ code.
   (Hint: for a binary code, the weight of the sum of two codewords of even weight is also even.)

6. Let G be a $k \times n$ matrix over $\mathbf{F}$ of full-rank and H be a $(n - k) \times n$ matrix over $\mathbf{F}$ of full-rank. If $GH^T = 0$, then the following two hold:
   1) The code with generator matrix G has H as a parity check matrix.
   2) The code with parity check matrix H has G as a generator matrix.

7. Find a generator matrix of the code C in Example 5.8.9.

8. Let C be a $[n, k]$ code over $\mathbf{F} = \mathbf{F}_q$, and G be a generator matrix of C. If none of the columns of G is zero-vector, then $\sum_{c \in C} w(c) = n(q - 1)q^{k-1}$.
   (Hint: Let $T = \sum_{c \in C} w(c)$, and $G = (G_1, \cdots, G_n)$, i.e. G_i be the i-th column of G. Any $c \in C$ is uniquely written as $c = yG = (yG_1, \cdots, yG_n)$ where $y = (y_1, \cdots, y_k) \in \mathbf{F}^k$. Thus yG_i contributes 1 to T if $yG_i \neq 0$; otherwise it contributes 0 to T. Since $G_i \neq 0$, the solution subspace of $yG_i = 0$ in variable vector y is $(k - 1)$-dimensional. When y runs over $\mathbf{F}^k$, the contribution of yG_i to T is $q^k - q^{k-1} = (q - 1)q^{k-1}$. Let i run over $1, \cdots, n$; the result follows.)

9. Prove Cor.5.8.7.

10. Complete the proof of Theorem 5.8.8.

11. Let $\mathbf{F}X$ be the group algebra of a finite group X over $\mathbf{F}$, and C be a code of $\mathbf{F}$. If C is neither the zero ideal nor the unity ideal of $\mathbf{F}X$, then $t(C) \geq 2$.   (Hint: in $\mathbf{F}X$, an element of weight 1 must be invertible.)

+12. If C is a cyclic code of length n and $(c_0, c_1, \cdots, c_{n-1}) \in C$ is a codeword, show that $(c_{n-1}, c_0, \cdots, c_{n-2}) \in C$ too.

(**Remark.** This is the original meaning of "cyclic codes".)

13. Take $\mathbf{F} = \mathbf{F}_3$, $n = 8$, u to be a root of $x^2 - x - 1$. Then u is a primitive 8-th root of unity by Example 5.7.15. Similar to Example 5.8.18, construct the trinary BCH code C over $\mathbf{F}$ with parity check matrix $\overline{H} = (1\ u\ u^2\ \cdots\ u^7)$ over $\mathbf{F}(u) = \mathbf{F}_9$; show a parity check $\mathbf{F}$-matrix of C and find the minimal distance of C.

$^+$14. Notations as in Theorem 5.8.15; and assume

$$h(x) = h_0 + h_1 x + \cdots + h_k x^k, \quad g(x) = g_0 + g_1 x + \cdots + g_{n-k} x^{n-k}.$$

Prove that $g(\xi), \xi g(\xi), \cdots, \xi^{k-1} g(\xi)$ form an $\mathbf{F}$-basis of C; and the matrices

$$G = \begin{pmatrix} g_0 & g_1 & \cdots\cdots & g_{n-k} & & \\ & g_0 & \cdots\cdots & g_{n-k-1} & g_{n-k} & \\ & & \cdots\cdots & \cdots & \cdots & \cdots \\ & & g_0 & g_1 & g_2 & \cdots & g_{n-k} \end{pmatrix}_{k\times n}$$

and

$$H = \begin{pmatrix} h_k & h_{k-1} & \cdots\cdots & h_0 & & \\ & h_k & \cdots\cdots & h_1 & h_0 & \\ & & \cdots\cdots & \cdots & \cdots & \cdots \\ & & h_k & h_{k-1} & h_{k-2} & \cdots & h_0 \end{pmatrix}_{(n-k)\times n}$$

are a generator matrix and a parity check matrix resp. of the cyclic code C. (Hint: For the first conclusion, see the proof of 5.8.15. Since $\xi^i g(\xi) h(\xi) = 0$, $0 \le i \le k-1$, there are $q_i(x) \in \mathbf{F}[x]$ such that $x^i g(x) h(x) = (x^n - 1) q(x)$; comparing the coefficients of $x^k, \cdots, x^{n-1}$ on both sides and noting that $\deg q_i(x) \le k-1$, one can have $\sum_{r+s=t} g_{r+i} h_s = 0$ for $k \le t \le n-1$ and $0 \le i \le k-1$; i.e. $GH^T = 0$; then apply Exer.6 above.)

5.9 p-adic Numbers

Let us recall how to construct the real field $\mathbf{R}$ from the rational field $\mathbf{Q}$. Besides the order relation "$\le$" on $\mathbf{Q}$, consider the usual absolute values $|a|$ for $a \in \mathbf{Q}$, which is referred to as a valuation in the following sense:

5.9.1 Remark. In general, a function $\varphi : F \to \mathbf{Q}$ from a field F to $\mathbf{Q}$ is said to be a *valuation* on F if the following hold (for all $a, b \in F$):

(N1). $\varphi(a) \ge 0$, and $\varphi(a) = 0 \iff a = 0$;

(N2). $\varphi(a \cdot b) = \varphi(a) \cdot \varphi(b)$;

(N3) (*Triangle Inequality*). $\varphi(a + b) \le \varphi(a) + \varphi(b)$.

To define valuations more generally and more precisely, the codomain of φ should be replaced by any *ordered field*; e.g. the real field $\mathbf{R}$ could be a candidate. For convenience and for easy understanding, however, here we prefer the concrete $\mathbf{Q}$. Furthermore, once a valuation φ is given on the field F, a function d on $F \times F$ is defined:

$$d(a, b) = \varphi(a - b), \qquad \text{for all} \quad a, b \in F\,;$$

which is called the *distance* on F associated with φ since the following axioms for distances are satisfied (for all $a, b, c \in F$):

(D1). $d(a, b) \geq 0$, and $d(a, b) = 0$ iff $a = b$;

(D2). $d(a, b) = d(b, a)$;

(D3) (*Triangle Inequality*). $d(a, b) \leq d(a, c) + d(c, b)$.

The proof of them is an easy exercise (Exer.2). For example, in $\mathbf{Q}$ the familiar distance of a and b is just defined as $d(a, b) = |a - b|$.

Return to the construction of $\mathbf{R}$ from $\mathbf{Q}$. One can prove that any *Cauchy sequence* of rationals has a unique "limit", though it may no longer be a rational. Of course, two Cauchy sequences have the same limit if their difference sequence is infinitesimal.

5.9.2 Remark. In a field F with a valuation φ (hence with a distance d), a sequence $\{a_i\} = (a_1, a_2, \cdots)$ is called a *Cauchy sequence* if for any rational $\varepsilon > 0$ there is an integer N such that $d(a_m, a_n) < \varepsilon$ for all $m, n > N$. Furthermore, We say that two Cauchy sequences $\{a_i\}$ and $\{b_i\}$ *have the same limit* (with respect to φ), and denoted by $\{a_i\} \sim_\varphi \{b_i\}$, if for any rational $\varepsilon > 0$ there is an integer N such that $d_p(a_m, b_m) < \varepsilon$ for all $m > N$.

As we remarked in (5.9.1), the values of d could be an arbitrary ordered field, e.g. $\mathbf{R}$ could be a candidate. But, for the moment during the construction of $\mathbf{R}$ from $\mathbf{Q}$, the distance takes values in $\mathbf{Q}$.

Then, let $\mathbf{R}$ be the set of all the limits (i.e. all the real numbers); there are an addition and a multiplication on $\mathbf{R}$ induced by the operations on $\mathbf{Q}$ such that $\mathbf{R}$ becomes a field containing $\mathbf{Q}$; and there is an order relation "$\leq$" induced by the order relation on $\mathbf{Q}$, hence the absolute value and the associated distance on $\mathbf{R}$ are defined. At last, $\mathbf{R}$ is *complete* and $\mathbf{Q}$ is a *dense subfield* of $\mathbf{R}$ in the following sense.

5.9.3 Definition. Let F be a field with a valuation $\varphi : F \to \mathbf{Q}$ (as remarked above, $\mathbf{Q}$ could be replaced by an ordered field).

1) F is said to be *complete* if any Cauchy sequence in F has a unique limit in F.

2) a subfield K of F is said to be *dense* if for any $u \in F$ and any $\varepsilon \in \mathbf{Q}$ with $\varepsilon > 0$ there is an $a \in K$ such that $d(a, u) < \varepsilon$.

A *remarkable fact* is that, however, the valuations (hence the distances) defined on $\mathbf{Q}$ are not unique; and, with each valuation on $\mathbf{Q}$, the above construction can be done similarly. Therefore, besides the real field $\mathbf{R}$, each valuation on $\mathbf{Q}$ gives rise to a field which is complete for the induced valuation from $\mathbf{Q}$, and contains $\mathbf{Q}$ as a dense subfield. They are just the so-called p-adic fields $\mathbf{Q}_p$; to some extent, their constructions are easier than $\mathbf{R}$. They are extensively applied to many mathematical and scientific areas, e.g. number theory, algebra, information and coding theory, and so on.

5.9.4 Definition. Let p be a prime number. For any $0 \neq a \in \mathbf{Q}$, there is a unique integer, denoted by $v_p(a)$ as it is of course dependent on a, such that $a = p^{v_p(a)} \cdot \frac{m}{n}$ with $p \nmid (mn)$. Setting $v_p(0) = +\infty$, we have a function $a \mapsto v_p(a)$. Let

$$\varphi_p(a) = p^{-v_p(a)}, \qquad \forall a \in \mathbf{Q} \; (\varphi(0) = 0 \text{ of course}),$$

which is called the *p-adic valuation*. Further, we define the *p-adic distance* on $\mathbf{Q}$ to be:

$$d_p(a, b) = \varphi_p(a - b), \qquad \forall a, b \in \mathbf{Q}.$$

5.9.5 Proposition. *Notations as above. The following hold ($\forall a, b \in \mathbf{Q}$):*

N1) $\varphi_p(a) \geq 0$; and $\varphi_p(a) = 0 \iff a = 0$;

N2) $\varphi_p(ab) = \varphi_p(a)\varphi_p(b)$;

N3') *(Strong Triangle Inequality).* $\varphi_p(a+b) \leq \max\{\varphi_p(a), \varphi_p(b)\}$.

Proof. Exer.4. □

5.9.6 Remark. The above (N3') is really stronger than (N3) in (5.9.1). In general, if a valuation φ on a field F satisfies (N3'), then we call it a *discrete valuation*. A discrete valuation $\varphi : F \to \mathbf{Q}$ has a further property:

$$(5.9.7) \qquad \varphi(a) \neq \varphi(b) \implies \varphi(a + b) = \max\{\varphi(a), \varphi(b)\}.$$

To see this, suppose $\varphi(a) > \varphi(b)$ and $\varphi(a+b) < \varphi(a)$; then, by Exer.1(1),

$\varphi(-b) = \varphi(b) < \varphi(a)$; however, $\varphi(a) = \varphi(a + b - b) \le \max\{\varphi(a + b), \varphi(-b)\}$; a contradiction.

Furthermore, associated with a discrete valuation φ on a field F, the distance d on F defined by $d(a, b) = \varphi(a - b)$ satisfies, besides (D1) and (D2), the following stronger condition (easy to derive from (N3')):

(**D3'**) (*Strong Triangle Inequality*). $d(a, b) \le \max\{d(a, c), d(c, b)\}$.

5.9.8 Notation. From now on, we fix the discrete valution φ_p and the distance function d_p on $\mathbf{Q}$ defined in (5.9.4). By $\mathbf{Q}(d_p)$ we denote the set of all Cauchy sequences in $\mathbf{Q}$, and abbreviate the relation $\{a_i\} \sim_{\varphi_p} \{b_i\}$ to $\{a_i\} \sim_p \{b_i\}$. Note that $\{a\} = (a, a, \cdots)$ for $a \in \mathbf{Q}$ means the constant sequence.

5.9.9 Proposition. *On $\mathbf{Q}(d_p)$, $\sim_p$ is an equivalence relation.*

Proof. The reflexity and the symmetry are obvious. Let $\{a_i\} \sim_p \{b_i\}$ and $\{b_i\} \sim_p \{b_c\}$. For any $\varepsilon > 0$, there is an N such that $d_p(a_m, b_m) < \varepsilon$ and $d_p(b_m, c_m) < \varepsilon$ for all $m > N$; hence

$$d_p(a_m, c_m) \le \max\{d_p(a_m, b_m), d_p(b_m, c_m)\} < \varepsilon.$$

That is $\{a_i\} \sim_p \{c_i\}$.   □

5.9.10 Definition. In $\mathbf{Q}(d_p)$, let $[\{a_i\}]$ denote the equivalence class containing $\{a_i\}$. Let $\mathbf{Q}_p$ denote the set of all the equivalence classes in $\mathbf{Q}(d_p)$. The elements of $\mathbf{Q}_p$ are called the *p-adic numbers*.

5.9.11 Lemma. *For any non-zero $[\{a_i\}] \in \mathbf{Q}_p$ there is a unique positive $a \in \mathbf{Q}$ such that, for any representative $\{a_i\}$ of $[\{a_i\}]$, there is an N such that $\varphi_p(a_n) = a$ for all $n > N$ (hence $a = 1/p^l$ for an integer l). In particular, any Cauchy sequence is bounded above.*

Proof. Since $\{a_i\} \not\sim_p \{0\}$, there is an $\varepsilon_0 > 0$ such that there is an infinite subsequence $a_{i_1}, a_{i_2}, a_{i_3}, \cdots$ of $\{a_i\}$ with every $\varphi_p(a_{i_k}) > \varepsilon_0$. On the other hand, by definition, there is an N such that $\varphi_p(a_m - a_n) > \varepsilon_0$ $\forall m, n > N$. Take $i_r > N$ and let $a = \varphi_p(a_{i_r})$. For any $n > N$, we have

$$\varphi_p(a_n - a_{i_r}) < \varepsilon_0 < \varphi_p(a_{i_r});$$

hence, by (5.9.7) we have

$$\varphi_p(a_n) = \varphi_p(a_n - a_{i_r} + a_{i_r}) = \max\{\varphi_p(a_n - a_{i_r}), \varphi_p(a_{i_r})\} = \varphi_p(a_{i_r}) = a.$$

Suppose $\{a_i'\} \sim \{a_i\}$ and $\varphi_p(a_n') = a' \in \mathbf{Q}$ for all $n > N'$; then for any real $\varepsilon > 0$, taking $n > \max\{N, N'\}$, by Exer.1(3) we have

$$|a - a'| = |\varphi_p(a_n) - \varphi_p(a'_n)| \leq \varphi_p(a_i - a'_i) < \varepsilon;$$

which forces that $a = a'$; i.e. a is uniquely determined. Setting $b = \max\{\varphi_p(a_1), \varphi_p(a_2), \cdots, \varphi_p(a_{i_r})\}$, we get $\varphi_p(a_i) \leq a + b$, $\forall\, i = 1, 2, \cdots$.

At last, if $\{a_i\} \sim_p \{0\}$, then, by the definition of "$\sim_p$", $\varphi_p(a_i)$ are clearly bounded above. □

We are ready to reach the aim of this section mentiond before (5.9.4); which are collected in the following theorem.

5.9.12 Theorem.

1) $\mathbf{Q}_p$ *is a field with the following operations:*

$$[\{a_i\}] + [\{b_i\}] = [\{a_i + b_i\}], \qquad [\{a_i\}] \cdot [\{b_i\}] = [\{a_i \cdot b_i\}].$$

2) *Define* $\overline{\varphi}_p([\{a_i\}]) = \varphi_p(a_i)$ *for large enough* i *whenever* $[\{a_i\}] \neq [\{0\}]$, *and define* $\overline{\varphi}_p([\{0\}]) = 0$; *then* $\overline{\varphi}_p$ *is a discrete valuation on* $\mathbf{Q}_p$, *hence a distance* $\overline{d}_p([\{a_i\}], [\{b_i\}]) = \overline{\varphi}_p([\{a_i\}] - [\{b_i\}])$ *is defined on* $\mathbf{Q}_p$; *the field* $\mathbf{Q}_p$ *with the discrete valuation* $\overline{\varphi}_p$ *is called the* p-*adic field.*

3) $\delta : \mathbf{Q} \to \mathbf{Q}_p$, $a \mapsto [\{a\}]$ *is a continuous injective homomorphism; and, identifying* $\mathbf{Q}$ *with its image in* $\mathbf{Q}_p$, *we have* $\overline{\varphi}_p|_{\mathbf{Q}} = \varphi_p$ *and* $\mathbf{Q}$ *is dense in* $\mathbf{Q}_p$.

4) *The* p-*adic field* $\mathbf{Q}_p$ *is complete.*

Proof. (1). We prove that the multiplication is well-defined; and it is similar for the addition. For $[\{a_i\}], [\{b_i\}] \in \mathbf{Q}_p$ we have

$$\begin{aligned}
\varphi_p(a_i b_i - a_j b_j) &= \varphi_p(a_i(b_i - b_j) + b_j(a_i - a_j)) \\
&\leq \max\{\varphi_p(a_i)\varphi_p(b_i - b_j),\ \varphi_p(b_j)\varphi_p(a_i - a_j)\};
\end{aligned}$$

since both $\{a_i\}$ and $\{b_i\}$ are bounded above by Lemma 5.9.11, for any rational $\varepsilon > 0$ there is an $N(\varepsilon)$ such that

$$\varphi_p(a_m b_m - a_n b_n) < \varepsilon, \qquad \forall m, n > N(\varepsilon).$$

So $\{a_i b_i\} \in \mathbf{Q}(d_p)$. Assume $[\{a_i\}] = [\{a'_i\}]$ and $[\{b_i\}] = [\{b'_i\}]$. Then

$$\begin{aligned}
\varphi_p(a_i b_i - a'_i b'_i) &= \varphi_p(a_i b_i - a_i b'_i + a_i b'_i - a'_i b'_i) \\
&\leq \max\{\varphi_p(a_i)\varphi_p(b_i - b'_i),\ \varphi_p(b'_i)\varphi_p(a_i - a'_i)\};
\end{aligned}$$

by the "bounded above" again, for any $\varepsilon > 0$ there is an $N(\varepsilon)$ such that

$$\varphi_p(a_n b_n - a'_n b'_n) < \varepsilon, \qquad \forall n > N(\varepsilon).$$

In other words, $[\{a_ib_i\}] = [\{a_i'b_i'\}]$; that is, the multiplication defined in (1) is independent of the choices of representatives.

It is routine to check that $\mathbf{Q}_p$ is a field; we just list the following information for the field $\mathbf{Q}_p$:

(5.9.13). The zero element $0 = [\{0\}]$; the unity element $1 = [\{1\}]$; and $-[\{a_i\}] = [\{-a_i\}]$; and, if $[\{a_i\}] \neq [\{0\}]$, then $a_i \neq 0$ except for finitely many a_i, and $[\{a_i\}]^{-1} = [\{a_i^{-1}\}]$ with ith-entry 0 for $a_i = 0$.

(2). By Lemma 5.9.11, $\overline{\varphi}_p([\{a_i\}])$ is well-defined; and it is positive if $[\{a_i\}] \neq 0$, i.e. (N1) holds. Assume $\overline{\varphi}_p([\{a_i\}]) = a$ and $\overline{\varphi}_p([\{b_i\}]) = b$. Then there is an N such that $\varphi_p(a_n) = a$ and $\varphi_p(b_n) = b$ for all $n > N$; hence $\varphi_p(a_nb_n) = ab$ for all $n > N$. That is,

$$\overline{\varphi}_p([\{a_i\}] \cdot [\{b_i\}]) = \overline{\varphi}_p([\{a_ib_i\}]) = ab = \overline{\varphi}_p([\{a_i\}]) \cdot \overline{\varphi}_p([\{b_i\}]);$$

so (N2) holds. The Strong Triangle Inequality (N3$'$) is proved similarly.

(3). By definition of the operations in $\mathbf{Q}_p$, δ is clearly a homomorphism; hence is injective (cf. 5.1.5). By the definition of $\overline{\varphi}_p$, we have

$$\overline{\varphi}_p([\{a\}]) = \varphi_p(a), \qquad \forall a \in \mathbf{Q};$$

i.e. $\overline{\varphi}_p|\mathbf{Q} = \varphi_p$; and δ is obviously continuous. For any $[\{a_i\}] \in \mathbf{Q}_p$ and any rational $\varepsilon > 0$ there is a positive integer N such that $d_p(a_m, a_n) < \varepsilon$ for all $m, n \geq N$. Then the constant sequence $\{a_N\} = (a_N, a_N, \cdots) \in \mathbf{Q}$, and by the definition of $\overline{\varphi}$, for large enough m we have

$$\overline{d}_p([\{a_i\}], [\{a_N\}]) = \overline{\varphi}_p([\{a_i - a_N\}]) = \varphi_p(a_m - a_N) < \varepsilon.$$

That is, $\mathbf{Q}$ is dense in $\mathbf{Q}_p$.

(4). Let $\{\alpha_i\}$ be a Cauchy sequence in $\mathbf{Q}_p$ where $\alpha_i = (a_{i1}, a_{i2}, \cdots)$ with $a_{ij} \in \mathbf{Q}$. Take positive integers $N(i)$ such that

(5.9.12.1) $d_p(a_{im}, a_{in}) < 1/i, \qquad \forall m, n \geq N(i).$

Adjusting recursively, we can assume that

$$N(1) < N(2) < \cdots < N(i) < \cdots.$$

Take $b_i = a_{iN(i)}$. First we claim that $\{b_i\} \in \mathbf{Q}(d_p)$.

Let rational $\varepsilon > 0$. Since $\{\alpha_i\}$ is a Cauchy sequence, there is an N' such that, if $m, n, k > N'$, then

$$
\begin{aligned}
(5.9.12.2) \qquad \varepsilon &> \overline{d}_p(\alpha_m, \alpha_n) = \overline{\varphi}_p(\alpha_m - \alpha_n) \\
&= \overline{\varphi}_p([\{a_{mj} - a_{nj}\}]) = \varphi_p(a_{mk} - a_{nk}) \\
&= d_p(a_{mk}, a_{nk}).
\end{aligned}
$$

Set $N = \max\{N', \frac{1}{\varepsilon}\}$ and $l \geq \max\{N(m), N(n), N'\}$. For any $m, n > N$, we have

$$d_p(a_{m,N(m)}, a_{m,l}) < \tfrac{1}{m} < \tfrac{1}{N} \leq \varepsilon \qquad \text{by } (5.9.12.1)\,,$$
$$d_p(a_{n,l}, a_{n,N(n)}) < \tfrac{1}{n} < \tfrac{1}{N} \leq \varepsilon \qquad \text{by } (5.9.12.1)\,,$$
$$d_p(a_{m,l}, a_{n,l}) < \varepsilon \qquad\qquad\quad \text{by } (5.9.12.2)\,;$$

hence, by the Strong Triangle Inequality, we get

$$d_p(b_m, b_n) = d_p(a_{m,N(m)}, a_{n,N(n)})$$
$$\leq \max\{d_p(a_{m,N(m)}, a_{m,l}), d_p(a_{m,l}, a_{n,l}), d_p(a_{n,l}, a_{n,N(n)})\} < \varepsilon\,.$$

Thus $\{b_i\}$ is a Cauchy sequence, hence $\beta = [\{b_i\}] \in \mathbf{Q}_p$. Furthermore, taking large enough $l > \max\{N(m), N(n), N'\}$, for $n > N$ the above argument also shows that

$$\bar{d}_p(\alpha_n, \beta) = d_p(a_{nl}, b_l) \leq \max\{d_p(a_{nl}, a_{ll}), d_p(a_{ll}, a_{l,N(l)})\} < \varepsilon\,.$$

That is, β is the limit of the sequence $(\alpha_1, \alpha_2, \cdots)$.   □

Similar to expressing a real number in an infinite decimal, we sketch a method to express a p-adic number in a "p-adic series". We begin with a remark on a number-theoretic fact.

5.9.14 Remark. Assume $v_p(a) \geq 0$. By definition we can write $a = n/m$ with $m, n \in \mathbf{Z}$ and $m > 0$ and $p \nmid m$. Then m is invertible modulo p^k for any $k > 0$, i.e. invertible in $\mathbf{Z}_{p^k}$; more practically, there are integers m', t such that $mm' + p^k t = 1$, i.e. $mm' = 1 (\bmod p^k)$ hence $1/m = m' (\bmod p^k)$. So there is a unique integer r with $0 \leq r < p^k$ such that $a \equiv nm' \equiv r (\bmod p^k)$; which means that $a = r + p^k b$ for $b \in \mathbf{Q}$ with $v_p(b) \geq 0$. This r is called the *principal residue* of a $(\bmod p^k)$.

Let $[\{a_i\}] \in \mathbf{Q}_p$. By Exer.6, we can assume:

$$v_p(a_n) = l \quad \text{and} \quad v_p(a_{n+1} - a_n) \geq l + n, \qquad \text{for all } n \geq 1\,.$$

In order to simplify the notation, we first assume that $l \geq 0$, that is, all a_i have denominators prime to p. Starting from a_1, by Remark 5.9.14 we have a unique s_0 such that

$$a_1 \equiv s_0 (\bmod p)\,, \quad \text{and} \quad 0 \leq s_0 < p\,;$$
$$a_1 \equiv r_1 (\bmod p)\,, \quad \text{where } r_1 = s_0\,.$$

As $a_2 \equiv a_1 (\bmod p)$, we see $a_2 \equiv r_1 (\bmod p)$, hence $v_p((a_2 - r_1)/p) \geq 0$; thus we have a unique s_1 such that

$$(a_2 - r_1)/p \equiv s_1 (\mathrm{mod}\, p), \quad \text{and}\ \ 0 \leq s_1 < p\,;$$
$$a_2 \equiv r_2 (\mathrm{mod}\, p^2), \quad \text{where}\ \ r_2 = s_0 + s_1 p\,.$$

As $a_3 \equiv a_2 (\mathrm{mod}\, p^2)$, we see $a_3 \equiv r_2 (\mathrm{mod}\, p^2)$, hence $v_p((a_3 - r_2)/p^2) \geq 0$; thus we have a unique s_2 such that

$$(a_3 - r_2)/p^2 \equiv s_2 (\mathrm{mod}\, p), \quad \text{and}\ \ 0 \leq s_2 < p\,;$$
$$a_3 \equiv r_3 (\mathrm{mod}\, p^2), \quad \text{where}\ \ r_3 = s_0 + s_1 p + s_2 p^2\,.$$

Recursively in this way, we get integers $s_0, s_1, s_2, \cdots$ such that each $0 \leq s_i < p$ and, letting $r_i = s_0 + s_1 p + \cdots + s_{i-1} p^{i-1}$, $v_p(a_i - r_i) \geq p^i$; in other words, $d_p(a_i, r_i) \leq 1/p^i$ for all $i = 1, 2, \cdots$; hence $[\{a_i\}] = [\{r_i\}]$. Instead of the sequence version, we express the p-adic number $\gamma = [\{r_i\}] \in \mathbf{Q}_p$ in series; then it is written as a p-adic power series:

$$\gamma = \sum_{i=0}^{\infty} s_i p^i = s_0 + s_1 p + \cdots + s_i p^i + \cdots, \qquad \text{with}\ 0 \leq s_i < p\,.$$

For general case that $l = v_p(a_i) < 0$, considering $\{a_i/p^l\}$, we reduce it to the above case.

Therefore, we get the existence of the following

5.9.15 Proposition.   *For any $\gamma \in \mathbf{Q}_p$ there is a unique p-adic power series $\sum_{i=0}^{\infty} s_{l+i} p^{l+i} = s_l p^l + s_{l+1} p^{l+1} + \cdots + s_{l+i} p^{l+i} + \cdots$ with integers $0 \leq s_i < p$ such that $[\{r_i\}] = \gamma$ where $r_i = \sum_{k=0}^{i-1} s_{l+i} p^{l+i}$ are the partial sums of the series.*

Proof. The uniqueness is left as Exer.7.   □

This is just why the name "*p*-adic number" is adopted. And, in fact the addition and the multiplication on $\mathbf{Q}_p$ can also be done in "*p*-adic way"; as shown in the following examples.

5.9.16 Example. Take $p = 3$; and consider $-1 \in \mathbf{Q}_p$; which is just the constant sequence $\{-1\}$. Using the above process we have the following calculation:

$$-1 \equiv 2 (\mathrm{mod}\, 3);$$
$$(-1 - 2)/3 = -1 \equiv 2 (\mathrm{mod}\, 3);$$
$$(-1 - (2 + 2 \cdot 3))/3^2 = (-1 - 2)/3 = -1 \equiv 2 (\mathrm{mod}\, 3);$$
$$\cdots \cdots \cdots;$$

thus, $-1 = 2 + 2 \cdot 3 + 2 \cdot 3^2 + \cdots$. In fact, for -1 we have an easier calculation:

$$-1 = 2 + 3 \cdot (-1) = 2 + 3(2 + 3 \cdot (-1)) = 2 + 2 \cdot 3 + 2 \cdot 3^2 + \cdots .$$

Moreover, the fact $1 + (-1) = 0$ appears as

$$
\begin{array}{cccccc}
 & 2 & 2 & 2 & \cdots & 2 & \cdots \\
+ & 1 & 0 & 0 & \cdots & 0 & \cdots \\
\hline
 & 1 & 1 & & \cdots & 1 & \cdots \\
\hline
 & 0 & 0 & 0 & \cdots & 0 & \cdots ;
\end{array}
$$

Next, consider $1/2$:

$$1/2 \equiv 2(\mathrm{mod}\, 3);$$

$$(\frac{1}{2} - 2)/3 \equiv 1(\mathrm{mod}\, 3);$$

$$(\frac{1}{2} - (2 + 1 \cdot 3))/3^2 \equiv 1(\mathrm{mod}\, 3);$$

$$\cdots \cdots \cdots;$$

thus, $1/2 = 2 + 1 \cdot 3 + 1 \cdot 3^2 + 1 \cdot 3^3 + \cdots$. In fact, we also have:

$$1/2 = 2 + 3 \cdot (-1/2) = 2 + 3(1 + 3 \cdot (-1/2)) = 2 + 1 \cdot 3 + 1 \cdot 3^2 + 1 \cdot 3^3 + \cdots .$$

Exercises 5.9

+1. Let φ be a valuation on a field in the sense of Remark 5.9.1. Then
   1) $\varphi(1) = 1 = \varphi(-1)$, and $\varphi(-a) = \varphi(a)$;
   2) $\varphi(a - b) \le \varphi(a) + \varphi(b)$;
   3) $|\varphi(a) - \varphi(b)| \le \varphi(a - b)$.

2. Derive (D1), (D2) and (D3) in 5.9.1 from the (N1), (N2) and (N3) in 5.9.1. (Hint: for (D2), cf. the above Exer.1(1).)

3. Let $v_p(a)$ be as in Def.5.9.4. For all $a, b \in \mathbf{Q}$ the following hold:
   1) v_p is a surjective map from $\mathbf{Q}^*$ (the non-zero rationals) onto $\mathbf{Z}$.
   2) $v_p(ab) = v_p(a) + v_p(b)$;
   3) $v_p(a + b) \ge \min\{v_p(a), v_p(b)\}$; and the equality holds if $v_p(a) \ne v_p(b)$.

4. Prove Prop.5.9.5.

5. Let $\varphi : F \to \mathbf{Q}$ be a discrete valuation and $d(a, b) = \varphi(a - b)$. Prove:
   1) The distance satisfies the Strong Triangle Inequality (D3').
   2) A sequence $\{a_i\}$ in F is a Cauchy sequence if and only if for any rational $\varepsilon > 0$ there is an integer N such that $d(a_{n+1} - a_n) < \varepsilon$ for all $n > N$.

+6. Let $v_p : \mathbf{Q} \to \mathbf{Z}$ be as in Def.5.9.4 and $[\{a_i\}] \in \mathbf{Q}_p$. Assume $\overline{\varphi}_p([\{a_i\}]) = a = 1/p^l$ where $\overline{\varphi}_p$ is defined in Theorem 5.9.12(2). Show that there is a representative $\{c_i\}$ of $[\{a_i\}]$ satisfies the following two condtions:

1) $v_p(c_n) = l$ for all $n \geq 1$;

2) $v_p(c_{n+1} - c_n) \geq l + n$ for all $n \geq 1$.

(Hint: any infinite subsequence of $\{a_i\}$ has the same limit as $\{a_i\}$.)

7. Prove the uniqueness of 5.9.15. (Hint: if $\sum_{i \geq 0} s_{l+i} p^{l+i} \neq \sum_{i \geq 0} s'_{l'+i} p^{l'+i}$ and $\{r_i\}$ and $\{r'_i\}$ are the corresponding sequences; let the first different coefficients be $s_k \neq s'_k$, then $\varphi_p(r_n - r'_n) = 1/p^k$ for all $n \geq k$.)

8. Find the p-adic power series of the following p-adic numbers:

   1) $p = 5$, $24 \in \mathbf{Q}_5$;

   2) $p = 3$, $-1/2 \in \mathbf{Q}_3$;

   3) $-1 \in \mathbf{Q}_p$ for arbitrary p.

9. Let $\gamma = \sum_{i \geq 0} s_i p^i \in \mathbf{Q}_p$. Find the p-adic power series of $-\gamma$.

10. Let $\gamma = \sum_{i \geq 0} s_{l+i} p^{l+i} \in \mathbf{Q}_p$. Prove that: $\gamma \in \mathbf{Q}$ if and only if it is a cyclic power series, i.e. from some position on, the coefficients satisfy $s_k = s_{k+N}$ for a fixed N. (Hint: $\gamma \in \mathbf{Q}$ iff $\gamma = p^l(b + \frac{a}{1-p^N} p^k)$, where $b = s_l + s_{l+1} p + \cdots + s_{l+k} p^k$ and $a = s_{l+k+1} + s_{l+k+2} p + \cdots + s_{l+k+N} p^{N-1}$.)

5.10   Quaternions

The complex field $\mathbf{C}$ is obtained by extending the real field $\mathbf{R}$ with the irreducible real polynomial $x^2 + 1$; and $\mathbf{C}$ is algebraically closed, this is just the Fundamental Theorem of Algebra in §5.5. Furthermore, this line of idea led to the discovery of quaternions when people went beyond the commutativity.

We begin with a concrete version with complex matrices.

5.10.1 Notation. Let $\bar{\alpha}$ denote the conjugate of $\alpha \in \mathbf{C}$. Let $M_2(\mathbf{C})$ denote the ring of all 2×2 complex matrices (in fact it is an algebra over $\mathbf{C}$, see Example 4.8.2). Consider the subset

$$\mathcal{H} = \left\{ \left(\begin{matrix} \alpha & \beta \\ -\bar{\beta} & \bar{\alpha} \end{matrix} \right) \middle| \alpha, \beta \in \mathbf{C} \right\} \subset M_2(\mathbf{C}).$$

5.10.2 Proposition. $\mathcal{H}$ is a skew field (i.e. a division ring, see Def.3.2.6) with the matrix addition and the matrix multiplication.

Proof. It is easy to see that $\mathcal{H}$ is a unitary subring of $M_2(\mathbf{C})$. For any $0 \neq A = \begin{pmatrix} \alpha & \beta \\ -\overline{\beta} & \overline{\alpha} \end{pmatrix} \in H$, the determinant $\det A = \alpha\overline{\alpha} + \beta\overline{\beta} \neq 0$; and

$$A^{-1} = (1/\det A) \cdot A^* = \begin{pmatrix} \overline{\alpha}/\det A & -\beta/\det A \\ \overline{\beta}/\det A & \alpha/\det A \end{pmatrix} \in \mathcal{H}.$$

Thus $\mathcal{H}$ is a skew field. It is non-commutative clearly, e.g.

$$\begin{pmatrix} i & 0 \\ 0 & i \end{pmatrix}\begin{pmatrix} 0 & -1 \\ 1 & 0 \end{pmatrix} \neq \begin{pmatrix} 0 & -1 \\ 1 & 0 \end{pmatrix}\begin{pmatrix} i & 0 \\ 0 & i \end{pmatrix}. \qquad \square$$

It is obvious that $\mathcal{H}$ is a 2-dimensional vector space over $\mathbf{C}$, hence is a 4-dimensional vector space over $\mathbf{R}$ (i.e. it is a 4-dimensional division R-algebra in notation of §4.8).

5.10.3 Proposition. *The following 4 elements form an $\mathbf{R}$-basis of $\mathcal{H}$:*

$$E = \begin{pmatrix} 1 & 0 \\ 0 & 1 \end{pmatrix}, \quad I = \begin{pmatrix} i & 0 \\ 0 & -i \end{pmatrix}, \quad J = \begin{pmatrix} 0 & -1 \\ 1 & 0 \end{pmatrix}, \quad K = \begin{pmatrix} 0 & i \\ i & 0 \end{pmatrix};$$

which satisfy the following relations:

$EE = E, \quad EI = I = IE, \quad EJ = J = JE, \quad EK = K = KE,$
$I^2 = J^2 = K^2 = -E,$
$IJ = K, \quad JK = I, \quad KI = J, \quad JI = -K, \quad KJ = -I, \quad IK = -J.$

Proof. For any $A = \begin{pmatrix} \alpha & -\beta \\ \overline{\alpha} & \overline{\beta} \end{pmatrix} \in \mathcal{H}$, letting $\alpha = a + bi$ and $\beta = c + di$ with $a, b, c, d \in \mathbf{R}$, we have $A = aE + bI + cJ + dK$. The relations can be checked straightforwardly. $\qquad \square$

Remark. Conversely, in fact, the operations on $\mathcal{H}$ are uniquely determined *only* by the vector space structure and the relations in Prop.5.3.10 as follows: for any $A_1 = a_1E + b_1I + c_1J + d_1K \in \mathcal{H}$ and $A_2 = a_2E + b_2I + c_2J + d_2K \in \mathcal{H}$, we have (for $A_1 \cdot A_2$, expand the brackets by distributivity):

(5.10.4)
$$\begin{cases} \begin{aligned} A_1 + A_2 \;=\; & (a_1 + a_2)E + (b_1 + b_2)I + \\ & (c_1 + c_2)J + (d_1 + d_2)K; \\ A_1 \cdot A_2 \;=\; & (a_1a_2 - b_1b_2 - c_1c_2 - d_1d_2)E + \\ & (a_1b_2 + a_2b_1 + c_1d_2 - c_2d_1)I + \\ & (a_1c_2 + a_2c_1 - b_1d_2 + b_2d_1)J + \\ & (a_1d_2 + a_2d_1 + b_1c_2 - b_2c_1)K. \end{aligned} \end{cases}$$

240   CHAPTER 5. FIELDS

With the operation rules above, one can check that

$$A^{-1} = (a_1^2 + b_1^2 + c_1^2 + d_1^2)^{-1} \cdot (a_1 E - b_1 I - c_1 J - d_1 K), \quad \text{if } A \neq 0.$$

Note that (5.10.3) and (5.10.4) are independent of the matrix version. Abstracting the structure of $\mathcal{H}$, we get the quaternions as follows.

5.10.5 Theorem.   *Let* **H** *be a 4-dimensional* **R**-*vector space with a distinguished basis* $\{e, i, j, k\}$. *Define a multiplication on* **H** *by the following relations and the distributive law (or "linearly extension" in some literatures):*

$$ee = e, \quad ei = i = ie, \quad ej = j = je, \quad ek = k = ke,$$
$$i^2 = j^2 = k^2 = -e,$$
$$ij = k, \quad jk = i, \quad ki = j, \quad ji = -k, \quad kj = -i, \quad ik = -j.$$

Then **H** *is a skew field (a division* **R**-*algebra), called the* skew field of the real quaternions.

Proof. The following is obviously an isomorphism of **R**-spaces:

$$\mathcal{H} \longrightarrow \mathbf{H}, \quad aE + bI + cJ + dK \longmapsto ae + bi + cj + dk.$$

As we showed in (5.10.4), the operations are determined by the vector space structure and the relations in 5.10.3; we see that 5.10.4 also holds for **H** provided E, I, J, K are replaced by e, i, j, k resp. Thus the above map preserves multiplications; hence it is an isomorphism of rings (in fact, of **R**-algebras). Since $\mathcal{H}$ is a skew field, so is **H**, cf. Exer.2.4.3. □

Exercises 5.10

1. Prove that the following subset of $M_4(\mathbf{R})$ is a skew field isomorphic to the **H** of the real quaternions:

$$\mathcal{K} = \left\{ \begin{pmatrix} a & b & c & d \\ -b & a & -d & c \\ -c & d & a & -b \\ -d & -c & b & a \end{pmatrix} \,\middle|\, a, b, c, d \in \mathbf{R} \right\}.$$

2. Show that the following subset is a division subring of **H** (which is the skew field of the rational quaternions):

$$\mathbf{H_0} = \{a + bi + cj + dk \mid a, b, c, d \in \mathbf{Q}\}.$$

3. Determine the center of the skew field **H** of the real quaternions.

Index